COMPUTER PROGRAMMING IN
QUANTITATIVE BIOLOGY

COMPUTER PROGRAMMING IN QUANTITATIVE BIOLOGY

R. G. DAVIES

*Department of Zoology and Applied Entomology
Imperial College of Science and Technology
London, England*

1971

ACADEMIC PRESS · LONDON AND NEW YORK

ACADEMIC PRESS INC. (LONDON) LTD.
24/28 Oval Road
London, NW1 7DX

U.S. Edition published by
ACADEMIC PRESS INC.
111 Fifth Avenue,
New York, New York 10003

Copyright © 1971 By ACADEMIC PRESS INC. (LONDON) LTD.

All Rights Reserved

No part of this book may be reproduced in any form by photostat, microfilm, or any other means, without written permission from the publishers

Library of Congress Catalog Card Number: 70-153536

ISBN: 0-12-206250-7

Printed in Great Britain by
ROYSTAN PRINTERS LTD.,
Spencer Court, 7 Chalcot Road,
London, N.W.1.

Preface

The availability of electronic digital computers is transforming the training of biologists in statistical and other quantitative and numerical techniques. The present book has therefore been written in an attempt to bridge the gap between an elementary acquaintance with Fortran coding and the more sophisticated programs now increasingly likely to be encountered by biologists. Its main purpose is two-fold: firstly to help the biologist to develop some facility as a Fortran programmer in the context of problems that arise naturally in his work, and secondly to provide him with a convenient set of fully explained and therefore easily adaptable programs covering many of the methods now widely used in biological research. Despite the existence of many recent books on other aspects of computer programming and on statistical and numerical methods in biology, there seem to be none that deal with the various topics in this integrated fashion and which discuss in detail the application of appropriate programming techniques to problems of major biological interest. The inclusion of chapters on Fortran coding and on some general principles underlying the programming and use of digital computers will, I hope, give the book something of the character of a self-contained and essentially practical introduction to computer science for biologists.

The choice of subjects for elaboration as programs has not been easy, but I have tried to meet the general needs of biologists and also to introduce a few of the more specialised fields of application. Only an elementary knowledge of mathematics is assumed initially and the treatment is generally at a level suitable for advanced undergraduate students of biology or for postgraduate and research biologists without special statistical experience. The book has been planned so that the first seven chapters contain material suitable for undergraduate courses, while the remaining sections deal with more advanced or specialised subjects. In both cases however, one hopes that the reader will always bear two points in mind: firstly, that to the biologist computation is always a means to an end, the real objective being some greater insight into a biological problem; secondly, that the ease with which computation is now possible places on the biologist an even greater responsibility than before to understand something of the principles on which his numerical or statistical analyses are based.

My thanks are due to Academic Press and the printers for so willingly meeting my wishes in the presentation of a rather complicated text. I am also very grateful to my wife and son for much help in the proof-reading and indexing.

London R. G. DAVIES
September, 1971

CONTENTS

Preface   v

Chapter 1

Introduction   1

Chapter 2

Electronic Digital Computers   5

Number Systems   5
   Decimal notation   5
   Binary notation   6
   Octal notation   8
   Interconversion of number systems   10
   Binary coded decimal   13

Computer Structure and Function   14
   Input devices   15
   Computer storage   17
   Arithmetic unit   23
   Output devices   25
   Control unit   27
Programming Languages   28

Chapter 3

Synopsis of Fortran Programming   32

Planning a Program   32
FORTRAN IV   36
   Fortran characters   36
   Statements and lay-out   36
   Constants   38
   Variables   39
   Arithmetic expressions   40
   Mathematical functions   42
   Other supplied functions   44

Arithmetic statements	45
STOP, PAUSE and END statements	46
Input—Output statements	46
Further FORMAT specifications	51
Transfer statements	57
Subscripted variables and the DIMENSION statement	61
The DO statement	64
Arithmetic statement functions	66
Function subprograms	67
Subroutine subprograms	69
Non-executable statements	74
Running a Program	80

Chapter 4

Three Simple Statistical Programs 82

PROGRAM 1: Computation of mean, variance, standard deviation and other statistics of a sample	82
PROGRAM 2: Correlation and regression	88
PROGRAM 3: Comparison of means of two samples	97

Chapter 5

Sorting, Tabulating and Summarising Data 105

PROGRAM 4: Sorting an array in ascending and/or descending order	106
PROGRAM 5: Ranking an array of numbers	111
PROGRAM 6: Tabulation of frequency tables	115
PROGRAM 7: Construction of two-way contingency tables	121
PROGRAM 8: Data-summarising Program 1	129
PROGRAM 9: Data-summarising Program 2	132
PROGRAM 10: Data-summarising Program 3	136
PROGRAM 11: Calculation of running means	141

Chapter 6

Analysis of Variance 145

PROGRAM 12: Two-factor analysis of variance without replication	149
PROGRAM 13: Incomplete three-factor analysis of variance (two factors with replication)	154
PROGRAM 14: Three-factor analysis of variance without replication	161

PROGRAM 15: Analysis of variance: Latin square design 166
PROGRAM 16: Analysis of variance: balanced incomplete block design 172

Chapter 7

Correlation and Regression Analysis 181

PROGRAM 17: Correlation and regression of data arranged in grouped frequency table 184
PROGRAM 18: Regression with tests of significance and linearity by analysis of variance 189
PROGRAM 19: Comparison of two regression lines by analysis of variance 193
PROGRAM 20: Autocorrelations in time-series 198
PROGRAM 21: Cross-correlations in time-series 202
PROGRAM 22: Best-fit linear functional relationship between two variables, each subject to error 208

Chapter 8

Matrix Methods 216

Definitions 216
Input and output of matrices 219
Elementary operations on a single matrix 224
Addition and subtraction of matrices 226
Matrix multiplication 226
Transpose of a matrix 230
Successive multiplication 230
PROGRAM 23: Dispersion and correlation matrices 232

Chapter 9

Further Matrix Methods 239

Determinants 239
PROGRAM 24: Determinant of a matrix 242
Matrix Inversion 247
PROGRAM 25: Matrix inversion 251
Latent Roots and Vectors 258
PROGRAM 26: Latent roots and vectors of a real symmetric matrix 260
PROGRAM 27: Latent roots and vectors of a real non-symmetric matrix 267

Chapter 10

Multiple Regression and Multivariate Analysis 273

PROGRAM 28: Multiple linear regression analysis 274
PROGRAM 29: Hotelling's T^2 and discriminant function for two groups 284
PROGRAM 30: Principal component analysis 291
PROGRAM 31: Multiple discriminant analysis 300

Chapter 11

Nonparametric Statistics 312

PROGRAM 32: Chi-squared test of association in two-way contingency tables 313
PROGRAM 33: Chi-squared test for goodness of fit 318
PROGRAM 34: Spearman's rank correlation coefficient 322
PROGRAM 35: Wilcoxon's signed ranks test 330
PROGRAM 36: Mann–Whitney test for two independent samples 333
PROGRAM 37: Kruskal–Wallis test for several independent samples 338

Chapter 12

Fitting Theoretical Distributions to Data 343

PROGRAM 38: Fitting a normal distribution and testing goodness of fit 344
PROGRAM 39: Fitting a Poisson distribution and testing goodness of fit 355
PROGRAM 40: Fitting a binomial distribution and testing goodness of fit 360
PROGRAM 41: Fitting a negative binomial distribution and testing goodness of fit 366

Chapter 13

Models and Simulation 376

Parameter studies 378
Numerical integration 380
PROGRAM 42: Numerical integration by Simpson's rule 380
Differential equations 387

PROGRAM 43: Solution of first-order ordinary differential equations by a Runge–Kutta method 388
Generation of random variables 394
Subroutines RAND1 to RAND7 396
Calculation of Probabilities from Statistic Distributions 403
PROGRAM 44: Computation of probabilities corresponding to the test-statistics $\chi^2 t$, z and F 406

Chapter 14

Three Special-Purpose Programs 410

PROGRAM 45: Probit analysis 410
PROGRAM 46: Single linkage cluster analysis 420
PROGRAM 47: Estimation of population-size by marking and recapture 430

Chapter 15

Efficiency in Programming 440

Choice of Efficient Algorithms 441
Economy in Computer Time 442
Conservation of Core Storage 447
PROGRAM 48: Dispersion and correlation matrices with compressed vector storage 456
Accuracy Considerations 458
Testing of Programs 466
Use of Available Package Programs or Subroutines 471
Conclusion 476

References 478

Author Index 483

Subject Index 486

Chapter 1

Introduction

One of the more striking developments in biology during the past fifty years has been the growth of quantitative and mathematical aspects of the subject (Bailey, 1967). Beginning with pioneers like Francis Galton and Karl Pearson, the statistical analysis of biological data received a great impetus during the nineteen-twenties from the work of R. A. Fisher and others, whose researches were soon applied widely in agricultural experimentation and genetics. A steady increase in the diversity and power of the methods available and in their fields of application has, in the last fifteen years, been enormously accelerated by the development of electronic digital computers (Ledley, 1965; Stacey and Waxman, 1965; Sterling and Pollack, 1966). The result has been a rapid invasion of still further branches of biology by statistical, numerical and mathematical techniques. A catalogue of the fields now involved to some degree or another would cover virtually every branch of biology and it is therefore more useful to indicate the general nature of the advances resulting from the use of digital computers.

Many widely known statistical methods such as the computation of correlation coefficients and regression equations, the analysis of variance and the performance of significance tests are relatively simple operations, usually requiring appreciably less time than the collection of the biological data. The execution of these analyses by computer is therefore a convenience rather than a fundamental improvement in methods, though the possibility of exploiting statistical techniques to the full, with very little computational tedium, has obviously encouraged a more searching analysis of many bodies of data. The same tendency has also promoted the development of large-scale surveys, previously forbidding on account of the sheer volume of data-processing required. Of much greater scientific interest is the very rapid development and application of methods which, while theoretically available at an earlier date, were hardly used in practice because of the great time needed to carry out the calculations. Techniques such as multiple regression analysis are extremely tedious to apply by hand calculation when there are more than four or five independent variables; with an electronic computer the purely computational procedures can be completed in literally a few seconds! Even more striking has been the growth of various methods of multivariate analysis, all particularly appropriate in biological work where many variables are often involved. Factor analysis—perhaps the best known of these methods and almost the only one to have been applied (though on a very limited scale) before the advent of digital computers—was formerly a major computational enterprise, even for unrealistically small bodies of data. A computer can carry out in a matter of minutes what might previously have required months or years of work with a desk-calculator. These and other powerful methods of multivariate analysis can now be applied without hesitation and new methods rapidly evaluated on extensive sets of empirical data. A comparable fundamental advance has also come about through the use of computers to manipulate and assess mathematical models that simulate various biological processes—ranging from evolutionary change to the behaviour of out-patient queues at hospitals. Multivariate methods and the construction of even mildly elaborate deterministic and probabilistic models are virtually unthinkable without a computer which gives the biologist great freedom in testing theories and building models while relieving him of the burden of immense computational routines.

While many biologists have welcomed the new opportunities for quantitative analysis of their problems, the inadequate mathematical background of others has made them reluctant to apply appropriate methods long available, or even leaves them ignorant of the techniques now at one's disposal. The dangers of this are clear when one realises the extent to which quantitative and numerical techniques are now transforming previously non-mathematical branches of biology. Plant and animal ecology have rapidly acquired a large repertoire of quantitative analytical methods (e.g. Greig-Smith, 1964; Pielou, 1970; Southwood, 1966; Williams, 1964), often applicable in nature-conservation and resource management (Watt, 1968). Population genetics is now almost a branch of applied mathematics, as any biologist will know who has tried to read such accounts as those by Kempthorne (1957), Li (1955) or Ewens (1969). Taxonomy—for two centuries the epitome of a biological science in its reliance on qualitative procedures and intuitive judgement—is undergoing an almost convulsive reform initiated by Sokal and Sneath (1963) and bound to leave a permanent numerical impress, however debatable were some of the principles first advocated. Inevitably, any quantitative treatment of classification and discrimination has had an impact on several other aspects of biology, from biogeography and floristic analysis to problems of medical diagnosis. Even physiology, perhaps the branch of biology which least needed the statistical advances of previous decades, is now appreciating the extent to which problems of metabolism, sensory perception and behaviour can be studied by simulation techniques. And, of course, the list of subjects profiting from computer techniques can be extended rapidly when one recalls the many restricted fields of biology in which numerical and statistical methods have long played an important role, such as the bioassay of pharmacological preparations, the evaluation of insecticides and fungicides, anthropometry, demography, the study of individual and sub-specific variation, and of crop yields or animal growth.

It is not suggested for a moment that all biological problems require a numerical or statistical approach; the point is simply that biologists need increasingly to be aware of the extent to which these methods are being applied and the role of digital computers in the analysis or formulation of hypotheses. The present book aims, therefore, to promote this awareness by providing the biologist with a guide to the preparation of representative computer programs in the widely used programming language Fortran IV. The programs developed are themselves capable of being applied directly to some of the more frequently encountered problems in quantitative biology. In particular, there seems to be a need for an account which bridges the gap between manuals on the general principles of Fortran programming (which rarely include examples of serious biological interest) and the highly sophisticated programs which the biologist will encounter once he embarks on the use of computer methods. Inevitably, all scientists who use computers in their research are heavily dependent on the work of professional programmers, but this relationship benefits greatly if the scientist has a sufficient understanding of programming techniques to translate at least the simpler of his statistical and mathematical methods into effective programs.

Finally, a few words on the plan of this book. It is assumed that the reader has an elementary knowledge of statistical methods and that he has perhaps also taken a short course in Fortran coding. For the sake of completeness, however, a synopsis of the major features of Fortran IV is presented in Chapter 3, and a short introduction to the statistical or mathematical techniques is given in each of the main sections under which programs are described (Chapters 4 to 14). Otherwise, the object has been to provide an introductory outline of computer structure and function, needed to appreciate many basic programming procedures (Chapter 2), to give an elementary account of some aspects of matrix algebra, which are essential for serious statistical programming (Chapters 8 and 9) and to offer a general guide to efficiency in programming (Chapter 15). All complete programs are accompanied by a flow-chart and a detailed discussion and have been machine-tested. The reader who follows the construction of these carefully should have no difficulty in devising further programs to suit his special requirements. He will also have available a selected set of over forty original programs covering a wide range of numerical methods useful in biological research. The emphasis throughout is on the needs of the individual biological research-worker who is, of necessity, an occasional programmer and perhaps also an amateur statistician. The bibliography will guide him to further sources of information—it is deliberately orientated towards topics and treatments likely to appeal to the biologist rather than to the statistician or computer-scientist.

Chapter 2

Electronic Digital Computers

Electronic computers are of two kinds: *analog computers* and *digital computers*. Analog computers—with which we are not concerned in this book—work by establishing a physical analogy to a mathematical operation in the form of some continuously varying mechanical or electrical quantity. Digital computers, on the other hand, operate with discrete numbers such as are used in the ordinary arithmetical calculations we perform with pencil and paper or with a desk-calculator. Digital computers, however, do not usually employ the same number system as we normally use and it is therefore desirable to introduce their operations by discussing briefly some of the ways in which numbers may be represented by numerical symbols or their electronic equivalents in a computer.

Number Systems

Decimal notation: The notation normally used in scientific work is the decimal system, in which there are ten digits—the numbers 0, 1, 2, ... 9, arranged so that the value of a number depends also on the position of the digits. It may therefore be described as a positional notation with the base 10. The meaning of this will be clearer if we recall that a number like 1,372 consists of 1 thousand + 3 hundreds + 7 tens + 2 units. This may be expressed by writing the digits, 1, 3, 7 and 2 in sequence, each in a column headed by the quantities 10^3, 10^2, 10^1, and 10^0 ($=1$), when the full significance of a positional system with base 10 becomes apparent:

$$
\begin{array}{cccc}
10^3 & 10^2 & 10^1 & 10^0 \\
1 & 3 & 7 & 2
\end{array}
$$

The column headings can, of course, be extended by multiples or sub-multiples of 10 in each direction, so that numbers like 13910.728 and 423.8714 can be understood as:

$$\begin{array}{ccccccccc} 10^4 & 10^3 & 10^2 & 10^1 & 10^0 & 10^{-1} & 10^{-2} & 10^{-3} & 10^{-4} \\ 1 & 3 & 9 & 1 & 0\ . & 7 & 2 & 8 & \\ & & 4 & 2 & 3\ . & 8 & 7 & 1 & 4 \end{array}$$

The advantages of a positional system over a non-positional system such as the Roman numeral system are obvious as soon as one tries to divide MCCCCXXV by XVII. There is, however, nothing sacrosanct about the base 10 and number systems depending on other bases can easily be constructed. Two are particularly important in computer science: the *binary* system to base 2 and *octal* to base 8. Hexadecimal (base 16) is also used.

Binary Notation: Just as decimal numbers were expressed above using columns headed by 10^3, 10^2, 10^1, 10^0 etc., so a binary number is expressed using powers of 2. Such a system requires only two digits, 0 and 1. For example the number fifteen is equal to $2^3+2^2+2^1+2^0$ ($=1$). Using the above method of headed columns we can therefore express fifteen in binary form as:

$$\begin{array}{cccc} 2^3 & 2^2 & 2^1 & 2^0 \\ 1 & 1 & 1 & 1 \end{array}$$

or more concisely, as 1111_2 where the subscript 2 denotes that we are using the base 2. Thus 15_{10} and 1111_2 are simply different ways of writing the same number—fifteen. A few other examples, which the reader can work out for himself, will perhaps make the principle clearer:

$$9_{10} = 1001_2$$
$$23_{10} = 10111_2$$
$$523_{10} = 1000001011_2$$

The same method can be used for numbers smaller than one by placing digits to the right of the binary point (the equivalent of a decimal point), remembering that $.1_2$ is equivalent to 0.5_{10}, i.e. 1×2^{-1} while $.01_2$ is equivalent to 0.25_{10} i.e. 1×2^{-2} and so on.

Binary numbers can, of course, be added and subtracted, multiplied and divided, bearing in mind that the only digits used are 0 and 1. For example:

$$\begin{array}{r} 10101 \\ +\ \ 1011 \\ \hline 100000 \end{array}$$

is the binary equivalent of $21+11=32$ in decimal notation. Or again:

$$\begin{array}{r} 1110 \\ \times \quad 101 \\ \hline 1110 \\ 0000. \\ 1110.. \\ \hline 1000110 \end{array}$$

is the equivalent of $14 \times 5 = 70$ in decimal notation.

At first sight the binary system appears unusually cumbersome, an average of about 3.3 binary digits being required for each decimal digit. Electrical and electronic devices are, however, well adapted to representing binary numbers since they frequently operate in one or other of two states. A switch or relay may be open or closed, a magnetic body magnetised in a given direction or the opposite one, and so on. It is for this reason that the binary system is used in electronic digital computers. A number such as seventeen can very easily be represented by a row of five electronic devices set so that the first and last are in the 'one' condition and the others in the 'zero' condition: 10001. This, in fact, typifies the way in which numbers are stored in the *computer registers.*

A register, whether it be a storage register or a working register used in arithmetical or control operations, is simply a linear array of electronic devices capable of storing the separate digits of one or more binary numbers. Each *binary digit* or *bit* represents the unit of information and a register capable of accommodating, say, 36 bits may be described as a 36-bit register. Alternatively one may speak of a computer with such registers as having a 36-bit word structure. Figure 1 illustrates schematically a register able to contain 36 bits, which we may number consecutively from 0 to 35. The first (leftmost in the diagram) is the so-called sign bit (0 for positive, 1 for negative is the usual convention) while the remaining 35 bit positions (numbers 1–35 inclusive) contain the binary number—in this case 1101, the leftmost numbers after the sign bit all being zeros. The register in Fig. 1 therefore contains the binary representation of the decimal number -13. Obviously a computer

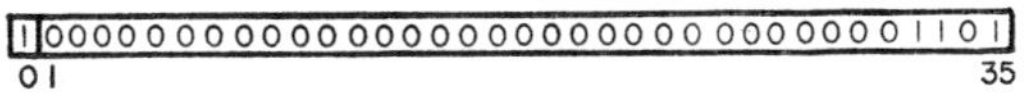

FIG. 1. Schematic representation of a computer register containing the number -13. There are 36 bit-positions, numbered 0 to 35 consecutively. The 0 position contains the binary digit 1: this is the 'sign bit' corresponding to a negative sign. The remaining 35 positions contain the binary number 1101 (preceded in this case by 31 zeros). Binary 1101 is the equivalent of decimal 13.

with a 36-bit word structure of this kind could accommodate integers ranging in the size from $-(2^{35}-1)$ when every bit is a 1 to $+(2^{35}-1)$ when the sign bit is zero and all the others are ones.

Binary representation of fractions is achieved by a somewhat different technique: they are stored in *floating-point* form, which means essentially that they are represented by a fraction and an exponent. Just as decimal 23.16 can be represented as 0.2316×10^2 with a fractional part, 0.2316, and an exponent 2 (the base 10 being understood), so a floating point binary number can be treated as a fractional part and an exponent of the base 2. These, together with the sign associated with the fractional part, can then be accommodated in, say, a 36-bit register on a principle such as that shown in Fig. 2.

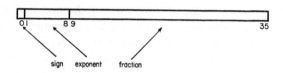

FIG. 2. Schematic representation of a computer register intended to accommodate a floating-point number. There are 36 bit-positions, of which the 0 position holds the sign bit, positions 1–8 hold the binary equivalent of the exponent and positions 9–35 hold the binary equivalent of the fraction.

Octal Notation: It is clear from the above that binary notation is deeply implicated in the physical functioning of a computer. Another form of notation—*octal*, to the base 8—also occurs frequently in computer science, but this is mainly for purposes of human communication: computer operations do not take place in octal form. Octal is used because binary numbers are long and difficult to remember or to write correctly. To convert them into decimals is not difficult but it cannot be done at sight. It is, however, possible to convert binary into octal and *vice versa* very easily, and since octal numbers are as convenient to remember or write as decimals they are commonly used in place of the latter. An octal number, applying the same principles as before, is written with the digits 0 to 7 inclusive, the position of each digit corresponding this time to a particular power of 8. Thus, with the column headings $8^2 \; 8^1 \; 8^0$ we can write 2 1 3 as an octal number equivalent to $(2 \times 64) + (1 \times 8) + (3 \times 1) = 139$. Now because $8 = 2^3$ there is a very simple relationship between binary and octal numbers. We first break up a binary number into groups of three digits, starting at the binary point and working in each direction, and we then give to each group of three digits its octal equivalent. The resulting octal digits, in a surprisingly simple way, immediately provide the octal equivalent of the whole binary number. Thus, taking the formidable-looking binary number 10,010,011,101.110110 we

NUMBER SYSTEMS

divide it into groups of three digits and write beneath each group its octal equivalent:

$$\begin{array}{cccccc} 010 & 010 & 011 & 101 & . & 110 & 110 \\ 2 & 2 & 3 & 5 & . & 6 & 6 \end{array}$$

Then 2235.66 is the octal representation of the very long binary number. Conversely, if one wished to convert, say, 237_8 into binary, one would take each octal digit in turn and write down its binary equivalent as a 3-digit group:

$$2_8 = 010_2$$
$$3_8 = 011_2$$
$$7_8 = 111_2$$

so that $237_8 = 10011111_2$ since the leading zero of the complete binary number can be dropped (as it normally is in any number system). All binary-octal conversions can therefore be carried out entirely through the following simple table:

Octal	Binary
0	000
1	001
2	010
3	011
4	100
5	101
6	110
7	111

Neither octal nor binary numbers occur to any extent in Fortran programming as such; the biologist can safely continue to think in decimal numbers virtually all the time! But the use of binary number systems accounts for many otherwise inexplicable features of computer operation and octal numbers will come to the programmer's notice in a few ways. It is, for example, occasionally necessary to use octal numbers in the control statements that accompany Fortran programs at large computer installations. They are also commonly produced by the computer when it divulges information about its internal condition. For instance, memory maps and "core dumps" which indicate how programs and data are stored in the computer (and which may be needed for diagnostic purposes when correcting a program) are often printed in octal. Even more commonly one finds that a program listed by a computer will be accompanied by a set of octal numbers indicating the numbers of the registers in which the separate instructions of the program are stored. For these reasons it is often desirable to convert numbers from one system to another. This can always be done from first principles as indicated above, but in practice it is usually quicker to use tables such as are published by some computer manufacturers or to employ the rules set out below.

Interconversion of Number Systems: As shown above, conversion from binary to octal and *vice versa* is so simple that no further rules are needed. Conversion from binary to decimal and the reverse require separate conversion of the integer and fractional parts of the number and the combination of the two results.

(a) *Binary to decimal integers.* The leftmost digit of the binary number is multiplied by 2 and the next binary digit added. This result is again multiplied by 2 and the next digit added and so on until all the binary digits have been included. The result is the decimal equivalent; e.g. to convert 11011 to decimal:

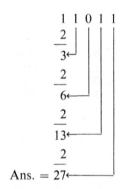

(b) *Binary to decimal fractions.* This may be done by adding a series of fractions: e.g. to convert binary .1011 to decimal:

$$.1011 = .5$$
$$+ .125$$
$$+ .0625$$
$$\overline{}$$
$$0.6875$$

(c) *Decimal to binary integers.* Divide the decimal number repeatedly by 2 until the quotient is zero. Note each remainder and then read them vertically upwards to give the binary number required, e.g. to convert 113_{10} to binary:

NUMBER SYSTEMS

```
2 | 113           remainder
2 | 56              1
2 | 28              0
2 | 14              0
2 | 7               0
2 | 3               1
2 | 1               1
    0               1
                        read thus
Ans. = 1110001
```

(d) *Decimal to binary fractions*: Multiply the decimal fraction repeatedly by 2, noting the carry-over into the units position. Read the carry-over numbers vertically downwards to give the binary fraction required: e.g. to convert decimal .061 to a binary fraction:

```
carry-over      .061
                   2
     0          .122
                   2
     0          .244
                   2
     0          .488
                   2
     0          .976
                   2
     1          .952
                   2
     1          .904
                   2
     1          .808
                   2
     1          .616
                   2
     1          .232
read              :
thus              :
```

$\therefore \quad .061_{10} = .0000111111_2 \ldots$

11

(e) *Octal to decimal integers*: Exactly the same rules apply as for binary to decimal, except that 8 replaces 2; e.g. to convert 123_8 to decimal:

$$\begin{array}{c} 1\ 2\ 3 \\ 8 \\ \hline 10 \leftarrow\!\!\rfloor \\ 8 \\ \hline 83 \leftarrow\!\!\!\rule{0pt}{0pt} \end{array}$$

$\therefore\ 123_8 = 83_{10}$

(f) *Octal to decimal fractions*. This may be done by adding multiples of the fractions $\frac{1}{8}, \frac{1}{64}, \frac{1}{512}$ etc.

$$\text{e.g.}\ .123_8 = \frac{1}{8} + \frac{2}{64} + \frac{3}{512} = \frac{64 + 16 + 3}{512}$$

$$= \frac{83}{512} = .1621\ldots$$

(g) *Decimal to octal integers*. As for decimal to binary, but with 8 replacing 2; e.g. to convert 83_{10} to octal:

$$\begin{array}{r|l|l} & & \text{remainder} \\ 8 & 83 & \\ 8 & 10 & 3 \\ 8 & 1 & 2 \\ & 0 & 1 \end{array}$$

$\therefore\ 83_{10} = 123_8$

(h) *Decimal to octal fractions*. As for decimal to binary, but with 8 replacing 2; e.g. to convert .16211 to octal:

$$\begin{array}{cc} \text{carry-over} & .16211 \\ & \underline{\quad 8} \\ 1 & .29688 \\ & \underline{\quad 8} \\ 2 & .37504 \\ & \underline{\quad 8} \\ 3 & .00032 \end{array}$$

$\therefore\ .16211_{10} = .123_8\ldots$

In practice, the integer conversions of octal to decimal and *vice versa* are the ones most likely to be needed (and then not very often) by the Fortran programmer. It is, however, worth noting that while integers in one system always have exact representation in another system, exact fractions in one may not have an exact representation in another, particularly if one is limited in the number of places after the point. This leads to the important problem of *round-off error*, which is further discussed on p. 460.

Binary Coded Decimal: The interconversion of binary and decimal numbers above is a replacement of one mathematical system of notation by another. It is, however, also possible to devise what are essentially number-codes, in which decimal numbers are coded in binary notation. For example, the number 21_{10} has its true binary equivalent 10101, but we could construct a code in which the digits 2 and 1 were converted separately into binary and the two binary groups placed one after the other. Since there are 10 decimal digits— 0 to 9 inclusive—we really require a code comprising groups of four binary digits. The most straightforward one—sometimes known as the 8–4–2–1 code —is as follows:

Decimal digit	Binary equivalent
0	0000
1	0001
2	0010
3	0011
4	0100
5	0101
6	0110
7	0111
8	1000
9	1001

Using this code the number 21_{10} would be written as 00100001, or 48_{10} as 01001000. Note once again that these *binary-coded decimal* numbers (BCD numbers) are not the true binary equivalents of 21 and 48, but are simply coded versions based on a binary system. Other similar BCD codes have been devised and are widely used. For Fortran programmers the exact nature of the code is not normally of any importance, but it may be very necessary to know whether information written on or read from magnetic tape or disk is in true binary or in binary coded decimal. As we shall see later, the magnetic tapes that normally provide the input and output of the central computer in a large installation may be in BCD, whereas tapes used by the programmer to store information in a form corresponding to its internal representation in the computer may be in true binary. It is also the case that some computers use a

BCD system for internal storage and manipulation of numbers—the so-called decimal computers—but these are not normally used in large-scale scientific computing.

Computer Structure and Function

The essential elements of a digital computer may be represented in a block diagram (Fig. 3). They are: (i) the input devices; (ii) the central storage unit; (iii) the arithmetic unit; (iv) the output devices, and (v) the control unit. Briefly, the *input devices* are intended to read information—data and programs—into the machine, the *central storage unit* or memory stores this information, the *arithmetic unit* operates on the numerical or logical data and the *output devices* make the results intelligible to the user. All these processes are controlled and co-ordinated by the *control unit*. It is worth emphasising one important feature: in the memory of a computer of this kind are stored both numbers and the program of instructions executed by the machine in manipulating the numbers. Both are stored in similar locations and it is the machine's ability to hold a full set of instructions and to execute them in the correct sequence which enables it to complete elaborate calculations without human intervention. Such a digital computer is referred to as a *stored-program computer* and since a virtually unlimited number of programs may be written the machine has enormous versatility. Further, since the execution of the program requires no interruption for human assistance, it may be carried out at very great speed. Ultimately, however, the computer can do no more than it is instructed to perform and ultimately all programs originate in the mind of a human programmer.

The major components of the computer may now be reviewed briefly and very simply, with special reference to their importance in programming.

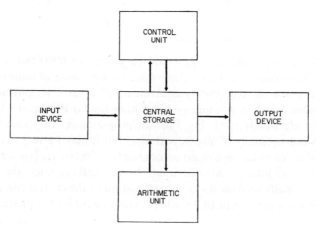

FIG. 3. Block diagram showing the essential components of an electronic digital computer.

COMPUTER STRUCTURE AND FUNCTION

Input Devices: These are electromechanical machines whose detailed structure and mode of operation depend on the input medium. Two input media are particularly important: *punched cards* and *punched paper tape*. A punched card of standard form is illustrated in Fig. 4. Numbers, alphabetical symbols, punctuation marks, etc. can be recorded on such cards by means of a *keypunch*—a hand-operated electrical machine with a keyboard resembling a typewriter. Each symbol is represented by one or more rectangular holes punched in the card in various positions along twelve horizontal rows and in 80 vertical columns. At the same time each symbol is usually typed out along the upper margin of the card, so that the contents of the card are immediately visible. Paper tape consists of a narrow band of tough, flexible paper on which

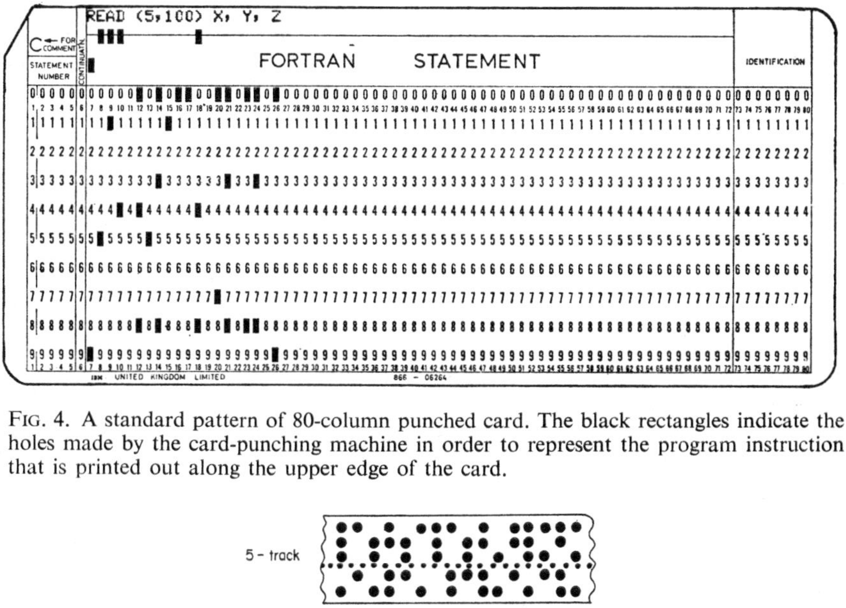

FIG. 4. A standard pattern of 80-column punched card. The black rectangles indicate the holes made by the card-punching machine in order to represent the program instruction that is printed out along the upper edge of the card.

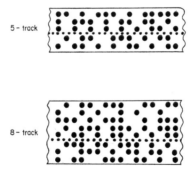

FIG. 5. Short lengths of punched paper tape. The continuous row of small holes near the middle of the tape are sprocket-holes and convey no information. Each numerical digit or other character is represented by a different combination of large punched holes in the 5 or 8 positions available in a row across the tape.

each symbol is recorded by a combination of circular holes punched across the tape (also by a machine with a typewriter-like keyboard). At the same time the data are typed out on a sheet of paper for the operator to read and check. Patterns of hole-punching vary both in punched cards and paper tape and a particular computer installation is usually able to cope with only a limited range of input media. The examples illustrated in Figs 4 and 5 are common but not general, and the programmer must obtain details of the precise characteristics of the input medium from the installation at which he works. In much of this book it is assumed that the reader will work with a punched-card system, but fortunately the writing of Fortran programs is not closely dependent on the detailed characteristics of the input medium (though the execution of the program is closely bound up with them). On both cards and paper tape the punched holes represent a coding of information, but this code is not based on the number systems discussed on pp. 5–13. Its ultimate conversion into one or other of the number systems is carried out by the computer.

Numerical data and programs punched on cards or tape can, of course, be stored indefinitely without special precautions and can be used repeatedly. When the information they bear is to be used they are fed into a *card-reader* or *paper tape reader* forming part of the computer system. These are relatively complicated machines which sense the pattern of punched holes. A *card-reader* holds the deck of cards in a hopper and by movements of a lever each card may be fed over a row of 80 READ brushes which can form electrical contacts only through the punched holes, or more commonly the positions of the holes are detected by photoelectric devices. The information on the cards is thus converted into a set of electrical signals which are subsequently transformed into binary or binary-coded decimal computer words. These are then stored temporarily in the in–out registers of the computer before being transferred to the high-speed central storage unit. Some such indirect method of transfer via a *buffer* is needed because of the discrepancy in speed between the relatively very slow card-reader (operating at perhaps 250 cards per minute) and the exceedingly fast operations within the computer. When a normal job is being run, the operation of the card-reader is under the control of the stored program(s) so that the appropriate parts of each card are read and interpreted. Punched paper-tape readers work in an analogous way, the tape being moved by sprocketted drive wheels, again under control of the program. The pattern of holes is detected by mechanical 'sensing pins' or by photoelectric devices which produce signals when a light source illuminates them through the holes.

The account given so far assumes that the input device is connected directly to the central computer. In some installations this may be so, the input operation being described as *on-line*. In other systems, however, *off-line* operation takes place, the cards or paper-tape for many separate jobs being read successively and the coded data transcribed on to magnetic disk or tape. When this has been done it is then read into the large central computer. Off-line operations of this kind may be associated with some form of *batch-processing*, whereby sets of jobs of a similar kind are processed together. It is also worth mentioning that card and paper-tape readers can cope with a variety of codes and that *relocatable binary* cards or binary tape (produced by the computer as a separate earlier operation) can also be read in.

Computer Storage: A computer's ability to store very large quantities of numerical data or instructions in its memory elements is one of the most intriguing features of its operation. In fact, a computer incorporates several different methods of storing information and making it available for control of execution, arithmetic manipulation, output, and so on. The information on punched cards or paper-tape is, in a sense, in storage, though it can be utilised only very slowly and is not physically 'inside' the computer except perhaps when it is being read. More realistically one can distinguish the following types of storage in a computer:

(a) The *central store* of very quickly available information, including the program and much or all of the data on which it operates, the intermediate quantities and the final results. This is sometimes known as *core-storage* and its size and other operational characteristics are of importance to the Fortran programmer. Its distinguishing features are its high operating speed, the fact that information can be obtained from any part of it equally easily (i.e. it has *random access*), and that it comprises a moderately large number (perhaps several tens of thousands) of static, erasable registers or locations.

(b) *Magnetic drum memories*, which form part of the auxiliary or *backing store* and are capable of holding very large quantities of information which is available moderately rapidly in a sequential fashion.

(c) *Magnetic disk memories*, which are very popular and intermediate in speed and cost between (b) and (d).

(d) *Magnetic tape memories*, which can also hold large quantities of information though it is not so quickly available. In addition to the above there are also a small number of registers in which information is stored temporarily while operations are being performed on it. These registers form part of the arithmetic and control units and are considered briefly on p. 24, but the four major types of large-scale storage require further discussion.

(a) *Central Storage Unit*

Several different kinds of central high-speed store are available, but the best known is the *magnetic core memory*. This consists of a large number of very small ring-shaped (toroidal) cores of ferrite, a magnetisable material with a characteristic, almost square, type of hysteresis curve (Fig. 6). Each core, measuring a millimetre or less in diameter, is threaded by a wire (the *input winding*). When a current is passed through this a magnetic flux is produced in the core, its direction depending on the direction of the current in the winding. The core will become magnetised either in a clockwise or a counter-clockwise direction, and this magnetism is very largely retained when the magnetising current ceases. Because it can exist in one or other of two directions of magnetisation, therefore, the core is capable of representing the binary digits 0 and 1. A second wire, the *sense winding*, is also threaded through the core and is used to "read" the value of the digit stored in the core. This it does in such a way that the core is always reset to zero after being read, so a third circuit is arranged to return the core to the conditions existing before it was sensed. The consequence of all this is that a binary digit can be stored in each core, that it is retained until needed and that it is not destroyed through the core being read. A digit can, of course, be replaced by another so that one may speak of this memory having *destructive read-in* and

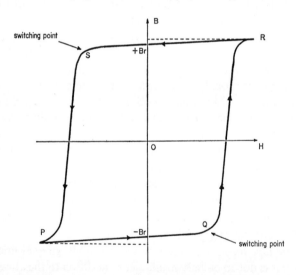

FIG. 6. The 'square loop' hysteresis curve of a ferrite core. The horizontal axis represents the intensity of the external magnetising field, H, due to the magnetising current. The vertical axis represents the magnetic flux density, B, induced in the ferrite. If the magnetising current is increased sufficiently, the flux density rises to a level corresponding to point R on the curve. When the current then drops to zero, the flux density falls only a little (i.e. to + Br). Further decrease in current causes only a little further fall in flux density until the 'switching point' S is reached, after which there is a rapid reversal of flux density to the point P. A converse situation holds from P, through the switching point Q, to R once again. The shape of the curve means that the ferrite core is normally in one or the other of two stable states of magnetisation, +Br and − Br, respectively close to positive or negative saturation and corresponding to zero or one in the binary number system.

non-destructive read-out. These latter properties are characteristic of all digital computer memories and underlie all programming operations.

The organisation of a large number of cores into a set of storage registers is, in principle, quite simple. The cores are arranged in a series of *core-memory planes*, each a flat, two-dimensional array in which any core may be selected for storage by a suitable ingenious arrangement of selection wires and sensed by the sense winding (Fig. 7). A plane may, for example, contain $128 \times 128 = 16{,}384$ cores and there may be, say, 36 such planes in the whole core-memory, making a total of over half a million cores, each able to store a single binary digit. The important point to notice is that each storage register consists of the series of similarly placed cores in successive planes. Thus the top leftmost core in the first plane accommodates the first bit of a given register, the top leftmost core in the second plane accommodates the second bit, and so on (Fig. 8). In the example given, therefore, the core-memory would comprise 16,384 registers or locations, each able to hold 36 bits. As we saw on p. 8 this would allow as a maximum the storage in each register of a sign bit followed by a binary integer equal to $2^{35} - 1$. Any integer of between $+(2^{35} - 1)$ and $-(2^{35} - 1)$—that is, of value $\pm 34{,}359{,}738{,}367$—could therefore be accommodated in any of the 16,384 storage registers of such a computer. In fact, of course, some numbers would probably be stored in floating-point fashion (p. 8) and some registers would contain coded

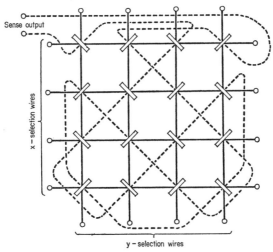

FIG. 7. Arrangement of cores and windings in a core-memory plane (schematic). Sixteen core are shown, arranged as a 4 by 4 array. If the current needed to saturate a core is denoted by I, then a current of $\pm I/2$ in any one of the *x*-selection wires and the same in any one of the *y*-selection wires will saturate only the core lying at their intersection. The state of a core may be sensed by applying a current of $+I/2$ to the appropriate pair of input lines and sensing the magnitude of the output voltage on the sense-winding.

instructions, the details of which we shall study later. In general, therefore, the number of planes making up the core-memory will determine the number of bits in each computer word, while the number of cores per plane will determine the number of words that can be stored. Each word is identified by a unique *address*, a number which defines the location at which the word is stored. The physical characters of the storage elements and associated circuitry also decide the speed at which a register can be written into or read from. The time taken to obtain a word from store—the so-called *access-time*, of the order of 1 μsec in modern computers—is also an essential factor in determining the speed at which the machine will operate.

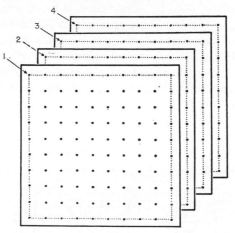

FIG. 8. Schematic representation of four core-memory planes forming a greatly simplified central storage unit. The large dots correspond to individual ferrite cores. Each plane contains 100 cores so that 100 'words' could be accommodated. The cores numbered 1, 2, 3 and 4 hold the four binary digits that make up a given word.

Faster than a magnetic-core memory, but more expensive and therefore less widely used is the *thin-film memory*, in which the basic unit is a very small, thin circle or rectangle (a "dot") of some magnetic material such as nickel-iron alloy. This acts as a small magnet in which the direction of magnetisation can be changed by an applied magnetic field so that a binary digit can be stored. Individual dots can be organised into planes as in the magnetic-core memory, so that the mechanism of storage is essentially similar in the two types.

Repeated reference has been made to the storage of instructions. The form in which these are stored depends on the computer, but a simple example is the arrangement in the IBM 7094 where the core-storage unit has a capacity of 32,768 words, each of 36 bits. An instruction in this machine consists of two essential parts, a binary-coded number specifying the operation to be performed—addition, multiplication, etc.—and a second binary number stating the address of the number to be operated on. These are stored in some such manner as Fig. 9. Bit positions 0–11 accommodate the binary

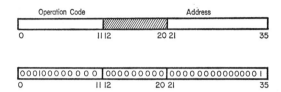

Fig. 9. Schematic representation of a computer instruction as held in the central storage unit of an IBM 7094 digital computer. The upper figure indicates the general instruction format, the lower one is the format for the operation ADD, performed on the contents of register 1.

code for the operation of adding an integer to one of the arithmetic unit registers (000100000000) while bit positions 21–35 inclusive hold the binary version of the address whose contents are to be added. In Fig. 9 this address is core-location 1 and it will be seen that the highest possible number for an address in this machine is $2^{15} - 1$ ($=32,767$), i.e. a 1 in each of the bit positions 21–35. The remainder of the 36-bit word in this computer (i.e. bit positions 12–20) is reserved for other purposes. Different computers may have different instruction formats—some, for example, may have two or three addresses—but the principle is similar and indicates how instructions are stored in essentially the same way as the numbers on which they operate (Fig. 10).

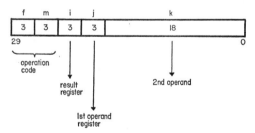

Fig. 10. Schematic representation of a 30-bit instruction format for the central processor of a CDC 6600 computer. Two such instructions comprise a 60-bit word in central memory. Other instruction formats also occur in this computer.

The three kinds of auxiliary storage devices listed on p. 17 are similar in several respects. In all cases they consist of movable surfaces coated with a material divided into tracks which can be subjected to local changes in the direction of magnetization. Each track is thus packed with a sequence of binary digits read or written by *read–write heads* placed in close proximity to the surface.

(b) *Magnetic Drum Storage*

These are cylindrical units on the surface of which are a series of parallel tracks, one of which is usually used to time the rotation of the drum while the others are available for storage. A drum of this kind might have 60 tracks (in addition to the timing track), each 60 inches in length and containing digits packed at 200 per inch. It would therefore be able to hold $60 \times 60 \times 200 = 720{,}000$ bits. The speed with which the drum operates depends on its rate of rotation. If it rotates at 10,000 rpm then any particular bit could be located under a read–write head in an average of 1/20,000 minutes = 3 msec. This time could be reduced if more than one head were placed above each track. Of greater importance to programmers, however, is the fact that access time can be greatly reduced if it can be arranged for the data to be read or written in a strictly sequential order, only the minimum movement of the drum then being needed between two successive bits. Under these conditions a drum with the physical characteristics we have indicated could allow 12,000 bits to be read in 6 milliseconds, the time of a single revolution. The same principle also applies to the other two auxiliary storage devices discussed below. A computer word consisting of, say, 60 bits can be read or written on drum storage in two ways. In parallel operation each of the 60 tracks is used to provide the 60 bit positions while in series operation each track is divided into consecutive sectors, each of 60 bits.

(c) *Magnetic Disks*

A magnetic disk unit comprises one or more rotating disks, rather like large gramophone records in appearance. When several disks are involved they are stacked one above the other, separated by spaces in which the movable read–write heads can be placed. The surface of each disk is divided into a series of concentric circular (not spiral) tracks, of which there may be several hundred, each accommodating binary digits at a density of perhaps 500–1000 b.p.i. (bits per inch). The disk rotates at speeds of the order of 1000 rpm and the storage capacity and access time can be estimated from these parameters, bearing in mind that the read–write heads need to be positioned above the correct track by a movable arm.

(d) *Magnetic tape*

Here the storage medium comprises a thin plastic ribbon, usually $\frac{1}{2}$–$1''$ wide and with some 7–14 tracks running along its entire length. Information is stored at various densities, commonly 200, 556 or 800 b.p.i. The tape is moved by a relatively elaborate *tape-drive unit* able to stop and start very quickly and to transport tape at speeds of perhaps 200 inches per second. The major drawback of tape is the relatively long access time—to locate data near the end of the tape all the earlier information must be passed over. To use the same data twice in a program means rewinding the whole length of tape on which it is recorded. Ingenious programming may minimize access-time, however, and the great convenience of readily portable reels of tape accounts for its popularity.

From an operational point of view, magnetic tape has a great advantage over cards or paper-tape as an input–output medium, since it can be read very much more rapidly. Large bodies of data, which may need to be run in conjunction with several different programs and are constantly being kept up to date, are therefore usually stored on magnetic tape, the transfer from punched cards to BCD tape being accomplished through a simple computer. Extensive files can also be stored on magnetic disks and the programming operations on these are often carried out in exactly the same way as on tape, using the concept of a "symbolic" input–output unit. According to the computer installation, this symbolic unit may be either tape or disk and there is a further distinction between a *user tape*, which can be dismounted and on which information is stored semi-permanently for repeated use, and a *scratch tape* or *scratch area* on disk which is used as a temporary backing store during the execution of a single program, beyond which the information is not preserved. Tapes on which data are recorded and which are used only as input media can be *file-protected*, i.e. they can be mounted in such a way that other information cannot inadvertently be written on to them; such an operation would, of course, obliterate the data they originally carried. Further details of the use of magnetic tape in programming are given on p. 449.

Arithmetic Unit: This comprises a small number of registers in which, under the control of the stored program, all the arithmetical operations of the computer are carried out: addition, subtraction, multiplication and division. The numbers on which these operations are carried out, known as operands, are obtained as required from the central storage unit and the results of the operations are returned to the storage unit. The arithmetic unit registers in which temporary storage or manipulation occurs usually take the form of arrays of electronic circuits known as "flip-flops", associated with which are further relatively complicated circuits. The details of these are available in books such as that of Bartee (1966) but are of little direct concern to the Fortran programmer. Of rather more interest to him is the organisation of the arithmetic unit into registers—which vary considerably in number and function from one computer to another—and the kind of basic operations performed in them. It is unfortunately difficult to generalize on the arrangement of registers, but the basic operations they perform are more uniform and include the following:

(a) a register may be *cleared*, i.e. set to zero.

(b) the contents of a register may be *complemented*. This is a process which may be used to represent a negative number. An example is the use of the "nines complement" in a machine operating on a binary-coded decimal system. The nines complement of a number is the result obtained when each digit is subtracted from nine, e.g., the nines complement of 23 is 76, that of 105 is 894. To illustrate its use, suppose we wish to subtract 23 from 98. This could be done by writing 23 as its nines complement, 76, *adding* 76 to 98 giving 174 and then transferring the carry figure (in this case 1), which is added to the least significant digit to give $74 + 1 = 75$. The advantage of this somewhat roundabout procedure is that subtraction can be carried out using only the same circuitry as is needed for addition. In a binary computer, of course, the equivalent of the nines complement is the ones complement, whose use is essentially similar. Other systems of complementation are also available.

(c) the contents of a register may be *shifted* to the left or right. Thus in a five-digit register the binary number 00101 ($=5_{10}$) would, on being shifted once to the left, become 01010 ($=10_{10}$). That is, it has in effect been multiplied by 2, the base of the binary system. Conversely a shift to the right would be equivalent to dividing by 2.

(d) the contents of a register may be *transferred* without change to another register.

(e) the contents of one register may be *added* to or *subtracted* from the contents of another register.

From these basic operations it is possible to construct methods, not only of adding or subtracting two operands from the central store, but also of carrying out multiplications and divisions. Other mathematical operations such as exponentiation (raising to a power) and the evaluation of logarithmic, exponential, trigonometric and other functions can also be reduced ultimately to addition and subtraction of numbers. The important point to grasp is the small number of basic operations and the fact that these can be combined and repeated in various ways to execute more elaborate instructions.

A very simple example will illustrate how two integers with the same sign are added in the IBM 7094 computer and at the same time introduce the important registers in the arithmetic unit of this and similar machines. The first number is located in central memory and transferred first to the storage register and then to the *accumulator*, replacing whatever may previously have been in the latter. The sign bit is, of course, transferred at the same time. The second number is then located, transferred to the storage register and thence to the *adders*, where it is added to the number from the accumulator, to which the sum is finally returned. From the accumulator the sum may be transferred to a location in central storage. It is worth noting that a register used to accumulate a result may be required to accommodate a number larger or smaller than it is able to. For example, if the number is represented in floating-point form, the greatest absolute value the exponent can have is determined by the number of bit positions in that part of the floating-point word. If in a particular computer the exponent can vary between $-m$ and $+n$, then any attempt to store an exponent smaller than $-m$ or larger than $+n$ will be signalled as an *underflow* or an *overflow* respectively. A message to this effect may then be produced by the computer, warning the programmer of what has happened.

When the arithmetic unit is required to carry out several operations to obtained a result, it is obviously advantageous for as many as possible of these to be performed concurrently ("in parallel") rather than in sequence, thus reducing the overall time for the computation. Similarly, in order to perform the relatively complex processes of multiplication there are many different ways in which the basic operations can be combined to produce a given result: the fastest is obviously the most desirable, but it is generally also the most expensive. These and many similar questions are the concern of the computer designer rather than the programmer, who must accept as it stands the electronic system which executes his programs.

Output Devices: Many of the results obtained in the course of computation are intermediate values, used for later stages of the calculations but of no interest in themselves. Other results are the final numerical answers which the user requires. Both kinds of results are stored in the central memory of the computer but the programmer will only normally require the second kind to be divulged. They are made available on a variety of output devices operating under the control of the program and carrying out processes which in many respects are the reverse of those proceeding in the input devices. Like the latter, output units may operate on-line or off-line (p. 16), and, as they are much slower than the central computing unit they are linked to it by some sort of buffering system. The main kinds of output devices are:

(a) the *high-speed printer*, which prints outs a permanent set of results on a long folded strip of paper;

(b) *high-speed card* or *paper-tape punching machines*;

(c) visual devices based on cathode-ray tubes, providing a temporary display of information; and

(d) magnetic tape equipment.

(a) *High-speed printer*

This is the mechanism through which output is usually made available to the user. Basically these machines resemble a typewriter, in that a set of metal type-faces press an inked ribbon against a paper surface. They operate at relatively high speeds, however, and may produce from a hundred to over a thousand lines of output a minute, using alphabetic and numerical characters as well as a range of common punctuation marks and simple mathematical symbols. The high speeds of operation are made possible in several ingenious ways. An example is the wheel-printer, in which the raised characters are distributed in bands around a constantly revolving drum. The number of bands is equal to the number of print-positions in a line, each band containing a complete set of characters which is rotated mechanically to the symbol required. Printers operate through binary-coded decimal information delivered from the computer; this is decoded and actuates the necessary machinery. The physical characteristics of the printed output vary from one machine to another, but most Fortran programs control a printer able to print 120 characters to a line 12 inches long. There are 60 such printed lines per page and the program must include all detailed instructions for the page lay-out.

(b) *Punched-card or paper-tape output*

Here the output device is a card-punch or paper-tape punching machine which is operated not by hand but through the receipt of coded information from a computer. The cards or tape issuing from such a device are, of course, similar to those which could be produced by hand, though they are produced more rapidly. If they are to be read they must be put through a listing machine. Punched-card or paper-tape output may be employed in a variety of situation, of which the following are typical: (i) the results may consist of normal decimal figures to be used as the input to another program, so that the user is saved the trouble of having to punch the data by hand. (ii) The output may consist of a Fortran program which has been stored in, say, BCD form on disk or magnetic tape and which is thus made available to a user requiring a deck of cards. In this case the output would be identical with that resulting from a hand-written program punched by an operator. (iii) The output may consist of a program converted into binary form exactly comparable to that in which it is represented in the central storage unit of a computer. Such *relocatable binary cards* can be used in place of the ordinary Fortran deck whenever the program is to be run again with different bodies of data. The significance of this important feature will be better appreciated after the reader has understood the process by which a Fortran program is "compiled" (p. 30).

(c) *Visual display devices*

These are less commonly employed, but can be used for displaying curves, the co-ordinates of which have been calculated by the computer. Apart from the fact that it is controlled by the output of a computer, such a device is essentially similar to the cathode-ray oscillograph used in electronic or electrophysiological work. A second type of visual output device employs a specially constructed tube that permits the visualization of alphabetic or numerical symbols and thus allows messages or sets of numerical results to be displayed. Rather different forms of visual display involve the mechanical plotting of graphs which are traced out by a pen on paper from co-ordinates calculated by the computer.

(d) *Magnetic Tape output*

In off-line operations the immediate output from the central computer may be on BCD magnetic tape which is later interpreted by a second, smaller machine in order to provide the printed output or punched cards that are normally required. More rarely the tape can be used to store data which then form the input to another program.

Control Unit: The function of the control elements is to ensure that the various operations described in the previous sections are carried out in a co-ordinated manner and in the correct sequence. In a general-purpose stored program computer this is done through the interpretation of the instructions that form the program and are stored in coded form in the central memory exactly as are the numerical data. The number of storage locations needed to run a program therefore depends on the number needed to accommodate the program plus those needed to hold the data, intermediate variables and results. If this total exceeds the capacity of the high-speed central memory, then some or all of the data will have to be accommodated in one of the auxiliary storage devices already described, leaving the central store for the instructions and intermediate results. The organisation of instruction words was outlined very briefly on p. 21 and varies appreciably from one computer to another. In a so-called single-address computer it must contain at least two sections, the *operation-code* (OP-code) defining the operation to be performed and the *address* part indicating the location whose contents are being operated on. In a two-address machine the instruction word contains the OP-code and two sections for addresses. One of these will contain the address of a number being operated on while the other may contain the address of a second operand or perhaps the address of the next instruction which the computer is to use. Three- and four-address computers also exist.

The number and nature of the instructions vary from one computer to another according to the design of the circuits and the operations which have to be carried out. In the CDC 6600, for example, there are 64 instructions for the central processor and another 64 peripheral and control instructions. Despite the great differences between the instruction sets of different computers, there are many basic instructions common to all, such as adding, multiplying, shifting, branching to other instructions and so on. Each instruction is commonly referred to by a mnemonic, alphabetic code-name, but is, of course, represented internally by the numerical OP-code.

The basic pattern of operations in many computers comprises an alternation of instruction cycles and execution cycles. In the *instruction cycle*, an instruction word is obtained from the central storage and interpreted, while in the *execution cycle* the operand is obtained and the instruction performed on it. In order to accomplish these processes a certain minimum number of registers is needed. An *instruction counter*, for example, records the addresses of the successive instruction words used, either incrementing the address by one during each instruction cycle or transferring to another instruction address as required by the program. An *operation-code* register stores the OP-code section of each instruction word, while a *memory address register* contains the location of the central memory from which a word is to be read or into which one is to be written. Finally, a *memory-buffer register* acts as an intermediate between the central memory and the other registers of the control and arithmetic unit. Inevitably, this is a very schematic outline of the arrangement of registers and the way in which instructions are handled and executed. To provide greater and more realistic detail, however, would entail going into the often very sophisticated design of individual computers and this is not necessary for the type of programming envisaged here.

Programming Languages

So far we have considered instructions only in terms of their internal representation as instruction words. We have also indicated how, on the one hand, each instruction forms a building block from which more complicated arithmetical process can be constructed while on the other it comprises a set of basic electronic operations in the circuits and registers of the machine. It is now time to trace how the instructions may be conveyed to the computer by a human programmer. The most direct way would clearly be for the programmer to prepare a sequence of instructions written in the same language as they are stored internally in the computer. Since computers differ in the number and design of their instruction words, this language will be peculiar to a particular model of computer. It is also a relatively long and tedious process to formulate complicated calculations in terms of the simple basic instructions, and it is particularly tiresome to keep track of the addresses of registers used to store variables. We can, however, imagine a very simple program intended to add two numbers, $c = a + b$, written in *machine language* somewhat as follows (except that for simplicity we shall use decimal notation for numbers which would be represented internally in binary form). The first column gives the address at which the instruction-word is stored, the second an invented OP-code number.

Address of instruction	OP-code	Address of operand	Operation
1	21	400	Clear accumulator and add to it the number stored in location 400
2	25	401	Add number stored in location 401 to contents of accumulator
3	32	500	Transfer contents of accumulator to location 500

Before such a program is executed location 400 must contain the number a, and location 401 the number b. When the program has been executed the sum of $a + b$ will be stored in location 500. Note especially that we can never refer to the variables a, b and c by name or by the actual values they may have in our computation; they are dealt with entirely in terms of the addresses of the locations where they are stored. Initially, of course, all instructions have to be made available to a computer in exactly this form. One of the striking features of the computer, however, is its capacity, when suitably programmed, to translate into this cumbersome machine language a program written in a more convenient form. The latter is then known as a *source program* and its machine-language equivalent as an *object program*.

The first step in this direction was the development of *assembly languages*. These resemble machine languages in that each statement of an assembly language generally corresponds to one instruction-word of the machine language. The assembly language statements are read into the computer and there translated automatically into machine language by a special program called the *assembler*. The latter has previously been stored in the computer and is, of course, itself written in machine language. The assembly language is much easier to use because of the following features:

(a) the binary OP-code is replaced by a mnemonic group of letters such as

CLA for 'clear and add' or FMP for 'floating multiply' or STA for 'store address'.

(b) Instead of giving the numerical address of each item of data, the programmer using an assembly language can refer to the variables by symbolic names—A, B, X, etc.—knowing that the correct addresses will be assigned automatically to each name by the assembler.

(c) Data can be written in a convenient form such as decimals and once again the assembler will translate them into binary form.

(d) The assembler will also make available a listing of the source and object programs so that they can be checked and it will detect certain errors and issue a diagnosis.

The very simple set of instructions on p. 29 could now be written as part of an assembly language program in some such form as:

```
CLA   A
ADD   B
STO   C
```

The assembler will assign addresses to the variables and see that the appropriate numerical values are stored in the correct locations when the data are read into the computer.

Despite their greater intelligibility, assembly languages are not easy to use and have the serious drawback that they are closely bound up with the instructions used by a particular model of computer. A great advance in programming techniques was therefore made possible by the development of *compiler languages*. These languages—of which FORTRAN is one—are much closer in appearance to English or to a mixture of English and algebraic notation. Like assembly languages they are translated into machine language by a special translating program which in this case is called a compiler program or, more simply, a *compiler*. One important difference between an assembly language and a compiler language is that each statement of the latter is no longer equivalent to a single instruction-word but may cause a number of machine instructions to be generated. The compiler language is therefore said to consist of *macro-instructions*. The result is that compiler languages are no longer dependent on the detailed structure of the computer. A Fortran program, for example, can be run on any suitable computer provided it has been equipped with a Fortran compiler.

The position is, therefore, that compiler languages are much easier to learn, have very wide applicability and are the medium in which the majority of scientific programs are written. When read into a computer in which a suitable compiler has first been stored, the compiler language program (source program) is first translated into a binary object program and the latter is then executed by the computer. Though FORTRAN in its various forms is one of the most important compiler languages it is by no means the only one. ALGOL is widely used in Europe for similar mathematical and scientific purposes while PL/1 (short for Programming Language One) has been developed more recently in the United States and may well become very popular. COBOL is used extensively for business and commercial purposes and there are a variety of other compiler languages adapted to rather specialized uses.

Thus, while the greater part of this book is concerned with the use of Fortran one must bear in mind that the versatility of this and other computer languages rests ultimately on the structure of the computer and the availability of an appropriate compiler. The compiler in turn is one of many programs forming part of the system which enables Fortran programs to be run. Such a system must be built up and maintained by the *systems programmer*, a great deal of whose work is concerned with assembly languages and machine structure and functioning. The programmer who works largely in a compiler language like Fortran is essentially an *applications programmer*, using his highly developed language for mathematical, statistical and scientific purposes and taking the existence of the system for granted.

Chapter 3

Synopsis of Fortran Programming

The present chapter includes an outline of the more important features of FORTRAN IV. It is obviously not intended to compete with the excellent Fortran IV manuals by Golden (1965), McCracken (1965), Organick (1966) and Dimitry and Mott (1966). These books—together with the many computer manufacturers' handbooks—must be consulted for comprehensive accounts of the language and some of its 'dialects'. The object here is rather to summarise only the essential techniques which all programmers need to have at their fingertips. Many of the refinements are used only rarely, even by professional programmers, and none of the non-standard features peculiar to certain machines will be mentioned, however useful some may occasionally be. There have been several distinct versions of Fortran since it was first introduced. Fortran II is still quite widely used but the later and generally current form of the language known as Fortran IV is adopted here, based as far as possible on the *Standard Fortran Programming Manual* of the National Computer Centre (1970).

Planning a Program

Before dealing systematically with Fortran IV it seems worth outlining the stages which lead to the production of a program. This involves decisions on the principal results required, the kinds of data on which the program will operate and the extent to which it might be generalised. The latter is an important aspect of planning since it extends the value of the program well beyond the immediate reasons for writing it. Secondly it is necessary to decide on the mathematical and statistical procedures which will be used. The choice of a computational method or *algorithm* is extremely important and is discussed in a little more detail elsewhere (p. 441). Having chosen a method it is necessary to consider how far it is suitable for use with a computer, how far it fulfils any accuracy considerations one might have in mind, and whether possible alternative procedures have been assessed with sufficient care. Biologists who begin programming with an inadequate mathematical or statistical background are obviously at a disadvantage at this stage and may well need to gain wider experience of the purely mathematical techniques before they can write adequate programs. While this book may help them, it is no substitute for a careful study of the many available works on mathematical methods, numerical analysis and statistical theory. Some of these are mentioned at appropriate points in the text.

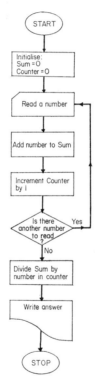

FIG. 11. Flowchart for finding the arithmetic mean of a set of numbers.

Having selected an algorithm suitable for use with a digital computer, it is now necessary to plan the detailed logical structure of the program. This operation is most conveniently carried out through the construction of *flowcharts* and is an essential preliminary that is to a considerable extent independent of the programming language used, though it naturally helps if one knows how the flowchart will eventually be coded. It is worth reiterating here that while most people can learn Fortran coding quite quickly, it is much more difficult learning how to set out in sequence the very simple logical steps from which a program must be constructed. For this reason detailed flowcharts are given and discussed for all the programs provided in later chapters of this book. Careful study of these will show what sort of problems arise repeatedly and how they may be solved. Two very simple illustrations will have to suffice at this point. Suppose first that we wish to find the arithmetic mean of a set of numbers. One way in which a program could be constructed is shown in the flowchart of Fig. 11. The points to notice are, first, that the process involves *initialization*, in this case setting two variables equal to zero. One of these will eventually provide the sum of all the numbers, the other will record how many numbers are being read. A *loop* is then set up so that numbers are read one by one, added to the variable 'sum' and the counter increased by 1 each time. The loop contains a diamond-shaped *decision box*

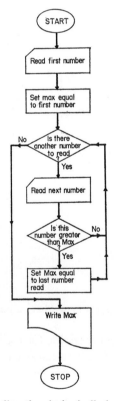

Fig. 12. Flowchart for finding the algebraically largest of a set of numbers.

where we ask each time whether there are any more numbers left to read. If the answer is 'yes' we complete the loop again; if it is 'no' then 'sum' now contains the total and dividing it by the number in the counter will give the required arithmetic mean. The arrows in the flowchart indicate the sequence in which the operations are carried out and from a flowchart of this kind a workable Fortran program can easily be written.

A slightly less obvious flowchart is that in Fig. 12 where we indicate how to find the largest number in a set of many numbers. We can decide to call this maximum number Max and we initialize by making it equal to the first number read. We then set up a rather more complicated loop involving two

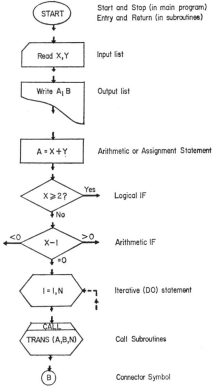

FIG. 13. Flowchart symbols used in this book. Further explanation of the operations denoted above is given in the text at various points in Chapter 3.

decision boxes. The second of these ensures that Max will always be increased to equal the largest number so far read, while the first shows where to terminate the looping process after the last number has been read.

The difficulty encountered by the beginner in constructing flowcharts arises mainly from uncertainty as to what constitutes an elementary logical step in the diagram. Only a full knowledge of what single operations are permitted by the programming language can eliminate this uncertainty, and for this reason facility in flowcharting will rapidly increase once one becomes familiar with Fortran or a similar compiler language. Indeed, an experienced programmer can usually dispense with the more obvious features of a full flowchart and will probably construct one in which only a few critical steps are set out explicitly. For the present, however, it is recommended that flowcharts should be as complete as possible. The logical structure of a program should be established *in detail* before coding can usefully begin.

Many elaborate conventions for flowcharts have been described but they need not be taken too seriously. In this book the main conventions are very simple and are indicated in Fig. 13. This is intended mainly for reference; the exact meaning of some of the procedures will be discussed later in the text.

FORTRAN IV

Fortran characters. Like any language, Fortran is written in only a limited number of *characters* or symbols, 47 in number. These are the ten numerical characters 0–9, the twenty-six alphabetical characters A to Z (all written as capital letters) and a group of punctuation and other characters $+ - * / . , ( ) = $ ' and \$. The numerical and alphabetical characters are together known as alphanumeric or *alphameric characters*, the remainder as *special characters*. Of these the apostrophe ' is used only under special circumstances (see H-specification, p. 51) and the \$ sign is rather rarely needed. Characters other than the above are invalid and their use will normally cause a program to be terminated, as may also the inappropriate use of certain valid characters. The numerals 1, 2 and 0 should be carefully distinguished from the letters I, Z and O. This may be done by writing the latter as I, $\mathcal{Z}$ and $\emptyset$ respectively. Apart from their occurrence in H-specifications, there are no rules governing the use of spaces—they can be employed freely to improve the legibility and appearance of a printed program.

Statements and lay-out. A Fortran program consists of a set of *statements*. These are constructed from the above characters in accordance with rules specified below and they should preferably be written out by the programmer on a standard Fortran coding form. Such a form comprises a series of lines, each divided into 80 columns. Each statement begins on a new line and normally occupies part or all of columns 7 to 72, one character per column as in Fig. 14. Columns 1–5 are reserved for the *statement number* which must accompany certain statements. Column 6 is the so-called *continuation column*. It normally contains nothing, but if a statement is too long to be accom-

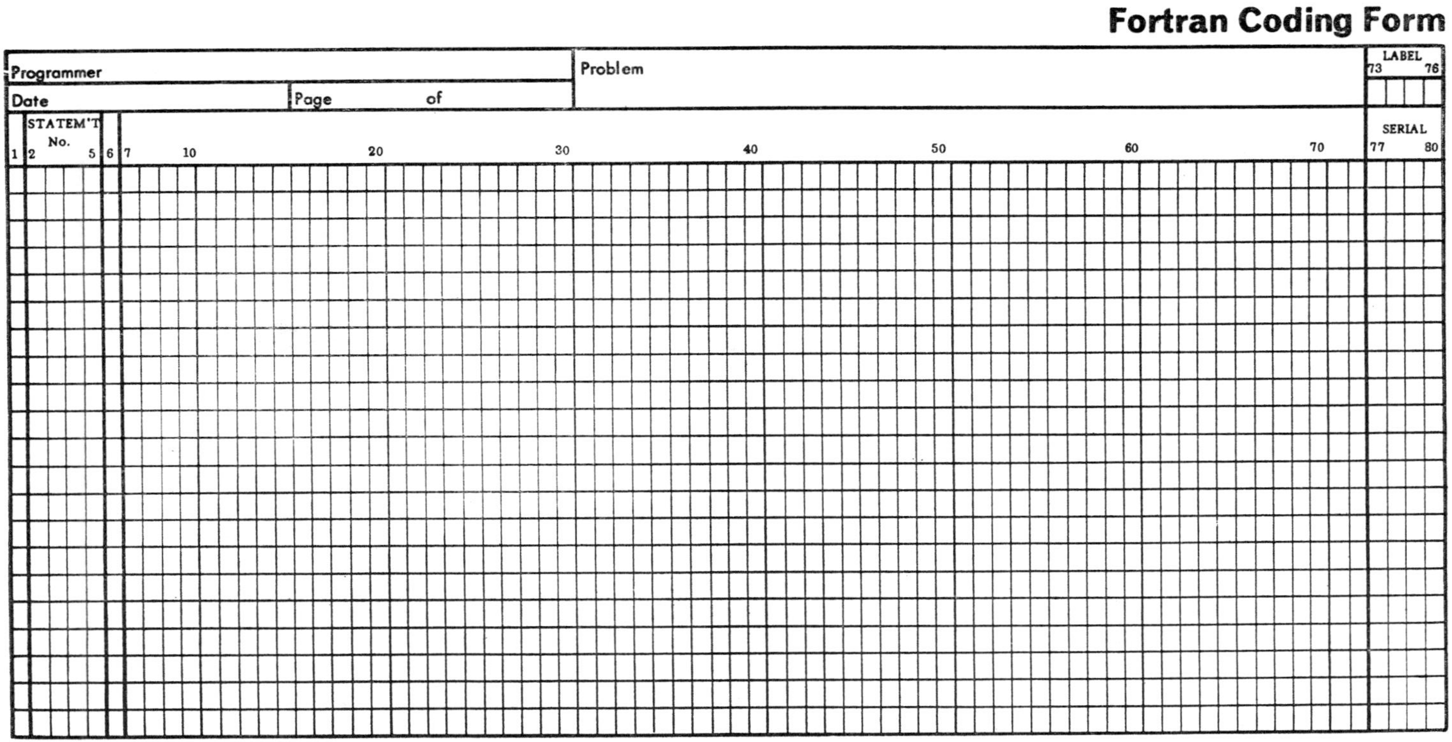

FIG. 14. A Fortran coding form. Each line on the form corresponds to a separate punched card, each column of the form to a separate card-column.

modated in a single line it may be continued on to columns 7 to 72 of one or more additional lines, each containing a non-zero character (e.g. 1, 2, 3 ... or A, B, C ...) in column 6. When a program is punched on to cards, each line of the written program is punched on to one card, the columns of the coding form corresponding to those of the punched card. The permitted number of continuation cards (i.e. those with a non-zero character in column 6) varies according to the compiler, but at least 9 are almost always permitted. Columns 73–80 of the punched card are not read by the Fortran compiler and may therefore be used by the programmer to identify the cards. As a rule they contain a sequence of numbers so that if the cards become accidentally shuffled they can readily be sorted into the correct order.

The statement number is an unsigned integer from 1 to 32767 (or even higher in some computers) and the numbered statements need not be in sequence. All statements requiring a number must have one and a given statement number cannot be used for more than one statement in a program. If a letter C occurs in column 1 the compiler will not read that card and the line is available for the programmer to write a *comment*, intended to make the program intelligible. These comments are, of course, printed out with the rest of the program when it is listed by the computer. Programmers are encouraged to make full use of intelligently worded comment cards—a program without them may be virtually unintelligible (even to the person who wrote it a few months previously!).

Constants. Constants are numerical quantities whose value is fixed throughout a program. They may be written in two forms, as *integer constants* or as *real constants*. Integer constants (also known as *fixed-point constants*) consist of a certain number of digits without a decimal point and without commas marking off the thousands, millions, etc. Examples of fixed-point constants are 21 or 21829 or 7. The maximum number of digits varies with the computer but up to 10 digits can be used on almost all large machines. Real constants (also known as *floating-point* constants) consist of 1 to 9 significant decimal digits with a decimal point, e.g. 21.35 or 0.0003. It is not necessary that a digit follows the decimal point—21. has the same significance as 21.0 has. Remember too that leading zeros do not constitute significant figures so that .00000123456789 is valid real constant, but 123456789000.0 is not (for most compilers) since the three zeros before the decimal point are significant figures. Integer and real constants can assume negative values and must then be preceded by a minus sign—e.g. −2.4 or −3, but the use of a plus sign before a positive constant is optional (e.g. +3.1416 is the same as 3.1416).

Floating-point constants can also be expressed in *exponent* form. This consists of an integer or real constant as defined above, followed by the letter E and then by a 1- or 2-digit integer constant denoting the exponent.

For example,

    2E2 means $2 \times 10^2 = 200$
    2.145E–3 means $2.145 \times 10^{-3} = .002145$
    $2.E+12 = 2 \times 10^{12} = 2,000,000,000,000$

Some further examples of valid constants are given below:

    Integer: 215713
                       –2
    Real: 2.
                       312.715862
                       –0.000013
                       +2.0
    Real (exponent form): 2.1735E02
                               2.128E+3
                               +1.0E–10
                               –1.0E+10
                               2E–3†

The following forms are invalid for the reasons given:

    2,000 (comma not allowed; should be 2000)
    21.0E1.5 (exponent must be an integer constant)
    7318596.21843 (exceeds 9 significant figures; valid only for some compilers)
    12.1E136 (exponent of more than 2 digits; valid only for some compilers).

Variables. A Fortran *variable* is represented by any combination of up to 6 alphameric characters, of which the first must be alphabetic. Examples are ALPHA, AX2, Y3, SUM, ROOT or NPROB. If a variable begins with the letters I, J, K, L, M or N it is an *integer variable* and will normally assume only integer values. If it begins with any other letter it is a *real variable* and will normally have only real values. This 'implicit typing' of variables by their initial letter can, if necessary, be over-ruled by the programmer with a special *type-declaration* (p. 74), in which case the variable is described as explicitly typed. It is desirable for reasons of intelligibility that variable

† This form is valid in Fortran IV. In Fortran II it would have to be 2.E–3 or 2.0E–3.

names should be mnemonic wherever possible, e.g. CHI2 for 'chi-squared' or SX for 'sum of x'. A variable name devoid of mnemonic significance is, however, quite valid provided it conforms to the rules above. Some examples of invalid names are:

VELOCITY (more than 6 characters)
3AB (begins with a numerical character)
LB/3 (contains a special character)

A special kind of Fortran variable, discussed below in greater detail, is the *subscripted variable*. Here a variable name (real or integer in type) is followed in parentheses by up to three integer constants or integer non-subscripted variables, separated by commas. Some examples of these are:

A (1), BETA (K), AN (I,J), X (J,K,L), ALPHA (1,2,3).

Arithmetic Expressions. The following six rules (a) to (f) must be observed:

(a) An arithmetic expression is a sequence of Fortran variables and/or constants joined by one or more of the following operational symbols, indicating the arithmetical process to be carried out:

+ denotes addition
− denotes subtraction
* denotes multiplication
/ denotes division
** denotes exponentiation (raising to a power)

Examples, with the corresponding algebraic expressions, are:

A + B	$a + b$
2.38 − ALPHA	$2.38 - \alpha$
X * Y * X − 3.0	$xyz - 3$
Y ** 3.21	$y^{3.21}$
Z/6.2832	$\dfrac{z}{6.2832}$

(b) When two or more operational symbols occur in an expression, they are evaluated in an order of priority that places exponentiation first, followed by multiplication or division (equal in priority), and then by addition or subtraction (also equal in priority). When two or more operations of the same priority occur, they are usually evaluated in turn from left to right. For example,

$$3.12 - \text{ALPHA} ** 3 + \text{X1} * \text{X2} - \text{A}/2.0$$

is evaluated as follows:

(i) the current value of ALPHA is raised to the power of 3.
(ii) X1 is multiplied by X2
(iii) A is divided by 2.0
(iv) From 3.12 is subtracted the value resulting from (i); to this is added the result of (ii); from this is subtracted the result of (iii).

(c) The above order of priorities and left-to-right evaluation can be varied by the use of parentheses. Expressions enclosed in brackets are evaluated first and several sets of parentheses can if necessary be 'nested' one within another. In the latter case the expression in the innermost parentheses is evaluated first, then the next innermost, and so on. An example is:

$$\text{A} * (\text{X} + 1.0)/(\text{B} - \text{C} ** 3) \quad \text{i.e.,} \quad \frac{a(x+1)}{b - c^3}$$

Here the expressions $(X + 1.0)$ and $(B - C ** 3)$ are first evaluated separately and then combined according to the multiplication and division operations. An example with nested parentheses is:

$$\text{A} * (\text{X} + 1.0 - 3.5 * (\text{B} + \text{C} ** 2))$$

Note that the brackets alone do not ensure multiplication—a * symbol must also separate the two expressions being multiplied, as in $3.5 * (B + C ** 2)$. Note also that the nested parentheses must balance, i.e. there must be equal numbers of right and left parentheses.

(d) All the quantities (variables or constants) in an expression must be of the same type (all integer variables and constants or all real variables and constants), e.g. M + N * IPQ/3 or 3.1416 * R * R (a suitable formulation of the expression πr^2). Invalid expressions would be M/2.3 or ALPHA $-2 * B1$. This rule is sometimes expressed by saying that mixed-mode expressions are not allowed. Some compilers do, in fact, allow them but they are undesirable for two reasons: they prevent the program from being run on less accommodating computers and even when allowed they take longer to evaluate than unmixed modes.

Note that when one integer variable or constant is divided by another, the result is truncated, i.e. if the answer includes a fractional part this is lost.

e.g. 6/4 is evaluated as 1 (the .25 of 1.25 being lost by truncation to give an integer result). 1/5 would be evaluated as 0. For this reason division by integer variables must always be considered with care. The result may depend on the order in which two or more operations are carried out. Attempts to divide by zero (whether in fixed-point or floating-point arithmetic) have various effects, some computers stopping execution after printing an error diagnosis. It must also be remembered when writing expressions involving division that division by zero will occur if the divisor is a variable that happens on occasion to assume a zero value.

(e) One exception to the rule concerning mixed modes occurs in relation to exponentiation. If A and B denote real variables and I and J denote integer variables, then the valid expressions are:

$$A ** B$$
$$A ** I$$
$$I ** J$$

but not I * A. That is, an integer variable can only be raised to an integer power, whereas a real variable can be raised to a real or an integer power. Further, such expressions as A ** (B + 2.0) or I ** (J + K − 2/M) are all valid. On the other hand, A ** B ** C is not a valid expression: it must be rewritten either as A ** (B ** C) or (A ** B) ** C, whichever is intended.

(f) No arithmetical expression can contain two successive operational symbols unless parentheses are employed. This rule is particularly important when operations involve negative quantities. Thus the algebraic expression $(a + 1)/-b$ cannot be rendered as (A + 1.0)/−B but only as (A + 1.0)/(−B). It must also be noted that an algebraic expression such as $a^{3.25}$ cannot be written in Fortran as A**3.25 if there is any likelihood of A assuming zero or negative values. This is because exponentiation involving a real power is normally carried out by first taking the logarithm of the base, and logarithms exist only for positive numbers. Similarly (−B**3.0) is impossible to evaluate for B ⩾ 0 though (−B**3) is a valid Fortran expression since the processor will compute it as (−B)*(−B)*(−B). Even when the base is positive an integer exponent should always be written as such (i.e. X**3 rather than X**3.0) since exponentiation by repeated multiplication is normally quicker and more accurate than the use of logarithms.

Mathematical Functions. The above five arithmetical operations often need to be supplemented by the use of logarithmic, exponential, trigonometric or other functions. Fortran makes this possible by the use of special function programs (considered further on p. 67). Many common mathematical functions are incorporated in the system in this way and can be used simply by writing the Fortran code-word followed in parenthesis by an arithmetic expression involving only real variables or constants. Thus, to express $\sin(\theta/2 + \theta^2)$ in Fortran we could write SIN (THETA/2.0 + THETA * THETA). The expression inside the parentheses is known as the argument of the function and a list of commonly available functions is given in Table 1.

Table 1 List of commonly available Fortran mathematical functions

Function	Operation	Fortran name
Napierian logarithm	$\ln x$ or $\log_e x$	ALOG
Common logarithm	$\log x$	ALOG10
Exponential (i.e., Napierian antilog.)	e^x	EXP
Sine	$\sin x$	SIN
Cosine	$\cos x$	COS
arc sine	$\sin^{-1} x$	ARSIN or ASIN
arc cosine	$\cos^{-1} x$	ARCOS or ACOS
arc tangent	$\tan^{-1} x$	ARTAN or ATAN
square root	$\sqrt{x}$	SQRT

This table requires a few comments. Many computer systems include several other common mathematical functions such as the error function, gamma function, log gamma function, hyperbolic functions and so on; others can readily be added through the use of special function sub-programs (p. 67). In some cases the Fortran names are not standardized and the programmer should therefore consult the list normally available at his computer installation. Biologists in particular may like to be reminded of a few mathematical peculiarities of these so-called basic external functions:

(i) The Fortran trigonometrical functions always employ an argument given in angular measure (radians) rather than in degrees. The programmer may therefore need to convert radians to degrees or *vice versa* when using these functions (1 radian = 57.2957795°; 1 degree = 0.01745329 radians).

(ii) Common antilogarithms are, of course, obtainable without a special function since the antilogarithm of x is simply 10^x or, in Fortran 10.0 ** X.

(iii) Certain functions only produce correct results when the arguments conform to the appropriate mathematical form. For example, the computer will usually stop execution if one tries to take the logarithm of zero or a negative number, or the square-root of a negative number, or to find an arc cosine with an argument greater in absolute magnitude than one. In some cases, however, the computer will make a recovery, e.g. it may, when required to take the square-root of a negative number, take the square-root of the corresponding positive number instead and print a warning message while continuing with the calculation. The results may, of course, be worthless in such a case. It should be remembered that the square-root function always provides the positive square-root of a number.

Other Supplied Functions. Certain other common mathematical operations can be carried out by the use of intrinsic Fortran functions that are supplied with the system and can be called in the same way as the mathematical functions listed above. They are given in Table 2.

Table 2   List of commonly supplied intrinsic Fortran functions

Function	Operation	Fortran name	Type	
			Argument	Function
Absolute value	$\lvert x \rvert$	ABS	real	real
		IABS	integer	integer
Truncation		AINT	real	real
		INT	real	integer
Remaindering	$x_1 \pmod{x_2}$	AMOD	real	real
		MOD	integer	integer
Maximum value	$\max(x_1, x_2, \ldots x_n)$	AMAX0	integer	real
		AMAX1	real	real
		MAX0	integer	integer
		MAX1	real	integer
Minimum value	$\min(x_1, x_2, \ldots x_n)$	AMIN0	integer	real
		AMIN1	real	real
		MIN0	integer	integer
		MIN1	real	integer
Float	convert integer to real	FLOAT	integer	real
Fix	convert real to integer	IFIX	real	integer
Transfer of sign	Sign of x_2 times x_1	SIGN	real	real
		ISIGN	integer	integer

Thus, to obtain the absolute value of a variable ALPHA one simply writes ABS(ALPHA) while to find the remainder when ALPHA is divided by BETA one would write AMOD(ALPHA, BETA). Note that the latter function has two arguments which are separated by a comma. Others, such as the minimum and maximum functions may have any number—e.g. MAX0 (L,M,N) will provide the largest of the three variables L, M and N.

Arithmetic Statements. These statements, which form the basis of all calculations, are of the form

$$a = b$$

where a is a real or integer variable, simple or subscripted, and b is any valid arithmetic expression (including one that incorporates functions such as those described in the two preceding sections). An example is:

$$\text{ALPHA} = A * A + \text{SIN} (2.0 * \text{PHI})/\text{ABS (BETA)}$$

The effect of this is to cause the computation of the expression on the right-hand side of the = sign and to store the result under the name of the variable given on the left of the = sign. Note that the = sign does not denote equality. It means, in effect, 'the current value of the variable on the left must be replaced by the value of the expression on the right'. Thus it is quite valid to write $N = N + 1$ or $A = A/2.0 + AK$. These would be nonsense in algebra, but in Fortran the first simply means that the current value of N is to be increased by 1, while the second means that the current value of A is to be replaced by a new one (A/2.0 + AK). In all such cases the original value of the variable on the left is destroyed. If, for some reason, one wished to preserve the original value as well as to obtain a new one, it would be necessary to define a new variable—i.e. to have two storage locations for the two quantities. For examples one could write

$$AA = A/2.0 + AK$$

so that the original value of A remained untouched and the new value would be stored as the separate variable AA. Writing a variable on the right-hand side of an arithmetic statement does not cause its current value to be destroyed. Note, too, that the variable on the left does not have to be of the same type as those in the expression on the right. Thus $N = A * A/2.0$ is quite valid. The result of carrying out the operation on the right in floating-point arithmetic is then stored in fixed-point form as the integer variable N. Conversely, BETA = $(2 \times 3)/4$ would store an integer result 1 as the real number 1.0 in the location BETA. Some examples of invalid arithmetic statements, with reasons, are:

ALPHA + 1.0 = A * B/C − D (left-hand side is an arithmetic expression—it should be a variable only)

Q = AMI/J − AM2 (right-hand side has mixed modes)

12AB = 12.0 * A * B (left-hand side is not a valid variable name).

STOP, PAUSE and END Statements. These are control statements with precisely defined functions. STOP may be placed at any point in the program and will cause the machine to cease execution. There may be more than one STOP statement in a program, though commonly a single one occurs as the penultimate statement. When a program is being run on a monitor system the STOP statement causes the monitor to proceed with the next available job. In such cases the statement CALL EXIT can usually be used in place of a STOP statement.

The statement PAUSE also causes execution to cease, but is used when some action is needed by the computer operator, who is subsequently able to start the machine again by operating the appropriate switch on the computer console. It might, for example, be necessary to mount a certain tape at that juncture. On some large computer installations, pauses of this kind are not allowed.

The END statement differs from the two preceding ones as follows:

(i) A program can contain only one END statement.

(ii) The END statement must always be present as the last statement in the program.

The END statement is not concerned with stopping execution but is a signal to the compiler that the program is complete and that no further statements will occur.

Input–Output Statements. The preceding Fortran rules allow one now to write a very simple program, provided that one can cause the computer to read data and to print out results. A simple account of input–output operations will therefore be interpolated at this point before proceeding to more complicated aspects of Fortran programming. Additional information on input–output techniques will be found in connection with matrix methods (p. 219) and with the storage of data on magnetic tape or disk (p. 449).

If, as is usually the case, one is reading data consisting of characters then the basic input statement is:

$$\text{READ } (i, n) \text{ } list$$

where i is an integer number allocated to an input tape or disk, n is the statement number of an accompanying FORMAT declaration (to be explained later) and *list* refers to a set of FORTRAN variables separated by commas. As a general rule the symbolic number of the input tape is 5 and this number is used in all the programs in this book. Thus, the statement

$$\text{READ } (5, 100) \text{ A, B, C, IJ, OPN, Z}$$

would cause the value of six variables A, B, C, IJ, OPN, and Z to be read from a card or cards where they are punched according to a specification given in an accompanying FORMAT statement which is numbered 100 and which must appear somewhere in the program (usually just after the relevant READ statement). The values of the six variables are then stored by the computer in locations referred to by the corresponding variable names.

Similarly, the basic output statement is

WRITE (i, n) *list*

where i refers to the output tape or disk, normally denoted by figure 6. For example

WRITE (6, 101) A, B, C

would cause the values stored in locations A, B and C to be printed according to a specification given by the accompanying FORMAT statement numbered 101.

The FORMAT statements are thus essential to both the input and output procedures described so far and must therefore be discussed in greater detail. A FORMAT statement has the general form

n FORMAT $(S_1, S_2, S_3, \ldots S_m)$

where n is the statement number and $S_1, S_2, \ldots S_m$ are specifications relating in turn to each of the variables in the list of an associated READ or WRITE statement. The FORMAT specifications are codes which indicate how the numerical data are to be read or written and we shall first indicate three such codes known as the I, F and E-conversions. Each of these consists of a letter followed by a specification of 'field width'. Field width indicates the number of columns of a data-card which are used to accommodate the numerical value being read, or the number of print-positions which are allowed for a number being written. Thus, for the I conversion the input statements

READ (5, 100) LAMBDA

100 FORMAT (I10)

would be interpreted as follows. An integer constant is to be read from a punched card and stored as the current value of the variable LAMBDA. This integer is punched in a field consisting of the first ten columns of the data card (as specified by the code I10 in the FORMAT statement). It is assumed that the number will be punched in a right-justified form, i.e. if the number to be read is, say, 105 it will be punched in the three right-most columns of the 10-column field, so that columns 1–7 are left blank, column 8 contains a 1, column 9 a zero and column 10 a 5. Again, for the statement

>   READ (5, 101) N1, N2, KAPPA
> 101 FORMAT (I5, I10, I10)

we have a list of three integer variables and therefore require a FORMAT statement with three I-specifications. These statements would cause the values of N1, N2 and KAPPA to be read from a single card on which the first number (corresponding to N1) would be punched right-justified in columns 1–5, the second (N2) in columns 6–15 and the third (KAPPA) in columns 16–25. If the I-conversion is used with a WRITE statement, a comparable output process takes place. For example the statements

>   WRITE (6, 102) MULT, NTRANS
> 102 FORMAT (I10, I5)

would cause the current value of the integer variable MULT to be printed out right-justified in the first ten print positions of a line of printed output and NTRANS to be printed similarly in the ensuing five positions (see, however, p. 52 for a qualification of this statement depending on the control of the printer carriage).

The F-conversion (of general form $Fw.d$) is used to specify the field width (w) and the number of digits (d) after the decimal point when a real number is read or printed out. For example F12.6 refers to a real number occupying a total field width of 12 columns (or 12 print positions) and having 6 digits after the decimal point. Another example,

>   READ (5, 103) ALPHA, BETA
> 103 FORMAT (F12.3, F12.4)

would mean that the values of the two real variables ALPHA and BETA were to be read from a single card; the first value would be punched in the first twelve columns, right-justified, and with the understanding that the digits in columns 10, 11 and 12 would be the three digits after the decimal point, while the value of BETA would be right-justified in columns 13–24 with digits in columns 21–24 corresponding to those after the decimal point.

The rather similar E-conversion (general form $Ew.d$) also handles real numbers but they are read or printed in the exponent form mentioned above for FORTRAN constants. Thus, if the statements

>   WRITE (6, 104) ALPHA, BETA
> 104 FORMAT (E10.4, E12.4)

were used, the value of ALPHA would be printed as, say, 0.2318E01 and that of BETA as, say, 0.6148E–3. In each case there are 9 printed characters (including digits, decimal point, sign and letter E) and these would be right-justified in a field width of 12 positions. The person reading the output would, of course, need to understand the convention by which Fortran expresses a constant in exponent form. Provided this is acceptable the E-conversion has the great advantage for output that one can specify a field-width knowing that a result will be accommodated within it. This, it will be noted, is not always true of the putput of quantities under the F-conversion. A programmer who specified that output should be printed under F5.2 would find, for example, that a result of 101287.31 could not be printed in full but might—depending on the computer—appear as 87.31. Output specifications for real numbers therefore require either the use of an E-conversion or of an F-conversion with a field width sufficiently large to cater for any reasonable result.

A few further points may be noted in connection with I, F and E FORMAT specifications:

(i) The separate field-specifications in the FORMAT statement must correspond in type with the variables in the input–output list.

(ii) The numerical values punched on a data card must agree exactly in type, position and field width with the field-specifications under which they are being read. If, for instance, the program contained the statements

    READ (5, 104) NTERM

104 FORMAT (I5)

and the data card being read had the required value punched in columns 10–12, the computer would not read it. Equally, if an integer value is to be read, then an I-conversion is needed while for real numbers an E- or F-conversion is required.

(iii) As explained above, a number being read under the F-conversion need not have a decimal point punched on the card. It is, however, permitted to punch a decimal point and in that case its position will over-ride the position indicated in the FORMAT statement. Thus we could use

    READ (5, 105) AX1, BX2

105 FORMAT (F10.0, F10.0)

yet if the values of AX1 and BX2 were punched (within the correct fields, of course) using decimal points, these values would be read correctly, the punched points over-riding the FORMAT. This applies only to input FORMAT statements, of course, and it is recommended that decimal points should always be punched on data cards.

(iv) When several variables are read, all with the same specification, we can write a multiple FORMAT specification. For example

    106 FORMAT (5F5.2)

has exactly the same effect as

    106 FORMAT (F5.2, F5.2, F5.2, F5.2, F5.2)

and is quicker to write or punch! If necessary a group of specifications can be repeated using *brackets*, e.g.

    107 FORMAT (I5, 2(I5, F10.0), 3(F5.2, I3))

is a quicker way of writing

    107 FORMAT (I5, I5, F10.0, I5, F10.0, F5.2, I3, F5.2, I3, F5.2, I3)

(v) It is possible for the number of field-specifications to exceed the number of variables. Thus in the statements

    READ (5, 105) A, B, C
    105 FORMAT (10 F7.2)

the computer would read A, B, C, each under the specification F7.2 and would then proceed to the next statement since there are no further variables to read. The fact that seven field-specifications were 'unused' does not matter. A comparable situation holds for WRITE statements.

(vi) If there are more variables to be read than there are field specifications, then the computer returns to the rightmost left parenthesis in the FORMAT statement and repeats the specification(s) within the parentheses. For example, in executing

    READ (5, 106) A, B, C
    106 FORMAT (F7.2)

the computer would read A under F7.2, then return to the left parenthesis and repeat for B and for C. This, however, means that the values of A, B and C must be punched on three successive cards, each in the first seven columns. A slightly more complicated example, this time based on output, is:

    WRITE (6, 107) N, B, C, D
    107 FORMAT (I5, (F7.2))

The value of N would first be printed in I5 format, followed—in the same line—by B in F7.2 format. Since C and D still remain to be printed, the computer would then return to the rightmost left parenthesis (i.e. the one before the letter F) and repeat the F7.2 specification, printing out C in the next line. Finally the same would be done again for D.

Further FORMAT specifications. The I-, E- and F-specifications are not sufficient for all purposes and others exist, of which we shall now consider the X-, and H-specifications, together with carriage control and slash formats.

The X-specification has the general form wX where w is an unsigned integer constant. It has the effect of skipping w columns of input or of creating w blank print-positions on output. Thus the statement:

READ (5, 108) A, B, C
108 FORMAT (F10.3, 10X, F10.3, 10X, F10.3)
or 108 FORMAT (2(F10.3, 10X), F10.3)

would cause the computer to read the value of A from columns 1–10 of a card, then to skip whatever might be punched in columns 11–20, to read B from columns 21–30, skip columns 31–40 and finally to read C from columns 41–50. Comparable spacing of output can be produced with the X-specification.

The H-specification (H for Hollerith, the American pioneer of punched-card data-processing) allows the programmer to read or write alphameric and special characters in order to write headings etc. Here we shall only indicate its use in output. The general form of the H-specification is nH $c_1 c_2 c_3 \ldots c_n$ where n is an unsigned integer constant and $c_1, c_2 \ldots$ are the corresponding n characters that are to be printed out. For example:

WRITE (6, 107)
107 FORMAT (21X, 20HANALYSIS OF VARIANCE)

would cause the heading 'Analysis of Variance' to be printed out. Note that the number of characters specified—20 in this case—must include the blanks between words. In this case there was no list of variables in the WRITE statement, but if we wish we can include a list, mixing H-specifications with I, E or F-specifications as in this example:

WRITE (6, 108) A, X, Y
108 FORMAT (11X, 3HA =, F10.3, 10X, 3HX =, F10.3, 10X, 3HY =, F10.3)

resulting in a line of output of the form:

$$A = \quad 103.718 \qquad X = \quad 109.214 \qquad Y = 1038.712$$

assuming, of course, that these happen to be the numerical values stored in the registers corresponding to A, X and Y.

The control of movement of the printer carriage is carried out by the program, using the *first character* of a line of output as follows:

> blank = begin new line (at left-hand side)
>
> 0 = skip a line then begin new line
>
> 1 = skip to top of a new page then begin new line.

Thus, suppose we had the output statements

> WRITE (6, 109) X
>
> 109 FORMAT (F10.3)

and that X was represented by the number 21.5. Now we have allowed an F10.3 specification for this number so that it would have the external representation bbbb21.500 where b signified a blank space. The first character in the field width of 10 would therefore be a blank and would ensure that X was printed at the beginning of a new line with *three* blanks preceding it (the first blank having been used to effect carriage control). Suppose, however, that X happened to have the value 456789.315. If we tried to print this using the same 109 FORMAT statement the first character would be a 4. This is not a standard carriage control character and something unpredictable would follow. The essential points are, therefore, that the first character of a line of output must be a blank, zero or 1 and that this character will not be printed out but is used to bring about the required movement of the printer carriage. A blank can be made to occupy the first position by using a sufficiently wide field length, but it is often safer to write 1X at the start of a FORMAT specification and thus ensure that a blank is created. To ensure that a zero or 1 is available it is necessary to use the H-specification. For example:

> WRITE (6, 109) X

could be associated with several different FORMAT statements depending on how one wished to print out the value of X. 109 FORMAT (1X, F10.3) would ensure that it was printed to occupy the first 10 print-positions of a new line immediately beneath the previous line of output. 109 FORMAT (1H0, F10.3) would ensure that it occupied the first 10 print positions of a new line with a blank line above it. Finally 109 FORMAT (1H1, F10.3) would ensure that it occupied the first 10 print positions of the first line of a new page. Remember that it is only the *first* character of a line of output that is used to control carriage movement in this way.

Slash-formats are those incorporating a slash or solidus / which indicates that an input or output record is to be skipped. Compare, for example, the following two input procedures:

>   READ (5, 99) A, B, C, D
> 99 FORMAT (4F10.3)
> and READ (5, 100) A, B, C, D
> 100 FORMAT (F10.3/F10.3/F10.3/F10.3)

This first one is of the pattern we have previously encountered and is an instruction to read A, B, C and D as four values punched respectively in columns 1–10, 11–20, 21–30 and 31–40 of a single card. The second pair of statements are instructions to read the values of the four variables as follows:

(i) Read A, punched in cols. 1–10 of *first* card
(ii) Read B, punched in cols. 1–10 of *second* card
(iii) Read C, punched in cols. 1–10 of *third* card
(iv) Read D, punched in cols 1–10 of *fourth* card.

The effect of the slash in each case, therefore, is to cause the computer to omit whatever may be in columns 11 onward of the first card and to proceed instead to the next card, and so on. A single slash at the beginning of a set of field specifications will cause the first card to be skipped completely. Two slashes here will cause *two* complete cards to be skipped and so forth. Between two field-specifications, however, *two* slashes must be inserted if a card is to be skipped completely. Thus,

>   READ (5, 101) A, B, C, D
> 101 FORMAT (/F10.3//F10.3/F10.3//F10.3)

would have the following rather complicated effect:

(i) The first card would be skipped completely.
(ii) The value of A would be read from columns 1–10 of the second card.
(iii) The remainder of the second card and all the third card would be skipped (two slashes).
(iv) B would be read from cols 1–10 of the fourth card.
(v) The remainder of the fourth card would be skipped.

(vi) C would be read from cols. 1–10 of the fifth card.

(vii) The remainder of the fifth card and all the sixth card would be skipped.

(viii) D would be read from cols. 1–10 of the seventh card.

As one may see, slash formats are extremely useful if one wishes to read only a few of the figures punched on a set of data cards, especially when they are combined with X-specifications that allow some of the columns on a particular card to be skipped. Briefly, then, for input n slashes at the start of a set of field specifications cause n cards to be skipped; n slashes between any two field specifications cause $(n-1)$ cards to be skipped while n slashes at the end of a set of field specifications usually cause n cards to be skipped ($(n-1)$ in some computers).

Similar conventions operate for slash formats in output records, the effect of a slash being to pass to the next line. Thus

WRITE (6, 110) A, B, C, D
110 FORMAT (//4F10.3)

means that two blank lines are printed before A, B, C and D appear in the next line, the first character of the first F10.3 being used as a carriage control blank. It should be noted that the usual carriage control symbols are always required in slashed output formats and that slashes incorporated in the literal message of a H-field are not used for output spacing.

*A-specifications and Variable Formats**

The A-specification is a device whereby alphameric data and special characters can be read into a computer and stored in coded form. They can then be printed out again, with or without previous manipulation. Thus, if one wished to carry out a statistical analysis involving 3 levels of a factor (called, in a particular analysis, LOW, MEDIUM and HIGH) it might be convenient to have these names printed out as part of the results. To do this, the names would have to be read in under an A-specification. The general form of this is Aw, where w is an integer denoting the number of characters (including blanks) to be dealt with, e.g.

READ (5, 100) A, B, C
100 FORMAT (3A6)

---

* This section is best deferred until the reader has some knowledge of input/output of arrays (p. 219).

The data card from which the variables A, B and C are read would bear 3 groups of characters, each occupying 6 successive columns starting with the first column. Thus the three fields might contain the characters

```
1 2 3 4 5 6   7 8 9 10 11 12   13 14 15 16 17 18
    L O W     M E D  I  U  M          H  I  G  H
```

The locations corresponding to the variables A, B and C would then contain coded equivalents of the words LOW, MEDIUM and HIGH (the first and last preceded by 3 and 2 blanks respectively). If at a later stage one then wished to arrange for these names to be printed out one would simply include in the program some such statements as

> WRITE (6, 101) A, B, C
> 101 FORMAT (1X, 3A6)

Note that we have assumed that the computer will accommodate 6 characters in each location, meaning that it has a 36-bit word-structure (p. 7). Not all computers do have this and the successful use of the A-specification requires a knowledge of the word-structure of the particular machine in use. Let us suppose that the computer word accommodates n characters, then with the A-specification Aw the following situations can arise:

$w = n$: The w characters will be stored in sequence and printed out in the same form.

$w < n$: The w characters will be stored left-justified in the register with $(n - w)$ blanks to the right. When printed out the w leftmost characters from the register will appear.

$w > n$: Only the n rightmost characters will be stored, the $(w - n)$ leftmost characters being lost from storage. Only the n rightmost characters will be printed, preceded by $(w - n)$ blanks.

Another use for the A-specification is to print a suitable title for the output, so identifying it as belonging to a particular problem or set of data. To do this a convenient method is to set up an array which will store, say, 72 characters (the number always available on a single punched card). On a computer with a 36-bit word-structure, this will require an array of twelve: say, NAME (1) to NAME (12), each element of the array being represented by 6 characters. One then writes:

> READ (5, 100) (NAME (I), I = 1, 12)
> 100 FORMAT (12A6)

and ensures that the READ statement operates on a data card bearing information such as:

PROBLEM 1 A.J.SMITH'S DROSOPHILA DATA

starting in column 1. This line contains 37 punched characters (including blanks between words) followed by 35 blanks, making a total of 72 'characters' stored in locations NAME (1) to NAME (12). Actually, of course, the essential information (37 characters) is stored in the first six elements of the array and in the single leftmost position of the seventh element, the remaining locations being 'filled' with blanks. At the appropriate place in the program one might then write:

WRITE (6, 102) (NAME (I), I = 1, 12)
102 FORMAT (// 1X, 12A6)

and the line

PROBLEM 1 A.J.SMITH'S DROSOPHILA DATA

would be printed out. Clearly the A-specification has a great advantage over the H-specification. The latter will only cause to be printed whatever is permanently written into the program, whereas the A-specification enables one to print out anything which has previously been read in under the same specification among the data.

Another of the important uses of the A-specification is that it allows one to vary the input and/or output formats in a program. This may be a very important consideration, especially where input is concerned. Imagine, for example, that one has a program that carries out an analysis of variance or some such frequently used statistical procedure. One may wish to employ it on several bodies of data, each of which has been punched on cards but using a different format. Anyone who ever handles other peoples' data knows that this happens repeatedly. If the input format were fixed one would have to run each set of data separately, altering the FORMAT statement and recompiling the program each time. It is obviously more efficient to arrange for the program to read in the appropriate format before each set of data. This can be done by punching a FORMAT statement (including parentheses but without a statement number or the word FORMAT) on a card and placing it before the data. The relevant statements in the program would then be something like the following:

DIMENSION INFMT (12), X(100, 100)
READ (5, 100) (INFMT (I), I = 1, 12)
100 FORMAT (12A6)
DO 1 I = 1, N
1 READ (5, INFMT) (X(I, J), J = 1, M)

The first data card would have punched on it the FORMAT statement, e.g.

(10F7.2)

and this information would be stored in the array INFMT (actually occupying only the first 8 out of 72 available positions). The loop incorporating the second READ statement would then read an array X, whose elements were punched row-wise according to the specification 10F7.2. A second set of data might be preceded by a card bearing

(12F6.0)

and an array punched in this format would be read, the process being repeated for each set of data with an appropriate variable format card ahead of it.

Note that in the preceding example the array INFMT had 12 elements. With a computer word that accommodated 6 characters, this would allow, as explained, a format specification of up to 72 characters. If it were necessary to cater for a longer format, then the array INFMT would have to be extended, say, to 24 or 36 elements, and the variable format would run on to a second or third card. It will be appreciated that if the array were dimensioned as INFMT (36) and the relevant READ statement were

READ (5, 100) (INFMT (I), I = 1, 36)
100 FORMAT (12A6)

then the data deck should *always* contain 3 variable format cards, even if the specification could be punched entirely on the first of these. Failure to observe such precautions will result in data for analysis being read as though it were part of a format, with consequent errors.

Transfer Statements. The statements in a program are normally executed one after the other, but it often happens that the programmer needs to deviate from this simple sequence in order to repeat certain statements or branch on two or more possible sequences and so on. These deviations are controlled by transfer statements of which we consider:

  (i) the unconditional GO TO
  (ii) the computed GO TO
  (iii) the arithmetic IF   and
  (iv) the logical IF

(i) *The unconditional GO TO statement.* This has the very simple general form GO TO n where n is a statement number, and it has the effect of transferring control to the statement bearing that number. By itself, the GO TO statement can, for example, be used to direct control to the next major section of a program after any one of several alternative sections has been executed. Used in conjunction with an IF statement, the GO TO statement can form a loop, the execution of which may be repeated as often as desired (p. 59).

(ii) *Computed GO TO statement.* This has the general form GO TO $(n_1, n_2, n_3 \ldots n_m)$, i where i is an unsigned non-subscripted integer variable and n_1, n_2 etc. are statement numbers. According as the value of i is $1, 2, 3 \ldots$ control is transferred to the statement numbers $n_1, n_2, n_3 \ldots$ respectively. For example, in the following statement:

GO TO (5, 10, 15, 20), INDEX

if the current value of INDEX were 1 then execution would pass to statement 5, if INDEX were 2 it would pass to statement 10, and so on. In all cases, therefore, the variable i of the general form is potentially able to assume the values $1, 2, 3 \ldots m$. Other values would be invalid in that context. The particular value that i assumes in a computed GO TO statement may be the result of it having been read in from a data card or it may be the result of an earlier computation or of an assignment made earlier in the program. The computed GO TO statement thus allows the programmer to select one out of any number of possible branches in his program. It is much used where the program aims to provide a variety of options, any one of which may be selected by reading in the appropriate value of the index i of a computed GO TO statement.

(iii) *Arithmetic IF statement.* This has the general form IF $(a) n_1, n_2, n_3$ where a is any arithmetic expression—including a single variable—and n_1, n_2, and n_3 are three statement numbers (which need not all be different). The effect of this statement is to transfer control as follows:

to n_1 if a is negative

to n_2 if a is zero

and to n_3 if a is positive.

It therefore allows the programmer to direct control in any of up to three possible ways according to the numerical values of some variable or function of variables. An example is:

IF (NTOT) 5, 10, 15

which would transfer control to statement 5, 10 or 15 according to whether the current value of NTOT were negative, zero or positive. Another example is:

IF (XBAR − XMAX) 5, 5, 10

This would have the effect of directing control to statement 5 if the value of XBAR were less than or equal to XMAX and to statement 10 if XBAR were greater than XMAX.

An arithmetic IF statement used in conjunction with a GO TO will allow the programmer to write a loop, as in the following simple program for finding the sum of the integers from 1 to 50:

```
   N = 0
   NSUM = 0
 1 N = N + 1
   NSUM = NSUM + N
   IF (N − 50) 1, 2, 2
 2 STOP
   END
```

(iv) *The logical IF statement.* This resembles the arithmetic IF statement but is more elaborate and allows a wider range of possibilities on which a decision can be taken. The general form of this statement is

$$\text{IF } (l) \ s$$

where *l* is a logical expression having various forms considered below and *s* is any executable Fortran statement other than another logical IF statement or a DO statement (p. 64). The logical IF operates by first evaluating the logical expression. If this is true then statement *s* is executed; if it is false then statement *s* is ignored and control passes directly to the next statement in the program. The logical expression *l*—which can also include arithmetical operations and is used here mainly in this way—has the following form:

$$a \ \ \text{.RO.} \ \ b$$

where *a* and *b* are Fortran variables or arithmetic expressions of the same mode and .RO. is one of the following relational operators:

.LT. less than
.LE. less than or equal to
.EQ. equal to
.NE. not equal to
.GE. greater than or equal to
and .GT. greater than

An example of the logical IF statement, then, would be:

IF (X .GT. Y) GO TO 5

The meaning is clear: if X is greater than Y, control passes to statement 5; if it is not greater control passes to the statement immediately following the logical IF statement. Another example is:

IF (A + B .NE. Y * Y) N = N + 1

If the statement in parentheses is true, the value of N is incremented by 1 and the program then continues with the next statement. If, however, the statement in parentheses is false, N remains unchanged and again the program continues with the next statement.

Two or more logical expressions may also be combined within the parentheses by the use of the following logical operators:

.NOT.
.AND.
.OR.

For example,

IF (X .EQ. Y .OR. X .EQ. Z) GO TO 10

The operator .NOT. is used mainly in logical operations rather than in the kind of mathematical operations considered here. The use of .AND. means that if both the logical expressions united by the .AND. are true, then the statement following the parentheses is executed. The .OR. is the so-called 'inclusive or' of symbolic logic; it means that if either one, or the other, or both of the expressions which it links are true, then the statement immediately following the parentheses is executed.

If a logical IF statement contains two or more of the relational and logical operators mentioned above, the component expressions are evaluated separately according to the following scheme of priorities:

Priority	Operation symbol
1	Arithmetic operators (p. 40)
2	.LT. .LE. .EQ. .NE. .GE. .GT.
3	.NOT.
4	.AND.
5	.OR.

If it is necessary to deviate from this scheme of priorities, additional parentheses can be nested within the outermost ones. The expressions within the innermost parentheses are then evaluated first as in the parenthesised arithmetical expressions discussed on p. 41.

One point that needs to be emphasized concerns the use of floating-point expressions in arithmetic and logical IF statements. For reasons indicated on p. 13 and discussed further on p. 460, round-off error in the computer representation of floating-point numbers may result in values which differ very slightly from those expected by the unwary programmer. The result may be that an expression like (X .EQ. Y) may be false when ordinary decimal arithmetic would suggest it to be true. It is therefore usually best instead of writing, say,

IF (X .EQ. Y) GO TO 10

to write something like:

IF (ABS (X−Y) .LT. 0.000001) GO TO 10

This allows a certain tolerance on the accuracy with which X and Y are represented or have been calculated by the computer. With integer variables no such danger exists as they can always be represented exactly in binary form.

Subscripted variables and the DIMENSION statement. It is often a great advantage to work with bodies of data arranged as an array. Arrays may be of 1, 2, 3 or more dimensions, though we shall consider only those of up to three. A one-dimensional array—often referred to as a vector (p. 217)—consists of an ordered set of numerical values (the elements of the array) such as 1.2, 1.5, 1.7, 1.8, 2.1 and so on. Algebraically the elements of such an array can be denoted as $a_1, a_2, a_3, \ldots$ and in Fortran as A(1), A(2), A(3), etc. The Fortran variable A is then described as a subscripted variable and the subscript may be of various general forms:

$$v$$
$$v + c$$
$$v - c$$
$$c * v$$
$$c * v + c'$$
$$c * v - c'$$

where v is a simple (i.e. non-subscripted) integer variable while c and c' are integer constants. The value assumed by a subscript must therefore always be an integer number and it must not be zero or negative. Examples are

X(1), X(J), ALPHA(MIN +1), K(3 * I −1). Note that integer subscripts can legitimately be used with real variables; such usage does not infringe the rule concerning mixed modes. Note also that a subscript of the form $c + v$ or $c - v$ is invalid.

A two-dimensional array or matrix (p. 216) is an ordered set of numbers arranged in rows and columns, e.g.

$$\begin{array}{cccc} 21.3 & 32.6 & 1.5 & 1.8 \\ 20.2 & 41.7 & 2.9 & 2.3 \\ 25.3 & 38.6 & 8.2 & 3.9 \end{array}$$

This array comprises 3 rows and 4 columns and is therefore known as a 3×4 array or 3×4 matrix. Algebraically the elements of such an array may be denoted as:

$$\begin{array}{cccc} a_{11} & a_{12} & a_{13} & a_{14} \\ a_{21} & a_{22} & a_{23} & a_{24} \\ a_{31} & a_{32} & a_{33} & a_{34} \end{array}$$

where the first subscript indicates the row (numbered from above downwards) and the second subscript denotes the column (numbered from left to right). Fortran denotes the elements of such a two-dimensional array as A(1, 1), A(1, 2), A(1, 3), etc. In general, a doubly-subscripted array element has its two subscripts separated by a comma. Each of the subscripts may be in any of the six valid forms listed above for singly subscripted arrays, e.g. X(I, J), ALPHA(M + 2, N −1).

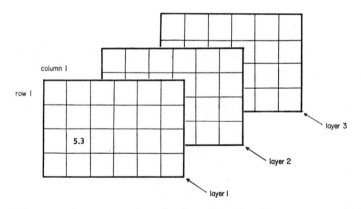

FIG. 15. Diagrammatic representation of a three-dimensional array, composed of rows, columns and layers. While it may be convenient to visualise the elements of a triply-subscripted array as stored in this way, the physical arrangement of storage locations in the computer has a different form.

A triply-subscripted array is one in which the elements are arranged in rows, columns and 'layers' (the third dimension) as suggested in Fig. 15. An individual element is referred to as, say, X(I, J, K) where I denotes the row number, J the column number and K the layer number. Thus the figure 5.3 in Fig. 15 would represent the element X(3, 2, 1).

In all the above cases a subscripted Fortran variable must be used consistently with 1, 2 or 3 subscripts according to the nature of the array in which it is an element. The same variable name may not be given to a non-subscripted variable in a given program and the name of a subscripted variable may appear without subscripts only in special circumstances. These are: in certain forms of input/output list (p. 221); as an argument in calling or defining a subroutine or function subprogram (p. 67); and in a type declaration, COMMON or EQUIVALENCE statement (p. 74). When the subscript is or contains a variable, this variable must be defined (i.e. given a numerical value) by some previous statement in the program.

When a non-subscripted variable first appears in a program, the compiler automatically reserves a storage location for it. When an element of an array is first mentioned, however, the compiler cannot know how many locations need to be reserved for accommodating all members of the array. This information must therefore be provided in a DIMENSION statement that must precede the first executable statement containing the array name. In practice it is usually placed at the beginning of the program. The general form of the DIMENSION statement is:

$$\text{DIMENSION } v_1, v_2, v_3, \ldots v_n$$

where v_1, v_2, v_3 etc. are the names of subscripted variables followed in parentheses by the maximum size of the array. For example,

$$\text{DIMENSION } A(100), B(20, 20), C(10, 10, 10)$$

reserves 100 consecutive storage locations for the singly subscripted array A, a total of 400 locations (20×20) for the doubly subscripted array B and a total of 1000 locations ($10 \times 10 \times 10$) for the triply subscripted array C. These numbers are therefore the maximum numbers of elements which arrays A, B and C will accommodate and if the program is to be used with data exceeding these dimensions, the values in the DIMENSION statement must be increased appropriately. The beginner will probably find it convenient to think of the reserved storage as having the same lay-out as a 1-, 2- or 3-dimensional array of printed numbers, though this is not physically the case.

The great advantages of using subscripted variables will become apparent when one considers the DO-statement below. The special problems of input and output of arrays are also discussed in connection with matrix operations on p. 219.

The DO statement. The DO statement is one of the most important in programming since it permits repeated execution of a particular set of statements. It is therefore used most in conjunction with the manipulation of array elements denoted by subscripted variables, though it is not confined to these. The general form of a DO-statement is

$$DO\ n\ i = m_1, m_2, m_3$$

where n, i, m_1, m_2 and m_3 are all integers or integer variables with the following significance:

- n is the number of the statement on which the DO-loop ends.
- i is the index whose value changes during the iterative cycle.
- m_1 is the initial value assumed by i.
- m_2 is the final value assumed by i.
- m_3 is the size of the step by which i is incremented at each pass through the DO-loop.

These definitions require some explanation which will be easier if we bear in mind an example of a DO-loop intended to sum an array of numbers stored as the singly subscripted variable X.

```
        DIMENSION  X (1000)
        SUM=0.0
        DO 1 I=1, N, 1
      1 SUM=SUM+X (I)
```

In this example the statement numbered 1 follows immediately after the DO-statement but generally there will be several intervening statements which, with the numbered statement, make up the *range* of the DO-loop. It is the statements included in the range of the loop which are executed repeatedly. The index variable here is I and on the first pass through the loop this is given the value 1. The statement numbered 1 therefore adds the value of X(1) to SUM and control then returns again to the DO-statement, in which I is now given the value 2, having been incremented by 1. The value of X(2) is then added to SUM and so on until I reaches the value N, when the last number in the array—X(N)—is added, the DO-loop is satisfied, and execution of the ensuing statements takes place. A few other examples of DO-loops are:

```
        SUMA=0.0
        DO 1 I=2, 20, 2
      1 SUMA=SUMA+X (I)
```

which sums X(2), X(4), X(6) ... X(20) of the array X.

$$\text{SUMNEG} = 0.0$$
$$\text{DO 3 I} = 1, \text{N}$$
$$\text{IF (X (I) .GE. 0.0) GO TO 3}$$
$$\text{SUMNEG} = \text{SUMNEG} + \text{X (I)}$$
$$3 \quad \text{CONTINUE}$$

which finds the sum of the negative numbers in array X. Note the CONTINUE statement on which the loop ends; this does nothing except to provide a statement on which the range of the loop ends. Note also that in the last example only the initial and final values of I were given: when this is done it is assumed that I is incremented by 1 at each pass through the loop. Subject to certain rules one DO-loop can be placed within another, when it is said to be 'nested'. An example of nested DO-loops is given in the following segment, which initialises all elements of the doubly-subscripted array X at zero:

$$\text{DO 2 I} = 1, \text{N}$$
$$\text{DO 2 J} = 1, \text{N}$$
$$2 \quad \text{X(I, J)} = 0.0$$

Here the two loops end on the same statement, the inner loop being satisfied on every pass through the outer one. It is, however, equally permissible for the inner loop to end on a statement that occurs before the one on which the outer loop ends.

The use of DO-loops is governed by a number of rules which may be summarised as follows:

(i) The last statement in a DO-loop must be an executable statement (i.e. it cannot be a statement such as FORMAT, DIMENSION or END).

(ii) The last statement cannot be any of the following: unconditional GO TO, computed GO TO, an arithmetic IF, another DO statement, STOP, PAUSE or RETURN. It may, however, be a logical IF.

(iii) When a set of DO statements are nested, all statements of the innermost loop must lie within the range of the outer loop. Two or more nested DO-loops may end on the same statement. Examples of valid and invalid nesting are depicted in Fig. 16.

(iv) When nested DO-loops are executed, the innermost loop is satisfied first and completion proceeds from the innermost to the outermost.

(v) Control may be transferred from within the range of a DO-loop to some other statement in the program which lies outside the range of the loop.

Control may not, however, be transferred from outside the range of a loop to a statement lying within the range.

(vi) The index i and the indexing parameters (m_1, m_2 and m_3) must not be redefined (i.e. given new values) within the range of a DO-loop.

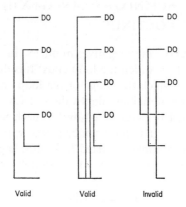

FIG. 16. Arrangement of valid and invalid nesting of DO-loops. The lines join the first and last statements in the loop. The programmer can draw lines of this kind on a program-listing in order to check the validity of the nested loops which it contains.

(vii) When a normal exit occurs from a DO-loop (i.e. through the loop being satisfied) the index i is no longer defined and must be given a new value before it can be used again. This may be done by a READ statement or by ensuring that the index variable appears to the left of an = sign in an assignment statement or by another DO statement. If a non-normal exit occurs—as when control is transferred out of the range of the loop before it is satisfied—the index retains the last value it had in the loop and is available for further computation.

Arithmetic Statement Functions. In certain programs a given function may be evaluated hundreds of times, each using a different variable. For example we may wish to evaluate $\sin \theta + \cos^2 \theta/2$ at one point, $\sin \phi + \cos^2 \phi/2$ at a second point, $\sin \beta + \cos^2 \beta/2$ at a third and so on. To simplify this procedure we may define an arithmetic statement function containing a dummy argument, say X, and *call* it repeatedly to operate on three different values of X (THETA, PHI and BETA in the above example). The general form of such an arithmetic statement is:

$$f(a) = e$$

where f is the function name (1–6 alphameric characters, the first alphabetic)

a is a dummy argument (i.e. a non-subscripted variable name)

and e is an arithmetic expression containing a but not involving subscripts. For the example discussed above we would write:

FUNCT (X) = SIN (X) + COS (X/2.0) * COS (X/2.0)

This statement is then placed ahead of all executable statements in the program. It is called by simply writing FUNCT—or whatever other name we have given it—follows in parentheses by the name of the argument we wish to use in place of the dummy argument. Thus our program could contain statements such as

Q = FUNCT (THETA)
A = X * ALOG (A) + FUNCT (PHI)
P = FUNCT (PHI)/FUNCT (THETA) * SQRT (FUNCT (BETA))

An arithmetic statement function may contain any number of dummy arguments (either integer or real). When a function containing two or more dummy arguments is called, the actual arguments must, of course, correspond to the dummies in number, order and type. A function statement may also include other variables which are treated as parameters and assume whatever values are assigned to them in the program. An example of a more complicated statement function is:

F(A, X, P) = COS (A)/SQRT (X * P) − ALPHA

Here A, X and P are dummy arguments, to be replaced by actual ones when the function is called, while ALPHA is a parameter whose value is assigned in the course of the program (as part of an input list or by occurring to the left of an = sign). A program may include several different statement functions and one may call another, provided that the calling function is preceded by the one being called. For example

PHI (X) = SIN (X) + ALOG (COS (X))
FUNCT (X, Y) = PHI (X) + ALOG (PHI (Y))

Function Subprograms. The arithmetic statement function, as we saw above, provides ('returns') a single value for a function that is expressed in a single arithmetic statement. It may be, however, that we frequently require the value of a function that can only be calculated by means of a series of statements. In this case we need to write a function subprogram. This is a complete program which may be of any length and which is compiled independently of the main program using it. The dummy arguments may include 1-, 2-, or 3-subscripted variables so that the function subprogram can accomplish many more useful things than the statement function. The function subprogram must begin with a statement of the general form:

$$\text{FUNCTION } a\,(x_1, x_2, x_3, \ldots x_n)$$

where a is the function name (1–6 alphameric characters, the first alphabetic) and $x_1, x_2, x_3 \ldots$ are the names of variables representing dummy arguments.

To call such a function (i.e. to use it in another program) one simply writes the name of the function where it is required. For example, one may imagine the subprogram FUNCTION SUM (X, M, N) which returns the value of the sum of all the elements in a doubly-subscripted array X, composed of M rows and N columns. If, in the course of a program requiring the sum of the elements of the array ALPHA, comprising NR rows and NC columns, one wrote

$$S = \text{SUM (ALPHA, NR, NC)}$$

then the necessary computation would be carried out.

The following points should be noted:

(i) The type of the function is implied by the first letter of the function name, unless one over-rides this implicit typing by a type declaration (p. 74). To do this one writes the first statement of the function subprogram as, say,

$$\text{INTEGER FUNCTION BETA (X, Y, Z)}$$

meaning that BETA, despite its first letter, is an integer variable. Similarly one might write REAL FUNCTION LAMBDA (A, B, C), or precede the word FUNCTION by one of the other type-declarative words (p. 74).

(ii) The function subprogram must include at one or more points the statement RETURN, indicating that control now returns to the program from which the function was called. If arrays are dealt with by the subprogram then the appropriate DIMENSION statement should be included, agreeing with the corresponding DIMENSION statement of the calling program.

(iii) The name of the function must appear at least once in the subprogram, either on the left-hand side of an assignment statement or in an input list.

(iv) A function subprogram can call another function subprogram but cannot call itself, nor can two function subprograms call each other.

(v) If a variable occurring in the calling program is used as the actual argument of a function subprogram, and if the latter assigns a new value to the variable, then this new value will, of course, be returned to the calling program and used in it thereafter. If under such circumstances one wished to preserve the original value of the variable, it would be necessary to duplicate it. This is most conveniently done at an early stage in the function subprogram by an assignment statement such as AA = A and thereafter using AA as the argument of the function.

(vi) Although it is often said that a function subprogram returns only a single value, it is, strictly speaking, more correct to say that it returns only a single value *associated with the function name*; other output values can be associated with each of the dummy arguments.

(vii) When a function subprogram has been written and is to be used in conjunction with a calling program, the card deck on which the function subprogram has been punched is placed after the deck comprising the main program. In some systems the function subprogram deck must also be preceded by one or more system control cards p. 80).

Subroutine Subprograms. The freedom to write Fortran subroutines is a great advantage of this method of programming. Essentially, a subroutine is a separate program which carries out all kinds of calculations and which can be called into action by a main program. A very large program is therefore often best conceived as a main program and a set of associated subroutines. Since many mathematical and statistical programs involve a number of common computational features, it is often possible to construct a very wide range of programs by linking together a set of standard subroutines with a calling program specially written for the circumstances involved. Thus we may be able to call on a subroutine which will carry out a standard set of transformations of raw data, another which will compute a linear regression between two variables and a third which will print out a set of points with known co-ordinates. By means of a relatively short calling program these can be linked into a coherent program capable of dealing with a large number of similar problems in succession. Many large computer installations have available a systems library of the more commonly used subroutines, stored on magnetic tape or disk in such a way that the programmer can call them directly, without having to incorporate a punched-card version of the subroutine in his program deck. It is, however, also usual for each programmer to build up a library of subroutines relevant to his own work (stored either as Fortran decks or in relocatable binary deck form). It is therefore necessary to know how a subroutine is constructed.

Essentially a subroutine subprogram is very similar to a function subprogram: it is a separate program written and capable of being compiled independently of the main program and the other subroutines with which it will be used. Like the function subprogram, it employs *dummy arguments* (also known as *formal parameters*), and these must be replaced in use by the actual arguments which must be transmitted between the calling program and the subroutine. The main points to be noted concerning subroutines are as follows:

(i) Each subroutine must begin with a statement of the general form

SUBROUTINE a $(x_1, x_2, x_3, \ldots x_n)$

or SUBROUTINE a

where a is a subroutine name (1–6 alphameric characters, the first alphabetic) and $x_1, x_2, x_3, \ldots$ are the dummy arguments which may be non-subscripted variable names, or the names of arrays (without subscripts) or the names of functions or other subroutines, or they may take the form of arithmetic expressions.

In the first case above, the dummy arguments are cited in parentheses, in which case they are said to be transmitted *explicitly*. In the second example the arguments have to be transmitted *implicitly* by a separate COMMON statement, whose use is discussed on p. 72.

(ii) Each dummy argument may be of any type, determined by its initial letter or by an explicit type declaration. If the arguments are functions or subroutines it is necessary to include in the main program a special EXTERNAL statement, discussed further on p. 75.

(iii) At least one of the dummy arguments will represent output from the subroutine, to be used by the calling program. As a rule there is also at least one argument representing input to the subroutine from the calling program. It is, however, also possible for the subroutine to operate on data which form part of an input list in the subroutine or to define a variable by including it on the left-hand side of an assignment statement. Unlike a function subprogram, a subroutine has no value associated with its name, so that a subroutine name can begin with any letter and is not typed. All the values are associated with the arguments. These are limited in number, but the limit is far larger than any normal program ever needs.

(iv) The subroutine must include at least one RETURN statement, indicating that control is to be returned to the calling program, i.e. that the subroutine has, for the time being at any rate, completed its task. Like a main program, too, the subroutine must include all the non-executable statements (e.g. DIMENSION, type-declaration statements, etc.). Where necessary these must agree with the corresponding statements of the calling program.

(v) In order to call the subroutine, a special statement is required in the main or calling program. This has the general form

CALL a $(x_1, x_2, x_3 \ldots x_n)$

when the actual arguments are being transmitted explicitly, or

CALL a

when they are being transmitted via a COMMON statement. A subroutine can call other subroutines but it cannot call itself, nor can two subroutines call each other.

(vi) Since subroutines are compiled quite independently of the main program with which they are associated, statement numbers and variable names in the subroutine can be identical with those in the main program. Two important consequences follow from this independence: (a) if there is a logical discrepancy between the subroutines and the main program this will not be detected by the compiler: the resulting program will not run properly, but the errors will not be diagnosed by the computer; (b) the actual arguments must be transmitted between the main program and the subroutine or between two subroutines. This second point is important enough to require more extended treatment.

(vii) Transmission of arguments explicitly is quite simple. Suppose one has a subroutine named VARI (A, N, V) which operates on a single-subscripted array A, containing N numbers, and computes their variance, V. Suppose also that we wished to use this subroutine in conjunction with a main program to compute the variance of an array X containing NVAL numbers whose variance was to be stored as a variable S2. We call the subroutine by the statement.

CALL VARI (X, NVAL, S2)

The actual arguments are X, NVAL and S2 and they are now set in correspondence with the dummy arguments A, N and V of the subroutine. The values X and NVAL must, of course, have been defined in the main program before the CALL statement, and the variance computed by the subroutine will, on return, be stored in S2. In other words, the actual arguments used in the CALL statement must agree in order, number and type with those in the list of dummy arguments in the SUBROUTINE statement. There is, of course, no objection to using the same variable names for the same quantities in both the main program and the subroutines. In such a case the actual arguments cited in the call are identical in name with those used in the subroutine.

(viii) Transmission of subroutine arguments by a COMMON declaration requires some explanation. A COMMON declaration may be placed ahead of executable statements in both the main program and the subroutines. Each variable mentioned in the COMMON statement of the main program is thereby set in correspondence with a similarly placed variable in the COMMON statement of the subroutine. Corresponding variables are then *assigned to the same storage registers in the computer*. It follows that by placing the actual arguments of a subroutine call in COMMON storage with the dummy arguments of the subroutine, one is in effect transmitting the variables between the two programs. For example, if the main program contains

    COMMON A, B, C
    ⋮
    CALL SUBRO

and the subroutine contains

    SUBROUTINE SUBRO
    COMMON X, Y, Z
    ⋮

then the variables referred to as A, B and C in the main program will be dealt with by the subroutine as though they were named X, Y and Z. Even if one were using the same variable names for the same quantities in both calling program and subroutine it would still be necessary to cite them in separate COMMON declarations, e.g.

    COMMON X, Y, Z
    ⋮
    CALL SUBRO   } In main program
    ⋮

    SUBROUTINE SUBRO
    COMMON X, Y, Z   } In subroutine
    ⋮

In all cases it is necessary that the variables in the common statement of the main program agree in number, order and type with those in the common statement of the subroutine.

A slightly more complicated situation arises when we wish to use COMMON declarations to transmit arguments between a main program and two or more associated subroutines, each operating on differently named variables. Let us suppose that the three actual arguments A, B and C of the main program are to be transmitted to the subroutine SUBA, whose corresponding dummy arguments are P, Q and R, while a further three actual arguments D, E and F are to be transmitted to a different subroutine SUBB with dummy arguments S, T and U. This could only be achieved by a combination of COMMON statements such as the following:

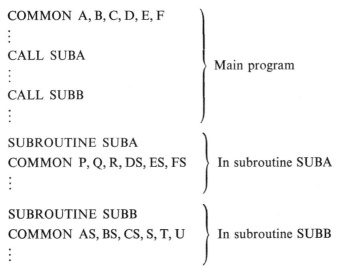

Since we are listing 6 actual arguments in the COMMON statement of the main program, we must include 6 in each of the associated subroutine COMMON statements. Three of these will, in each case, however, be fictitious variables whose only function is to 'pad out' the COMMON statement so that a correspondence is forced between A, B, C and P, Q, R on the one hand, and between D, E, F and S, T, U on the other. The dummy variables are DS, ES, FS, AS, BS, and CS in the above example. More complicated situations of this kind can arise which are best solved through the use of several blocks of COMMON storage that are labelled separately. Details of this procedure are available in the general Fortran manuals.

(ix) The arguments of a subroutine may be arrays, in which case the variables must be included in a DIMENSION statement. Many compilers do actually allow dimensions to be passed in the COMMON statement, but separate DIMENSION statements are preferable in many ways. The dimensions so quoted in the subroutine must agree with those given for the corresponding variable in the main program, unless one employs the device of *adjustable dimensions* (p. 76).

Non-executable Statements. Repeated reference has been made to statements like the DIMENSION, FORMAT and COMMON statements, which are not executed in the sense that an arithmetic statement or a DO instruction is, but which specify certain conditions within which execution proceeds. These non-executable statements (or declarations as they are sometimes called) include the following:

Type declarations
DIMENSION
COMMON
EQUIVALENCE
DATA
FORMAT
NAMELIST (not a Standard Fortran statement)
Statement Functions

The places at which these may be inserted in a program varies with the compiler in use, but it is a completely safe guide to include them in the order given above and at the start of the program before any executable statements. Only FORMAT declarations are commonly found scattered about the program close to the READ and WRITE statements in which they are cited. We shall not deal with DATA and NAMELIST statements here, and Statement Functions have already been discussed on p. 66 while FORMAT statements are dealt with in connection with input and output (pp. 47–57). It remains to extend our acquaintance with Type declarations, DIMENSION, COMMON and EQUIVALENCE statements.

(i) *Type declarations.* These are intended to over-ride the implicit typing of variables (first letter I to N denoting integer variables, others real) or to indicate other types. Type declarations have the form:

INTEGER $a_1, a_2, a_3 \ldots a_n$
REAL $m_1, m_2, m_3 \ldots m_n$
DOUBLE PRECISION $a_1, a_2, a_3 \ldots a_n$
LOGICAL $a_1, a_2, a_3 \ldots a_n$
COMPLEX $a_1, a_2, a_3 \ldots a_n$
EXTERNAL $b_1, b_2, b_3 \ldots b_n$

where $a_1, \ldots a_n$ and $m_1, \ldots m_n$ are subscripted or non-subscripted variable names and $b_1, b_2, b_3 \ldots b_n$ are FUNCTION or SUBROUTINE subprogram names. The significance of the type declarations are:

INTEGER denotes that $a_1, a_2, a_3 \ldots a_n$ are integer variables, irrespective of their initial letter e.g. INTEGER SAMPLE is used if, for some reason, one wishes the variable SAMPLE to take integer values. All values assigned to SAMPLE and all operations involving it must then treat the variable as though it were integer.

REAL denotes that $m_1, m_2, m_3 \ldots m_n$ are to be treated as real variables, e.g. REAL LAMBDA, NUM1, NUM2.

DOUBLE PRECISION specifies that $a_1, a_2, a_3 \ldots a_n$ are double-precision quantities as specified on p. 464. These are not used in any of the programs listed in this book, but any program can be modified to a double-precision version through the use of this statement and of double precision functions.

LOGICAL specifies that the following variables are logical quantities which may only assume the values .TRUE. and .FALSE. This type of variable is of very limited use in statistical programs and is not discussed further in this book. Full details are given in general Fortran manuals.

COMPLEX indicates that the variables are complex numbers of the form $x + iy$ where $i = \sqrt{-1}$. Each such variable is indicated by two numbers, a real part, x, and an imaginary part, y. Complex numbers are not widely used in programs of biological interest and are not discussed further in this book. Again, full details of the coding of COMPLEX variables is given in Fortran manuals but their effective use presupposes the requisite mathematical knowledge.

EXTERNAL specifies that the variables listed are the names of subprograms. This ensures that when a main program includes a call statement that has as one or more of its arguments a subprogram name, then the compiler will recognise it as such and not treat it as an ordinary variable. For example, a main program containing

EXTERNAL COS
⋮
CALL SUBRO (A, COS, X)

ensures that COS in the call to SUBRO is treated as the cosine function. The names listed in an EXTERNAL statement may be either the Fortran mathematical library functions available in all computer systems with a Fortran compiler (e.g. COS, SQRT, ALOG10) or FUNCTION subprogram names, or SUBROUTINE subprogram names but not the built-in 'open' subroutines (e.g. ABS, AMAX1, FLOAT, etc.), or statement functions specially written by the programmer.

(ii) *DIMENSION statements.* The use of these in reserving storage space for subscripted variables denoting 1-, 2- or 3-subscripted arrays has already been dealt with briefly on p. 63 but a few additional comments are needed.

(a) Dimensional information can also be given in COMMON and in some type-declaration statements, e.g.

> REAL NX, LAMBDA (100), NUMBER (50,50)

indicating that the arrays LAMBDA and NUMBER have the respective dimensions 100 and 50 × 50. Again,

> COMMON A, B(100), C

conveys dimensional information regarding B. If a variable is dimensioned in a COMMON statement or a type declaration, the same information must not be repeated in a DIMENSION statement. In many respects it is better to restrict all such information to DIMENSION statements.

(b) FUNCTION and SUBROUTINE subprograms must contain DIMENSION statements in respect of all arrays that they deal with. The dimensions given in these should agree in size and number with the dimensions of corresponding arrays in the calling program (unless one uses the device of adjustable dimensions mentioned below). Note that when corresponding arrays are dimensioned in both the calling program and a subprogram, the required number of locations is only reserved once (by the calling program).

(c) The need for agreement between the dimensions of a calling program and a FUNCTION or SUBROUTINE subprogram limits somewhat the flexibility of the subprograms. If the programmer has written a subroutine with, say, DIMENSION A(50,50) B(50), C(50) and wishes to use it with a main program involving arrays of 100 × 100 for A and 100 for B and C, he must alter the DIMENSION statements of the subroutine, which must then be recompiled. This is inefficient and a method of adjustable dimensions is therefore available on some computers for specifying the dimensions of arrays in SUBROUTINE and FUNCTION subprograms. The DIMENSION statement *in the subprogram* then has a form such as:

> DIMENSION A(N,N), B(N), C(N)

where the parenthesised character is an integer variable and not an integer constant as is normally the case. Furthermore, all the variables involved (i.e. both the array names and the dimensioning variable) must be defined, that is conveyed to the subprogram as arguments. These may be transmitted explicitly or through a COMMON statement.

It may happen that as written a subprogram contains arrays that are not dummy arguments. If these arrays are to be adjustably dimensioned they must also be included as arguments of the subprogram. A DIMENSION statement in a subprogram may, however, include both fixed and adjustable dimensions, e.g. DIMENSION A(N,N), B(100), C(N) is valid. The method of adjustable dimensions is very valuable but should be applied with care since its inappropriate use may cause a lack of agreement between arrays which the programmer wishes to correspond.

(iii) *COMMON statements.* A COMMON statement has the form:

$$\text{COMMON } V_1, V_2, V_3, \ldots V_n$$

where $V_1, V_2, V_3, \ldots V_n$ are variable names, e.g. COMMON A, B, C. The statement has the effect of reserving a special area of storage—*common storage*—for the variables mentioned. If these variables represent arrays, then they must be dimensioned, either in a preceding DIMENSION statement or by enclosing the array dimensions in parentheses in the COMMON statement:

e.g. DIMENSION A(10,10), B(100)
 COMMON A, B

or COMMON A(10,10), B(100)

Variables dimensioned in a COMMON statement must not be dimensioned elsewhere as well. Storage of simple variables or arrays in common storage locations means that they are accessible to a main program and any number of subprograms with corresponding COMMON statements, so that the latter form an efficient method of transmitting arguments between one program and another, as discussed on p. 72. A variation of this principle, involving the establishment of two or more separately labelled blocks of common storage, is also used but for details of this the reader must consult a general Fortran manual. A single common storage area as employed in this book is sometimes referred to as 'blank common' to distinguish it from the more complicated 'labelled common'.

(iv) *EQUIVALENCE statements.* These are used mainly to conserve storage space by employing the same physical locations to store two or more arrays. It is, of course, assumed that these arrays are not required simultaneously during the program, so that the first can be dealt with completely before the second, third, etc. array is set up in the same registers. Fundamentally, however, the EQUIVALENCE statement is simply a device for storing two variables in the same location. Thus, if A and B are two non-subscripted variables, the statement

EQUIVALENCE (A, B)

ensures that values of A and B will be stored in a single location. If B represented a subscripted array, however, one would write, say,

EQUIVALENCE (A, B(1))

to ensure that the value of A is stored in the same location as B(1). Alternatively one might write

EQUIVALENCE (A, B(5))

which would ensure that the values of A and B(5) were stored in the same location.

More often one wants to store whole arrays so that they utilise, partially or entirely, the same set of locations. The coding to achieve this varies slightly from one computer to another, but the basis of the method is to put one element of one array in equivalence with one element of the other array. Corresponding elements are then automatically made equivalent throughout both arrays (or as much of them as overlap). For example,

DIMENSION X(50), Y(50)
EQUIVALENCE (X (1), Y (1))

would put X(1) into equivalence with Y(1) and therefore ensure automatically that X(2) was equivalent to Y(2), X(3) to Y(3) and so on. This might be represented diagrammatically as in Fig. 17 where the boxes denote locations and the symbols indicate the number they contain. The shaded box denotes

Fig. 17. Relationships of elements of the singly-subscripted variables X and Y when stored in accordance with the statement EQUIVALENCE (X(1), Y(1)). The squares denote successive storage locations, occupied in turn by the elements listed left and right.

the location specifically identified in the EQUIVALENCE statement. If one had written EQUIVALENCE (X(1), Y(5)) the storage relationships would be as shown in Fig. 18.

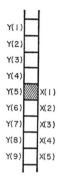

FIG. 18. Relationships of elements of the variables X and Y when stored in accordance with the statement EQUIVALENCE (X(1), Y(5)). Compare with Fig. 17.

When two-dimensional arrays are placed in equivalence one must remember that the physical storage locations are always sequential, with the first subscript of the Fortran variable changing most rapidly (i.e. a two-dimensional array is stored columnwise). The following statements would thus have the storage lay-out indicated in Fig. 19.

    DIMENSION A (2,3), B (3,4)
    EQUIVALENCE (A (1,1), B(1,1))

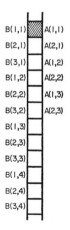

FIG. 19. Relationships of elements of the double-subscripted arrays A and B with dimensions and equivalence declaration as given in text.

In practice the commonest situation is one in which one wants to make equivalent two double-subscripted arrays of the same dimensions. In such cases the appropriate statement would be one like:

EQUIVALENCE (A (1,1), B (1,1))

though some compilers require that double or triply subscripted arrays should be indexed as though they were singly subscripted. Note that in general one should not make equivalences between arrays of different types.

Running a Program

After a program has been written and is being tested, or when a reliable program is to be used for carrying out a computation, certain practical considerations arise. These will be dealt with briefly and in sequence, though it must be emphasised that different computer installations employ different methods of handling jobs. The following information may therefore need revision in your own case, depending on the procedure laid down in the instructions to users that are always available at each computing unit.

1. Write the program and data on a Fortran coding form and check visually for errors.

2. Get the program and data punched and, if possible, verified (i.e. checked on a verifying machine).

3. Obtain a listing of the program and data and check this again visually if it has not been verified.

4. Make up the deck. This entails inserting the appropriate system control cards. These vary so much from one system to another that there is no point in attempting to describe them in detail here. Essentially they represent input to the monitor or supervisory program which will control the execution of your job and probably many others which are being run with it on some kind of batch processing system. Commonly there are a few system control cards at the beginning of the deck which specify such things as: the system to be used; the compiler required; the time and probable output expected; whether tapes or disk auxiliary storage is needed; whether the deck includes binary or assembly language cards; and a name or code identifying the job and the user. Often the different components in the deck (main program function, subprograms, etc.) require to be separated by control cards and there is also one before the data cards. When several sets of data are being processed in sequence there may also be separator cards between them. Finally there is some sort of end-of-file card which ensures that control returns to the monitor so that the next job after yours can be handled.

5. The complete deck (between backing boards and held together by rubber bands, or in a card tray) is usually handed to the receptionist at the computer unit. Further responsibility for the running of the program then passes to the personnel of the reception centre and the computer operators.

6. When processing is being carried out the normal arrangement at an installation of any size is that the decks for each job are usually assembled in batches so that all those of a particular kind are handled together, e.g. all those requiring less than 1 minute of processing time, or all those employing special users' tapes that need to be mounted by hand as required, etc. The decks are then placed in a card-reader often connected to a small computer which is used only to prepare a BCD magnetic tape or disk version of the programs and data.

7. In due course the BCD tape or disk area is read by the central processing unit. All programs in Fortran are then compiled, i.e. translated into binary object code by the compiler program resident in the computer.

8. The output from the central processor is again produced on disk or BCD magnetic tape. This is used by a high-speed printer to produce the normal printed output or, when requested, to operate a card punch producing punched cards. It is normally in these forms that the output is returned to the user via the reception centre.

Chapter 4

Three Simple Statistical Programs

We shall begin with an attempt to produce Fortran programs for three simple and frequently recurring statistical procedures: (i) the calculation of the mean, variance and other statistics of a sample, (ii) the computation of the correlation coefficient between two variables, and (iii) the significance tests used to compare the means of two samples. In each case the discussion of the program is preceded by a very brief summary of the statistical method. Further details of these methods can be found in any standard text book of statistics (e.g., Bailey, 1959; Campbell, 1967; or Dixon & Massey, 1969).

Program 1: Computation of mean, variance, standard deviation and other statistics of a sample

Much quantitative work in biology involves sampling a population (in the statistical sense) and summarising the properties of the sample in terms of a few statistics. The two most important of these are the *arithmetic mean*, which provides a measure of the central value of the sample and the *variance*, which estimates the extent to which the observations are dispersed. The mean of $\bar{x}$ of a set of n observations $x_1, x_2, x_3, ..., x_n$ is defined as

$$\bar{x} = \frac{\sum_{i=1}^{n} x_i}{n} \qquad (1)$$

where $\sum_{i=1}^{n} x_i$ denotes the sum of $x_1, x_2, x_3, ..., x_n$. Σ is the capital Greek letter sigma and the summing operation is sometimes denoted more briefly as

Σx, it being assumed that summation is carried out over all values of x_i from $i = 1$ to $i = n$.

The sample *variance*, s^2, is defined by the expression

$$s^2 = \frac{\Sigma(x - \bar{x})^2}{(n - 1)}. \tag{2}$$

That is, the variance is equal to the sum of the squares of the differences between each value of x (i.e., $x_1, x_2, x_3, ..., x_n$) and the sample mean, $\bar{x}$, divided by one fewer than the number of observations in the sample. As defined above, $\bar{x}$ and s^2 are also unbiassed estimates of the mean and variance of the population from which the sample was drawn. The formula (2) for the sample variance is not always very convenient for computation and is often better expressed in a somewhat different form, i.e.,

$$s^2 = \frac{\Sigma x^2 - (\Sigma x)^2/n}{(n - 1)}. \tag{3}$$

The remaining three statistics to be computed by this program are derived from the variance. The first of these is the *standard deviation*, defined as the positive square root of the variance, i.e.,

$$s = \sqrt{\frac{\Sigma x^2 - (\Sigma x)^2/n}{(n - 1)}}. \tag{4}$$

Like the variance, this provides a measure of the variability of the observations. It is expressed in the same units as were used to measure the original observations and for this reason it is unsuitable should we wish to compare the variation in two samples measured in different units. We can, however, derive from the standard deviation a dimensionless number, the *coefficient of variation*, which provides a measure of variability that is independent of the units of measurement used. This is defined as

$$c = \frac{s}{\bar{x}} \times 100, \tag{5}$$

that is, the coefficient of variation is the ratio of the standard deviation of the sample to its mean, expressed as a percentage. Finally, we wish to compute the *standard error*. This provides us with a measure of the reliability of the sample mean; it is, in fact, the standard deviation of the sample mean and is defined as

$$e = \frac{s}{\sqrt{n}}. \tag{6}$$

Although the calculation of means and the estimation of dispersion by variance, standard deviation and so on, is a simple process, biological work often requires that these calculations be performed on a large number of samples, each containing various numbers of observations. The program has therefore been devised to allow any number of samples to be analysed in succession, each sample composed of any number of observations. As the flow-chart shows, this is achieved by the use of two indices, ISTOP and ISAMP, which are represented by integers in the first and second card-column of each data card. ISAMP remains zero (i.e. the card is not punched in this column) until the last card of the sample is reached, when it is given a non-zero integer value (e.g. 9). ISTOP remains zero (i.e. unpunched) until the last card *of the last sample*, when it too is given a non-zero value. The value of ISAMP is then tested after each card has been read and control transferred to the computation as soon as it is clear that the last card of a sample has been processed. Similarly, ISTOP is tested after the results of each sample are printed out, and control transferred to STOP only after it is known that there are no further samples to analyse. The two decision points ISAMP = 0? and ISTOP = 0? are clearly indicated in the flow chart.

The program begins with a set of comments and continues by dimensioning X to allow for the fact that several values will be printed on each data card. We shall assume that not more than 10 values will occur on each card and therefore write DIMENSION X(10). So long as the number is 10 or less the program can be used without change; if more than 10 values per card are to be accommodated, then the DIMENSION statement (and the FORMAT declaration accompanying the first READ statement) will require alteration. Having written a title for the analysis, the program initialises at zero the variables for which accumulated sums are later required, i.e. NS, the sample number; N, the number of observations in the sample; SM, the sum of the various values of X; and SSQ, the sum of the squares of these values. These initial settings at zero are essential since if that is not done the locations used by the computer for these variables may contain from the start whatever figures were left over from previous calculations.

It is now time to read in the contents of the first data card. In addition to ISTOP and ISAMP (which will both be zero unless the card happens to be the only one in the sample), the card contains a value for M, which gives the number of observations of X that follow on the card. Normally, if there are 10 observations for X, then M will be 10. But if one chooses to put more or fewer than 10 observations per card, then M will be amended accordingly. And, of course, when the card is the last one of a sample it will probably contain fewer values of X than the previous cards of the same sample. By giving a value to M, therefore, the subsequent values of X on the card can be read in by an implied DO-loop and a DO-loop (terminating on statement 3) is also used to accumulate the sums for N, X and SSQ relevant to the data on the card.

The first decision point is now reached; if ISAMP is zero, control returns to the READ statement (numbered 2) and the next card will be read and its contents processed; if ISAMP is not zero (i.e. if the end of the sample has been reached), control proceeds to the series of statements used to compute the following variables: XBAR, the arithmetic mean of the sample; VAR, the variance; SDEV, the standard deviation; COEFF, the coefficient of variation; and SERR, the standard error. These computations are self-explanatory, but it is worth noting that before undertaking them we must convert N, an integer number, into its real equivalent, AN. Had this not been done our expressions for XBAR, VAR, etc., would have involved mixed modes and many computers would have cancelled the execution of the program.

We are now ready to print out the values of XBAR, VAR, etc., but before doing so we increment the value of NS so that the number of the sample may be printed out first. Provided the statement NS = NS + 1 is executed before the results are printed out, it does not matter whether the statement is placed here or at some earlier point in the computations for that particular sample.

Once the results for a sample have been printed, it is necessary to determine whether there are further samples to be analysed. This is done by the test ISTOP = 0? shown in the last decision box of the flowchart. So long as ISTOP is zero—i.e. so long as the last card of the last sample remains unread —control will be transferred to statement 1, the variables representing accumulated sums will be reset to zero and the analysis of a new sample begins. It should be noted that when control is transferred from the last decision box to a point where a new sample is begun (i.e. to statement 1), NS is not included in the process of initialisation: once set to zero at the very beginning of the program, NS must continue to be increased by 1 as each successive sample is dealt with.

Finally, if at the last decision box ISTOP is not zero (i.e. the last card of the last sample has been read), then control passes to STOP and the execution of the program is complete.

Along with the listing of this program is a sample set of data and the results which were printed out by the computer when data and program were run together. From the standpoint of programming tactics (as opposed to the mechanics of Fortran coding), three points should be noted:

(i) the use of the variables ISTOP and ISAMP to enable the computer to 'know' when the end of each sample and of all the samples is reached.

(ii) the device of initialisation at zero followed by a DO-loop to accumulate sums and sums of squares.

(iii) the use of formula (3) to calculate the variance, in place of the more explicit formula (2). In this way the variance can be calculated without know-

Flowchart for Program 1

Computation of mean, variance, standard deviation, standard error and coefficient of variation.

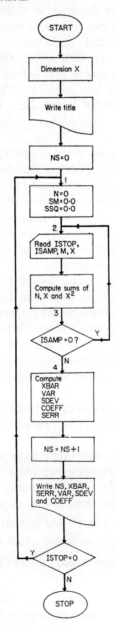

ing the sample mean in advance. A little thought will show that formula (2) would work only if we read all the data twice (once to calculate the mean and a second time for the variance), or if the whole set of observations were stored centrally in an array of their own. The method adopted is preferable to either of these alternatives when a large body of data is being processed. It can, under some circumstances, lead to a reduction in accuracy, however, and formula (2) is preferable when all the data are held in the central memory.

Program 1

Computation of mean, variance, standard deviation, standard error and coefficient of variation; with sample data and output.

```
C
C............................................................................
C
C   THIS PROGRAM COMPUTES THE MEAN, STANDARD ERROR, STANDARD DEVIATION,
C   VARIANCE AND COEFFICIENT OF VARIATION FOR ANY NUMBER OF SAMPLES,
C   EACH OF ANY SIZE. INPUT PARAMETERS ARE AS FOLLOWS –
C       ISTOP = 9 ON LAST CARD OF LAST SAMPLE, OTHERWISE LEAVE BLANK
C       ISAMP = 9 ON LAST CARD OF EACH SAMPLE, OTHERWISE LEAVE BLANK
C       M     = NUMBER OF OBSERVATIONS ON CARD (MAXIMUM = 10)
C       X     = VECTOR OF OBSERVATIONS ON CARD (MAXIMUM OF 10)
C
C............................................................................
C
        DIMENSION X(10)
        WRITE (6, 100)
  100   FORMAT (16H1SUMMARY OF DATA///)
        NS = 0
    1   N = 0
        SM = 0.0
        SSQ = 0.0
    2   READ (5,101) ISTOP, ISAMP, M, (X(I), I = 1, M)
  101   FORMAT (2I1, I3, 2X, 10F6.2)
        DO 3 I = 1, M
        N = N + 1
        SM = SM + X(I)
    3   SSQ = SSQ + X(I)*X(I)
        IF (ISAMP .EQ. 0) GO TO 2
    4   AN = N
        XBAR = SM/AN
        VAR = (SSQ – SM*XBAR)/(AN – 1.0)
        SDEV = SQRT(VAR)
        COEFF = SDEV/XBAR* 100.0
        SERR = SDEV/SQRT(AN)
        NS = NS + 1
        WRITE (6,102) NS, XBAR, SERR, VAR, SDEV, COEFF
  102   FORMAT (1X, 10HSAMPLE NO., I6//1X, 6HMEAN = , F9.4/
       A11X, 16HSTANDARD ERROR = , F9.4/11X, 10HVARIANCE = , F9.4/
       B11X, 20HSTANDARD DEVIATION = , F9.4/11X,
       C26HCOEFFICIENT OF VARIATION = , F9.4//)
        IF (ISTOP .EQ. 0) GO TO 1
        STOP
        END
```

Program 1 (*continued*)

DATA

```
 9   7   7.28   1.10    5.27   4.29    7.00   5.82    4.41
    10  10.20  12.10    9.35  11.70   11.31  10.80   12.27  10.71  14.72  11.90
    10  14.18   9.28   19.60  15.22    9.11  10.02   11.72  14.80   9.00   8.28
99   4  12.10  12.20   13.73  16.10
```

OUTPUT

SUMMARY OF DATA

SAMPLE NO. 1

      MEAN =   5.0243
      STANDARD ERROR =   0.7868
      VARIANCE =   4.3336
      STANDARD DEVIATION =   2.0817
      COEFFICIENT OF VARIATION =   41.4335

SAMPLE NO. 2

      MEAN =   11.8083
      STANDARD ERROR =   0.4388
      VARIANCE =   4.6219
      STANDARD DEVIATION =   2.1499
      COEFFICIENT OF VARIATION =   18.2063

Program 2: Correlation and regression

It is often necessary to test whether there is an association of some kind between two variables. We might, for example, measure the heights of fathers and of their eldest sons and investigate whether the two sets of measurements were related. A measure of the association is provided by the *covariance*, defined as follows. If one set of observations consists of the quantities $x_1, x_2, x_3, \ldots x_n$ and the other of the quantities $y_1, y_2, y_3, \ldots y_n$ then the covariance of x and y is denoted by

$$C_{xy} = \frac{\Sigma\,(x - \bar{x})(y - \bar{y})}{(n - 1)} \qquad (7)$$

which may also be written in a form that is sometimes more suitable for computation as:

$$C_{xy} = \frac{\Sigma\,xy - \Sigma\,x\,\Sigma\,y/n}{(n - 1)}. \qquad (8)$$

There is an obvious analogy with the variance (see formulae (2) and (3) above) but whereas the variance is always a positive quantity (being based on the square of $(x - \bar{x})$ as shown in formula (2)), the covariance may be positive or negative. When the cross-product $(x - \bar{x})(y - \bar{y})$ is positive the two variables are associated in the same sense (i.e. tall fathers have tall sons and short ones have short sons); when it is negative the opposite is true. Like the variance, the covariance depends on the units in which the original measurements are made. It can, however, be standardised by dividing it by the product of the separate standard deviations of x and y. The resultant number is the familiar *correlation coefficient*, more formally known as the product–moment correlation coefficient to distinguish it from other measures of correlation. This may be written

$$r = \frac{C_{xy}}{s_x s_y} \qquad (9)$$

or as either of the following equivalent formulae:

$$r = \frac{\Sigma (x - \bar{x})(y - \bar{y})}{\sqrt{\Sigma (x - \bar{x})^2 \Sigma (y - \bar{y})^2}} \qquad (10)$$

$$r = \frac{\Sigma xy - \Sigma x \Sigma y/n}{\sqrt{(\Sigma x^2 - (\Sigma x)^2/n)(\Sigma y^2 - (\Sigma y)^2/n)}} \cdot \qquad (11)$$

The correlation coefficient may take any value from -1 to $+1$. A value of zero denotes that there is no correlation between the two variables while a value of $+1$ denotes complete positive correlation and of -1 denotes complete negative correlation. In practice, values of, say, $+0.7$ or -0.52 are the sort encountered. A positive value denotes that on the whole the two variables tend to vary together, a large value of x being accompanied by a large value of y and *vice versa*. A negative value denotes the opposite: large values of x are accompanied by small values of y and *vice versa*. Knowing the number of observations on which r is based, its statistical significance can be determined from published tables (e.g. Appendix 4 in Bailey, 1959).

Despite its most formidable appearance, expression (11) above may be the most useful for many computational purposes. It does not require the previous computation of the covariance and the two variances. These are, however, easily obtainable because many of the steps in the calculation are the same. The program below does in fact calculate the covariance and variances (as well as the means of the two sets of measurements) at the same time as the correlation coefficient is computed.

The correlation coefficient is not the only measure of association between two variables and in certain circumstances it may be necessary to compute the *regression* of one variable on the other. The meaning of this relationship becomes clear if we consider a simple scatter diagram in which each point represents the associated values of x and y. The value of x is plotted on the horizontal axis (abcissa) and the corresponding value of y on the vertical axis (ordinate) (Fig. 20). In this diagram x represents the so-called independent variable and y the dependent variable. These are commonly interpreted as having a causal relationship in the sense that a particular value of x determines a corresponding value of y. For example a particular temperature determines the speed of development of a cold-blooded animal. It is, however, best not to confuse causal relationships and statistical associations. A more appropriate interpretation would be that the independent variable x has some arbitrary and perhaps irregular distribution while for any value of x the dependent variable is normally distributed with a mean independent of its constant variance.

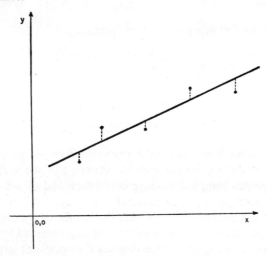

FIG. 20. Scatter diagram and regression line of y on x. Each point represents a pair of observations. The line is placed so that the sum of the squares of the distances from the points to the line is a minimum. The distances concerned are those indicated by broken lines.

Returning to our scatter-diagram, if x and y are linearly related then the points representing the different associated values will tend to fall on a straight line, the *regression line*, which we can draw by eye if we wish. When the association is a direct one (i.e. corresponding to a positive correlation coefficient) the line has a positive slope (i.e. slopes upwards to the right as normally plotted). When the correlation is negative the slope of the line is also negative (i.e. slopes down to the right). Figure 21 illustrates these relationships.

As is well-known, a straight line of this kind can be represented by an equation

$$y = bx + a. \qquad (12)$$

This is the so-called regression equation of y (the dependent variable) on x (the independent one). The equation contains two *parameters* (i.e. constants for a particular set of values of x and y): b is the slope of the line and a is the

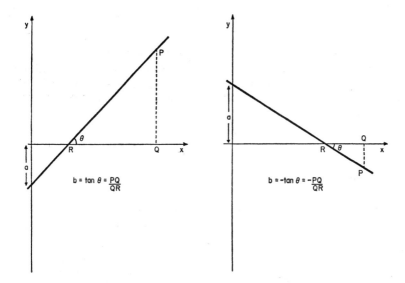

Fig. 21. Diagrams illustrating the geometrical meaning of the parameters a and b in the regression equation $y = bx + a$. The left-hand figure shows a regression line of positive slope ($b > 0$), the right-hand one a line of negative slope ($b < 0$).

intercept it makes on the y-axis. Computing the equation to a regression line therefore involves computing the best estimates of b and a from the available pairs of observations (x, y) (Fig. 20). This can be done from the following two formulae:

$$b = \frac{\Sigma xy - \Sigma x \, \Sigma y / n}{\Sigma x^2 - (\Sigma x)^2 / n} \tag{13}$$

$\left(\text{which is simply a convenient way of writing } b = \dfrac{C_{xy}}{s_x^2}\right)$ and

$$a = \bar{y} - b\bar{x} \tag{14}$$

The question immediately arises as to whether a computed value of b differs significantly from zero. To answer this one needs to know the standard error of the regression coefficient b, which can be computed from the formula:

$$e_b = \frac{\sqrt{\dfrac{\Sigma y^2 - (\Sigma y)^2/n - \dfrac{(\Sigma xy - \Sigma x\, \Sigma y/n)^2}{\Sigma x^2 - (\Sigma x)^2/n}}{(n-2)}}}{\sqrt{\Sigma x^2 - (\Sigma x)^2/n}} \qquad (15)$$

This quantity is quite easy to evaluate since the sums of squares and cross-products which it contains are all calculated in the course of computing the variances and covariance. To make use of this standard error one carries out a t-test as described in any statistical textbook: the ratio of the slope b to its standard error is distributed as Student's t with $(n - 2)$ degrees of freedom. From tabulated values of t (e.g. in Bailey, Appendix 2) it is then possible to decide the probability that the true regression coefficient estimated by b is different from zero.

In biological work it is often desirable to calculate values of r, b and a for a series of samples of different sizes. The following program has therefore been constructed so as to allow any number of samples, each of any size, to be dealt with in a single run. This is accomplished by the two integer variables ISTOP and ISAMP which are used as described for Program 1 on p. 84. A non-zero value (e.g. 9) is punched for ISAMP in column 2 on the last card of each sample and one for ISTOP in column 1 on the last card of the last sample.

Apart from the above two columns, each data card will also contain an integer M, denoting the number of pairs of observations (X and Y) on the card. In the program given, M is supposed to have a maximum of 5, i.e. there will be a total of up to 10 observations on each card, each of the five X values being followed by the corresponding Y value. Obviously if the data are such that more than 5 pairs of observations are to be accommodated on a card, the DIMENSION statement and the FORMAT declaration accompanying the first READ statement must both be changed.

As the flow chart shows, the program proper starts with a DIMENSION statement for X and Y and the writing of a title. The counter NS, denoting sample number, is set at zero as is also N, the number of pairs of observations in the sample. The other variables initialised at zero are SMX and SMY (the sums of X and Y respectively), SMXY (the sum of the cross-products XY), together with SSQX and SSQY (the sums of squares of X and Y). The first data card is then read, using an implied DO-loop for the M paired values of X and Y. A DO-loop ending on statement 3 then accumulates the totals of N, SMX, SMY, SMXY, SSQX and SSQY for the M values on the card.

The whole process, starting with the reading of another card, is continued until the contents of the card with ISAMP $\neq$ 0 have been dealt with. Looping

Flowchart for Program 2

Computation of covariance, correlation coefficient, and regression equation. Note that the basic structure of this program is identical to that shown in the flowchart for Program 1 (p. 86).

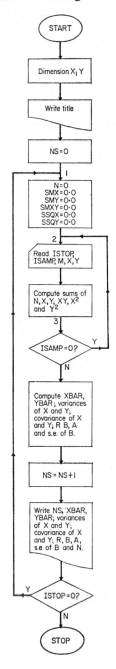

then stops and the means (XBAR and YBAR), the variances (VARX, VARY), the covariance (COVAXY), the regression coefficients (BY and AY) and the standard error of BY (SEB) are computed in succession. The sample counter N is next increased by 1 and the results for this sample printed out. So long as further cards are available (i.e. ISTOP = 0), control returns to statement 1 and the next sample is processed. When the last card has been read and its data dealt with, the index ISTOP is no longer zero and the execution of the program stops.

The print-out lists the values calculated for each sample. It should be noted that the FORMAT statement gives an E-conversion with field-width 16.8 for the variances and covariances, thus enabling any correct value to be printed. The FORMAT codes for the means and for BY and AY are, however, F9.4 and will require amendment if values outside this range are likely to be encountered.

Program 2

Computation of covariance, correlation coefficient and regression equation; with sample data and output.

```
C
C.........................................................
C
C   THIS PROGRAM COMPUTES THE CORRELATION COEFFICIENT BETWEEN X AND Y
C   AND ESTIMATES THE PARAMETERS IN THE REGRESSION EQUATION OF Y ON X.
C   OUTPUT ALSO INCLUDES MEANS AND VARIANCES OF X AND Y, COVARIANCE
C   OF X AND Y AND STANDARD ERROR OF SLOPE OF REGRESSION LINE.
C   INPUT PARAMETERS ARE AS FOLLOWS –
C       ISTOP = 9 ON LAST CARD OF LAST SAMPLE, OTHERWISE LEAVE BLANK
C       ISAMP = 9 ON LAST CARD OF EACH SAMPLE, OTHERWISE LEAVE BLANK
C       M     = NUMBER OF PAIRS OF OBSERVATIONS ON CARD (MAXIMUM = 5)
C       X     = INDEPENDENT VARIABLE
C       Y     = DEPENDENT VARIABLE
C
C.........................................................
C
        DIMENSION X(5), Y(5)
        WRITE (6,101)
   101  FORMAT (36H1CORRELATION AND REGRESSION ANALYSIS///)
C
C   INITIALISE FOR ACCUMULATION OF SUMS.
        NS = 0
      1 N = 0
        SMX = 0.0
        SMY = 0.0
        SMXY = 0.0
        SSQX = 0.0
        SSQY = 0.0
C
C   READ DATA AND ACCUMULATE SUMS AND SUMS OF SQUARES AND PRODUCTS.
      2 READ (5,102) ISTOP, ISAMP, M, (X(I), Y(I), I = 1,M)
    102 FORMAT (2I1, I4, 10F6.2)
        DO 3 I = 1,M
```

Program 2 (*continued*)

```
      N = N + 1
      SMX = SMX + X(I)
      SMY = SMY + Y(I)
      SMXY = SMXY + X(I) * Y(I)
      SSQX = SSQX + X(I) * X(I)
    3 SSQY = SSQY + Y(I) * Y(I)
      IF (ISAMP .EQ. 0) GO TO 2
C
C COMPUTE STATISTICS.
      AN = N
      XBAR = SMX/AN
      YBAR = SMY/AN
      A = SSQX - SMX* SMX/AN
      B = SSQY - SMY* SMY/AN
      C = SMXY - SMX* SMY/AN
      VARX = A/(AN - 1.0)
      VARY = B/(AN - 1.0)
      COVAXY = C/(AN - 1.0)
      R = C/SQRT (A*B)
      BY = C/A
      AY = YBAR - BY*XBAR
      SEB = SQRT((B - C*C/A)/(AN - 2.0))/SQRT(A)
      NS = NS + 1
C
C OUTPUT RESULTS.
      WRITE (6,103)NS, XBAR, YBAR, VARX, VARY, COVAXY, R, BY, AY, SEB, N
  103 FORMAT (1X, 10HSAMPLE NO.,I6//11X, 11HMEAN OF X = , F9.4/
     A11X, 11HMEAN OF Y = ,F9.4/11X, 15HVARIANCE OF X = , E16.8/
     B11X, 15HVARIANCE OF Y =, E16.8/11X, 23HCOVARIANCE OF X AND Y =,
     CE16.8/11X, 25HCORRELATION COEFFICIENT = , F9.4/
     D11X, 16HY = BX + A, B =, F9.4, 5X, 3HA = ,F9.4/
     E11X, 21HSTANDARD ERROR OF B = ,F9.4/11X, 14HNO OF POINTS = ,I6//)
      IF (ISTOP .EQ. 0) GO TO 1
      STOP
      END
```

DATA

```
    5  7.10  2.93 10.41  5.29  8.11  4.00 16.12  9.28  3.05  1.37
 9  3  8.22  5.08  6.12  3.00  8.00  4.12
    5  8.00  4.12  6.12  3.00  8.22  5.08  3.05  1.37 16.12  9.28
99  3  8.11  4.00 10.41  5.29  7.10  2.93
```

OUTPUT

CORRELATION AND REGRESSION ANALYSIS

SAMPLE NO.   1

```
           MEAN OF X =    8.3912
           MEAN OF Y =    4.3837
           VARIANCE OF X =   0.14203614E 02
           VARIANCE OF Y =   0.55152841E 01
           COVARIANCE OF X AND Y =  0.86972094E 01
           CORRELATION COEFFICIENT =    0.9826
           Y = BX + A,  B =    0.6123     A =  - 0.7544
           STANDARD ERROR OF B =    0.0472
           NO OF POINTS =     8
```

Program 2 (*continued*)

SAMPLE NO.   2

    MEAN OF X =   8.3912
    MEAN OF Y =   4.3837
    VARIANCE OF X =   0.14203613E 02
    VARIANCE OF Y =   0.55152841E 01
    COVARIANCE OF X AND Y =   0.86972094E 01
    CORRELATION COEFFICIENT =   0.9826
    Y = BX + A,   B =   0.6123   A = − 0.7544
    STANDARD ERROR OF B =   0.0472
    NO OF POINTS =   8

From a programming point of view there is little in this program which does not occur in Program 1. ISTOP and ISAMP are used to control the reading of the correct number of cards and the accumulation of sums is brought about by a DO-loop after initialisation to zero. It is perhaps also worth noting that the computation of the covariance by summation of cross-products follows formula (8) rather than the more explicit (7).

A less obvious aspect of the program involves a potential weakness, of a kind which the programmer needs always to bear in mind. In computing the correlation coefficient we make use of relation (9) and divide the covariance by the standard deviations of x and y. Now it could happen that a set of data consisted of identical numbers for all values of y (or of x, or of both). If such a thing happened, the variance and standard deviation of y (and/or x) would be zero and the computer would, in effect, be required to carry out a division by zero. Some compilers will return a zero value for such an operation, thus giving a value $r = 0.0$, which is the statistically correct answer. Other compilers, however, will cease execution when required to divide by zero. A possibility like this can be guarded against by suitable programming (as we shall see on p. 236). The likelihood of any realistic data consisting of identical values is, however, remote and most programmers would probably feel it unnecessary to modify the program given above. The point is discussed because it illustrates how an apparently satisfactory program might work repeatedly, yet fail when a quite unusual set of data were presented to it.

A final warning is necessary for users of correlation and regression programs. While it is always possible to calculate a correlation coefficient and the equation to a regression line, it does not follow that the problem being investigated yields data which are statistically suitable for such purposes. The question of when such statistics provide a proper measure of association, and what conditions must be satisfied before they can be meaningfully interpreted falls outside the scope of this work (but see p. 181). Like all statistical procedures, they are based on mathematical models whose relevance should be assured before they are used. It would be very unfortunate if the ease with which a computer carries out analyses like this led to their being used inappropriately. Far from freeing the biologist from statistical problems, the computer really forces him to inquire even more closely into the assumptions on which statistical methods are based.

Program 3: Comparison of means of two samples

One of the most frequently required statistical procedures in biological work is a test of significance of the difference between the means of two samples. Let us suppose that one of the samples takes values $a_1, a_2, a_3, \ldots a_{n_a}$ and the other one $b_1, b_2, b_3, \ldots b_{n_b}$ so that the numbers of observations in the two samples are respectively n_a and n_b. The means are $\bar{a}$ and $\bar{b}$ and the test to be carried out depends on several things. In the first place we must consider a test for large samples, where n_a and n_b are both greater than, say, 30. To compare means in this case we compute a quantity d from the formula:

$$d = \frac{|\bar{a} - \bar{b}|}{\sqrt{\dfrac{s_a^2}{n_a} + \dfrac{s_b^2}{n_b}}} \quad \text{where } s_a^2 \text{ and } s_b^2 \text{ are the sample variances.} \quad (16)$$

To understand this formula we note first that the numerator is the absolute value of the difference between the two means. The denominator is the square root of the variance of the difference between the two means, i.e. the standard deviation of the difference. We are therefore expressing the difference in units of one standard deviation. Provided the samples are drawn from two normally distributed populations, d represents a normal deviate with zero mean and unit standard deviation. The probability that a departure as great or greater than the computed value of d can occur by chance is then available from published tables. In fact the only values of d and P usually required are:

d	P
1.960	0.05
2.576	0.01
3.291	0.001

Thus the probability that the means are significantly different is less than 0.05 if d is greater than 1.960, less than 0.01 if d is greater than 2.576 and so on.

If the samples are small (i.e. n_a or n_b less than, say, 30), the above normal deviate test must be replaced by a t-test. The exact test required will depend on whether the two samples are drawn from populations with the same or different variances. It is therefore first necessary to compare the sample variances by a variance-ratio test, i.e. to compute F, the ratio of the larger to the smaller variance. This will be

$$F = \frac{s_a^2}{s_b^2} \tag{17}$$

or

$$F = \frac{s_b^2}{s_a^2} \tag{18}$$

according to which variance is the larger. Associated with F are two degrees of freedom: n_1 (corresponding to $(n_a - 1)$ or $(n_b - 1)$ according to whether s_a^2 is greater than s_b^2 or *vice versa*) and n_2, which is one less than the number of observations in the other sample.

If the two variances are unequal then the computed value of F (for n_1 and n_2 degrees of freedom) will exceed the critical values of F tabulated for the probabilities $P = 0.05$ and smaller. If, on the other hand, the computed value of F is less than the critical value, then it can be assumed that the two sample variances are not significantly different.

We now return to the question of whether the means of the two small samples differ significantly. If the variances of the two samples can be assumed to be equal, then we compute:

$$t = \frac{|\bar{a} - \bar{b}|}{s\sqrt{\frac{1}{n_a} + \frac{1}{n_b}}} \tag{19}$$

where s, the best estimate of the common variance, is given by

$$s = \sqrt{\frac{\Sigma (a - \bar{a})^2 + \Sigma (b - \bar{b})^2}{n_a + n_b - 2}}. \tag{20}$$

The quantity t, computed from (19) has $(n_a + n_b - 2)$ degrees of freedom and the corresponding probability can be found from tables of the t-distribution.

If, however, the variances of the two samples are unequal, then we require a different form of the t-test. We compute the quantity

$$t' = \frac{|\bar{a} - \bar{b}|}{\sqrt{\frac{s_a^2}{n_a} + \frac{s_b^2}{n_b}}} \tag{21}$$

which is distributed approximately like t, with a number of degrees of freedom given by the formula:

$$\text{d.f.} = \cfrac{1}{\sqrt{\cfrac{u^2}{n_a - 1} + \cfrac{(1-u)^2}{n_b - 1}}} \quad (22)$$

where

$$u = \frac{s_a^2/n_a}{s_a^2/n_a + s_b^2/n_b} \quad (23)$$

Formula (22) does not in general give an integer value for the number of degrees of freedom, so the value is rounded off to the nearest whole number and used with t' in a table of the t-distribution.

Before discussing details of the program which will carry out the above tests a few general points may be noted. Reference has been made to tabulated values of d, t and F, which are required at various points. It is not difficult to write a computer program which enables one to dispense with tables and to calculate, with sufficient accuracy, the probabilities required. (See p. 403). This would, however, lead us too far aside at present so the following program will print out value of t, d and F with the appropriate degrees of freedom. Where small samples are concerned, the perfect program would first carry out an F-test and then—depending on the associated probability—carry out the appropriate form of t-test. Since we do not wish to halt the execution of the program while we (or a computer operator) consults a set of statistical tables, we shall arrange for the F-test to be followed by computations for both forms of the t-test for small samples. The user will then use tables to assess the probability level corresponding to F and will choose the appropriate value of t from the two given.

Reference to the flowchart and program listing shows that after the appropriate introductory comments we dimension two arrays A and B, to accommodate the observations forming the two samples. The dimensions of 1000 enable all reasonable sets of data to be handled, but can obviously be increased if very large samples are being dealt with. The next step is to read a program specification card (or parameter card as it is sometimes called). This has punched on it the values of NPROB (a reference number given by the user) and NA and NB, the numbers of observations in the two samples. NPROB also has the function of providing a sentinel card, since if it is given a zero value it will cause execution to stop. A card punched with a zero in column 5 is therefore placed at the end of the data deck (a blank card could be used, but is less desirable). Assuming, however, that NPROB has some non-zero value, the program then provides two READ statements, causing the observations punched on data cards for the first and second sample (A and B, respectively) to be read under the FORMAT 10F7.0. This method of reading in the data needs a little explanation. Each sample is read in by an implied DO-loop, but we use two READ statements because it is convenient to begin each sample on a new card. Had we written

READ (5,101) (A (I), I = 1, NA), (B (I), I = 1, NB)

we could still have read in two arrays but the individual observations would have had to be punched in a continuous string. This is often inconvenient because each sample usually has a very definite identity of its own and it is desirable to be able to handle it as a separate group of cards.

The program now proceeds to carry out the major computational steps, the calculation of the mean and variance for each of the two samples separately. In each case this involves initialising at zero the variables which will be used to accumulate the sums and sums of squares and then accumulating them by DO-loops. These terminate on statements 2 and 3 respectively. The variances are then stored in VA and VB respectively and at the same time we take the opportunity of computing the standard errors of the two samples as SEA and SEB. These are not required for the significance tests but they are useful quantities if the data are being presented in summary form and they can be used to compute confidence limits for the means of the two samples.

After completing these calculations we reach the decision-box in which we ask whether the samples are large or small. Here, as earlier in the decision-box where NPROB is used, we employ a logical IF statement. This directs control to statement 4 if the samples are small, otherwise we proceed with the significance test for large samples. A single arithmetic statement computes D, the normal deviate, using the previously calculated means and variances. Here—as also when calculating the means in the previous section—we make use of the variables ANA and ANB, which are simply the floating-point equivalents of NA and NB. Having computed D we then execute a WRITE statement which prints out all the information we require for the large samples. Control then returns to statement 1 and we read the problem specification card for the next two samples.

If, however, our samples were small and the decision box had directed us to statement 4, the program would carry out the small-sample significance tests. These begin with the computation of the variance ratio F, followed by a decision box catering for the possibility that the variance of sample A is less than that of B. In such a case the reciprocal of the previous ratio is used and the degrees of freedom are interchanged accordingly. The computation of the two values of t and their degrees of freedom is then followed by a WRITE statement covering all the small sample tests. Note that the degrees of freedom calculated from formula (22) above are printed out under F5.0 so that rounding to the nearest whole number is automatically ensured. Finally

Flowchart for Program 3

Comparison of means of two samples by normal deviate or *t*-test.

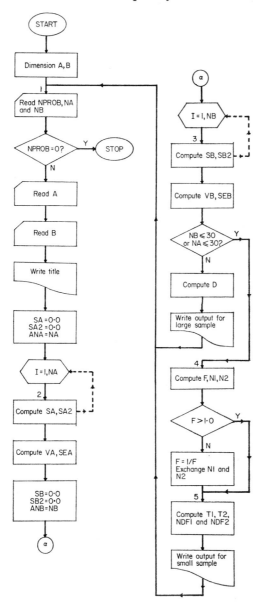

control returns to statement 1 so that the next problem specification card can be read.

From a programming standpoint, the most interesting feature of this program is the way decision boxes are used to control several possible types of test depending on the nature of the data. The use of a problem specification card is also worth noting. This can be a very useful feature of large programs, providing a quick method of defining a number of essential parameters as well as offering the opportunity of specifying a variety of options through the use of index numbers.

Program 3

Comparison of means of two samples by normal deviate or *t*-test; with sample data and output.

```
C
C.........................................................................
C
C   THIS PROGRAM COMPARES THE MEANS OF TWO SAMPLES BY A NORMAL DEVIATE
C   OR T-TEST AS APPROPRIATE. FOR COMPUTATIONAL DETAILS SEE BAILEY (1959)
C   STATISTICAL METHODS IN BIOLOGY, PP. 171-174.
C   INPUT PARAMETERS ARE AS FOLLOWS -
C       NPROB = NUMBER OF PROBLEM
C       NA    = NUMBER OF OBSERVATIONS IN FIRST SAMPLE
C       NB    = NUMBER OF OBSERVATIONS IN SECOND SAMPLE
C       A     = VECTOR OF OBSERVATIONS IN FIRST SAMPLE
C       B     = VECTOR OF OBSERVATIONS IN SECOND SAMPLE
C
C.........................................................................
C
        DIMENSION A(1000), B(1000)
C
C   READ SAMPLE SPECIFICATION AND DATA.
      1 READ (5,100) NPROB, NA, NB
    100 FORMAT (3I5)
        IF (NPROB .EQ. 0) STOP
        READ (5,101) (A(I), I = 1,NA)
        READ (5,101) (B(I), I = 1,NB)
    101 FORMAT (10F7.0)
        WRITE (6,102) NPROB
    102 FORMAT (////8H PROBLEM, I4)
C
C   COMPUTE MEANS, VARIANCES AND STANDARD ERRORS.
        SA = 0.0
        SA2 = 0.0
        ANA = NA
        DO 2 I = 1,NA
        SA = SA + A(I)
      2 SA2 = SA2 + A(I) * A(I)
        ABAR = SA / ANA
        SA2 = SA2 - SA * ABAR
        VA = SA2 / (ANA - 1.0)
        SEA = SQRT(VA/ANA)
C
        SB = 0.0
```

Program 3 (*continued*)

```
      SB2 = 0.0
      ANB = NB
      DO 3 I = 1,NB
      SB = SB + B(I)
    3 SB2 = SB2 + B(I) * B(I)
      BBAR = SB / ANB
      SB2 = SB2 - SB * BBAR
      VB = SB2 / (ANB - 1.0)
      SEB = SQRT(VB/ANB)
      IF (NA .LE. 30 .OR. NB .LE. 30) GO TO 4
C
C   EXECUTE SIGNIFICANCE TESTS FOR LARGE SAMPLES.
      D = ABS(ABAR - BBAR) / SQRT(VA/ANA + VB/ANB)
      WRITE (6,103) ABAR, SEA, BBAR, SEB, D
  103 FORMAT (/6X, 22HMEAN OF FIRST SAMPLE = , F10.3/
     A  6X, 32HSTANDARD ERROR OF FIRST SAMPLE = , F10.3/
     B  6X, 23HMEAN OF SECOND SAMPLE = , F10.3/
     C  6X, 33HSTANDARD ERROR OF SECOND SAMPLE = , F10.3/
     D  6X, 16HNORMAL DEVIATE = , F10.3)
      GO TO 1
C
C   EXECUTE SIGNIFICANCE TESTS FOR SMALL SAMPLES.
    4 F = VA / VB
      N1 = NA - 1
      N2 = NB - 1
      IF(F .GT. 1.0) GO TO 5
      F = 1.0 / F
      N1 = NB - 1
      N2 = NA - 1
    5 V = (SA2 + SB2) / (ANA + ANB - 2.0)
      T1 = ABS(ABAR-BBAR) / SQRT(V) / SQRT(1.0/ANA + 1.0/ANB)
      NDF1 = NA + NB - 2
      T2 = ABS(ABAR-BBAR) / SQRT(VA/ANA + VB/ANB)
      U = (VA/ANA) / (VA/ANA + VB/ANB)
      DF2 = 1.0 / (U*U / (ANA-1.0) + (1.0-U)**2 / (ANB-1.0))
      WRITE (6,104) ABAR, SEA, BBAR, SEB, F, N1, N2, T1, NDF1, T2, DF2
  104 FORMAT (/6X, 22HMEAN OF FIRST SAMPLE = , F10.3/
     A  6X, 32HSTANDARD ERROR OF FIRST SAMPLE = , F10.3/
     B  6X, 23HMEAN OF SECOND SAMPLE = , F10.3/
     C  6X, 33HSTANDARD ERROR OF SECOND SAMPLE = , F10.3/
     D  6X, 16HVARIANCE RATIO = , F10.3/ 6X, 4HN1 = , I5, 10X, 4HN2 = , I5/
     E  6X, 29HT (VARIANCES ASSUMED EQUAL) = ,F10.3,6H (WITH,I5,6H D.F.)/
     F  6X, 33HT (VARIANCES NOT ASSUMED EQUAL) = , F10.3, 6H (WITH, F5.0,
     G  6H D.F.))
      GO TO 1
      END
```

DATA

```
  1   10    8
  2.0   3.0   1.0   3.0   2.0   4.0   1.0   3.0   1.0   2.0
  1.0   1.0   3.0   4.0   4.0   1.0   1.0   1.0
  2   31   31
  2.0   3.0   3.0   4.0   1.0   2.0   3.0   2.0   2.0   3.0
  3.0   4.0   4.0   3.0   2.0   4.0   1.0   2.0   3.0   3.0
  2.0   4.0   1.0   2.0   3.0   2.0   3.0   1.0   1.0   2.0
  2.5   5.0   5.0   6.0   3.0   4.0   5.0   4.0   4.0   5.0
  4.0   6.0   6.0   5.0   4.0   6.0   3.0   4.0   5.0   5.0
  4.0   6.0   3.0   4.0   5.0   4.0   5.0   3.0   3.0   4.0
  4.5
  0
```

Program 3 (*continued*)

OUTPUT

PROBLEM 1

    MEAN OF FIRST SAMPLE =   2.200
    STANDARD ERROR OF FIRST SAMPLE =   0.327
    MEAN OF SECOND SAMPLE =   2.000
    STANDARD ERROR OF SECOND SAMPLE =   0.500
    VARIANCE RATIO =   1.875
    N1 =   7   N2 =   9
    T (VARIANCES ASSUMED EQUAL) =   0.347 (WITH   16 D.F.)
    T (VARIANCES NOT ASSUMED EQUAL) =   0.335 (WITH   12. D.F.)

PROBLEM 2

    MEAN OF FIRST SAMPLE =   2.500
    STANDARD ERROR OF FIRST SAMPLE =   0.172
    MEAN OF SECOND SAMPLE =   4.500
    STANDARD ERROR OF SECOND SAMPLE =   0.172
    NORMAL DEVIATE =   8.224

Chapter 5

Sorting, Tabulating and Summarising Data

Unlike the previous chapter, this one includes very little new statistical or mathematical material and incorporates only a few of the statistical concepts that have already been introduced. Instead, it aims at indicating a few techniques for handling numerical data so as to summarise or display some of their properties in more convenient or precise form. It might therefore be regarded as an introduction to some elementary Fortran methods of *data processing* rather than statistical analysis. Biological data are usually relatively time-consuming to collect and most biologists—unless they are working as a team on some kind of survey—are unlikely to collect vast quantities of numerical information. They may, nevertheless, find some of these methods useful as a preliminary to more highly analytical techniques. It has been assumed in writing some of the programs that the data may be accumulated over a relatively long period and that it is stored on punched cards. Processing therefore requires programs with some flexibility in the following respects. Firstly it is an advantage if the input and output formats can be decided by the user according to the special requirements of the data; we shall accordingly introduce variable FORMAT techniques. Secondly, it is often desirable that records should be counted automatically while being processed. Thirdly, we shall arrange for any number of similar problems to be dealt with in succession. The first and third of these requirements are closely related: there is little point in arranging for successive problems to be handled unless one is free to change formats. At this stage we shall not consider the storage of data on magnetic tape. Simple records can be handled quite easily in this way once the normal tape manipulation statements (pp. 449–451) are understood, but the more elaborate problems of file-processing and updating are not considered in this book. Anyone who needs to store very extensive or complex records needing periodic revision will have to seek special guidance. It is, however, surprising how much data can be handled efficiently by quite simple methods.

Eight programs are discussed in this chapter. The first two will sort and rank an array of numbers, the third provides a set of tabulated frequency distributions and the fourth enables one to construct contingency tables. All four show how Fortran can carry out the processes of sorting, selecting and arranging numbers. These are not commonly thought of as arithmetical operations, but they can ultimately be reduced to processes like addition, multiplication and division. The next three programs produce summaries of simple statistical quantities—mean, variance, range, etc. for three different kinds of primary data, while the last program constructs the running means sometimes used in smoothing data. It is worth bearing in mind that many of the sorting, selecting, scanning and arranging procedures occur repeatedly as incidental features of more complicated Fortran programs. Ranking, for example, plays a large part in non-parametric statistics (Chap. 11). An important feature of the art of programming—perhaps its most important aspect—is the capacity for reformulating in arithmetical terms the processes of manipulation and recognition that one carries out intuitively when doing a calculation with pencil and paper. To find the maximum number in an array, or to record the positions of all the zeros in a two-way table, or to allocate numbers to previously defined classes represent exercises in simple, logical thought to be translated into a set of instructions for the computer. It is the object of this chapter to indicate a few such 'tricks of the trade' and to develop the reader's capacity for evolving others, at the same time producing a set of useful data-processing programs.

Program 4: Sorting an array in ascending and/or descending order

The object of this program is to read in a linear array of numbers and to sort them in order of size. This is done either in ascending order (the algebraically smallest number coming first), or in descending order (the largest number first), or in both orders. The user will have the opportunity of choosing which arrangement he needs. Basically the method consists first of sorting the numbers in ascending order and then reversing the sorted sequence should a descending order by required. Sorting is carried out by an 'exchange sort' method described below.

After some explanatory comment statements the program begins with a type-declaration INTEGER OUTFMT. This is intended to ensure that the array OUTFMT, which will accommodate the variable output format specifications, is treated as an integer variable. This is because it will be read and stored under an A-specification (p. 54) and some computers require that only integer variables can be so stored. An alternative way of coping with the situation would have been to use an implicitly typed integer variable, such as NFMT.

The DIMENSION statement allows the array being sorted to consist of up to 1000 numbers and also provides storage space for the variable input and output formats. A problem specification card is then read, giving the reference number of the problem (NPROB), the size of the array (N) and a code number MODE which is set at 1, 2 or 3 depending on whether one needs ascending, descending or both sequences to be worked out. A test on NPROB enables the program to stop execution after all problems have been handled, using a sentinel card (the last data card) on which NPROB is punched as zero.

The variable input and output formats are then read and stored under an A-conversion that allows up to 80 characters for each format. The two data cards carrying these format specifications will be placed immediately after the problem specification card and will carry such information as

$$(10F7.0)$$

punched starting with the left parentheses in column 1 of the card. The input specification is stored as part of the 80 characters available in the array INFMT and the output specification in the array OUTFMT. The 20A4 specification under which INFMT and OUTFMT are read assumes a computer word-structure of 4 or more characters per word. Needless to say, the data cards for the problem must be punched in accordance with the INFMT specifications while the OUTFMT specifications must provide for suitable print-out. The array A is next read in from the data cards by an implied DO-loop under control of the specifications stored in INFMT and the sorting process then begins.

Essentially this consists of comparing each number in the array in turn with the number above it. If the latter number is greater than or equal to the previous one, the two numbers are left as they were. If the lower number in the array is bigger, however, it is interchanged with the one above. The process begins with the first and second numbers in the original array and ends with the last but one and the last. This first scan will ensure that the largest number in the array is moved to the end, though it is also possible that several of the numbers may happen to be in correct ascending order near the end. This possibility is used to reduce sorting time by noting the point at which the last interchange took place. A second run through the array up to the point of last interchange is now made, resulting in a new point of last interchange, and so on until this point is reduced to 1, when the array will be completely sorted in ascending order.

The above explanation may help in following the program segment running from the comment

C BEGIN ASCENDING SORT

up to but not including statement 4. To begin, an upper limit for the scan is placed at the penultimate position of the unsorted array by setting

$$LIMIT = N - 1$$

while the point of last interchange (INTER) is initialised at 1 (for reasons we shall see later). A DO-loop ending on statement 3 now begins the comparison of number-pairs from A(1) up to A(N). At each pass through the loop, INTER is increased to I, the loop index, whenever an interchange of numbers occurs. The final value of INTER when the loop is satisfied therefore indicates the point of last interchange.

If, by a fortunate change, the numbers were already in ascending sequence before the sorting program started, INTER would have remained at the initial setting of 1. This possibility is therefore tested. If, as is far more likely, INTER is now some value greater than 1 the sorting loop is repeated by the GO TO 2 statement after an adjustment of LIMIT to one less than the current value of INTER. This sequence continues until eventually all the necessary interchanges have been made and INTER remains at 1. At this point the program heading is printed and a computed GO TO statement prints out the ascending series if MODE was set at 1 or 3 but transfers control to statement 6 if MODE was set at 2 (i.e. only a descending sequence was required).

The numbers are then arranged in a descending sequence by exchanging the lowest and highest numbers, then the second lowest and second highest, and so on. These exchanges, like the previous ones used to sort the numbers in ascending order, are accomplished by a DO-loop and a special temporary storage location TEMP that stores one of the numbers being exchanged so that it is not lost when the other number is read into its location. Another feature of this DO-loop is that its upper index parameter NHALF is defined by the previous instructions:

$$NHALF = N/2$$

This statement is valid whether N is an even or an odd number. In the latter case truncation in an integer mode division gives a value to NHALF which represents only the first $(n - 1)/2$ numbers out of the n present in the series. But the central number out of an odd-numbered total does not need to be reversed and the program is therefore perfectly effective.

Having reversed the ascending sequence to produce a descending one, the program prints it out and returns to statement 1 to read the specification

Flowchart for Program 4

Sorting an array of numbers in ascending and/or descending order.

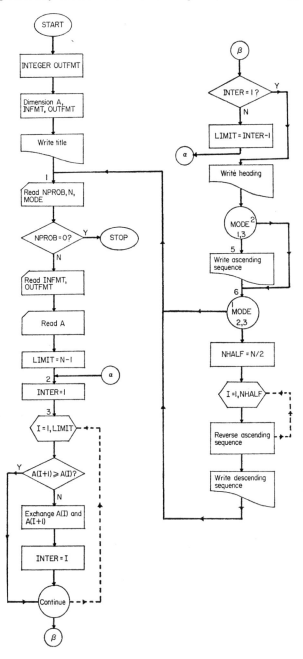

card for the next sample (or the sentinel card if no further problems remain). A flowchart for the whole program is given on p. 109.

Program 4

Sorting an array of numbers in ascending and or descending order.

```
C
C............................................................
C
C   THIS PROGRAM ARRANGES A SET OF NUMBERS IN ASCENDING AND/OR DESCENDING
C   ORDER OF SIZE. INPUT PARAMETERS ARE AS FOLLOWS –
C       NPROB  = NUMBER OF PROBLEM
C       N      = NUMBER OF OBSERVATIONS IN SAMPLE
C       MODE   = 1 FOR ASCENDING ORDER ONLY
C                2 FOR DESCENDING ORDER ONLY
C                3 FOR BOTH
C       INFMT  = VECTOR HOLDING INPUT FORMAT
C       OUTFMT = VECTOR HOLDING OUTPUT FORMAT
C       A      = VECTOR OF OBSERVATIONS
C
C............................................................
C
        INTEGER OUTFMT
        DIMENSION A(1000), INFMT(20), OUTFMT(20)
C
C   WRITE TITLE.
        WRITE (6,100)
  100   FORMAT (1H1,/// 40X, 27HSORTING AN ARRAY OF NUMBERS)
C
C   READ PROBLEM SPECIFICATION AND VARIABLE FORMATS.
    1   READ (5,101) NPROB, N, MODE
  101   FORMAT (3I5)
        IF (NPROB .EQ. 0) STOP
        READ (5,102) (INFMT(I), I = 1,20)
        READ (5,102) (OUTFMT(I), I = 1,20)
  102   FORMAT (20A4)
C
C   READ IN ARRAY A
        READ (5,INFMT) (A(I), I = 1,N)
C
C   BEGIN ASCENDING SORT
        LIMIT = N – 1
    2   INTER = 1
C
C   COMPARE EACH NUMBER WITH NEXT AND INTERCHANGE IF NECESSARY.
        DO 3 I = 1,LIMIT
        IF (A(I + 1) .GE. A(I)) GO TO 3
        TEMP = A(I)
        A(I) = A(I + 1)
        A(I + 1) = TEMP
        INTER = I
    3   CONTINUE
C
C   ALL NUMBERS ABOVE INTER ARE NOW IN ORDER. REPEAT WITH NEW VALUE
C   OF LIMIT IF NECESSARY.
        IF (INTER .EQ. 1) GO TO 4
        LIMIT = INTER – 1
        GO TO 2
    4   WRITE (6,103) NPROB, N
  103   FORMAT (///8H PROBLEM, I5, 13H   (CONTAINING, I5, 9H NUMBERS))
```

Program 4 (*continued*)

```
C
C   SELECT WHETHER ASCENDING SERIES IS TO BE PRINTED.
        GO TO (5,6,5), MODE
      5 WRITE (6,104)
    104 FORMAT (/11X, 15HASCENDING ORDER/)
        WRITE (6,OUTFMT) (I, A(I), I = 1,N)
C
C   SELECT WHETHER DESCENDING SEQUENCE IS TO BE COMPUTED
      6 GO TO (1,7,7), MODE
C
C   REVERSE SEQUENCE AND PRINT.
      7 NHALF = N / 2
        DO 8 I = 1,NHALF
        N1 = N - I + 1
        TEMP = A(I)
        A(I) = A(N1)
      8 A(N1) = TEMP
        WRITE (6,105)
    105 FORMAT (/11X, 16HDESCENDING ORDER/)
        WRITE (6,OUTFMT) (I, A(I), I = 1,N)
C
C   REPEAT FOR NEXT SAMPLE.
        GO TO 1
        END
```

The main features to be noted in the program are the use of variable formats, the method of exchanging numbers using a temporary storage location, and the whole strategy of the exchange sort. It is worth remembering that the method used to create first the ascending array and then the descending one does actually move the numbers to new storage locations so that the order of the original array is completely destroyed and replaced at the end of the problem by *either* an ascending array *or* a descending one. It sometimes happens that the original order of an array needs to be retained and in such cases a sorting program must allow for an additional array or arrays to hold the sorted numbers. It is also possible, given an ascending array, to print out the values in descending order without moving them to new storage locations. The reader in need of practice is invited to develop these simple alternatives to Program 4. It will remind him that a single basic algorithm—in this case the exchange sort—can often be implemented in several different ways to cater for different requirements. Careful consideration of the range of possible requirements should always precede the construction of a program.

Program 5: Ranking an array of numbers

This short program assigns to each number in an array the corresponding rank, starting with the smallest number ranking 1 for an ascending series and with the largest number ranking 1 for a descending series. The method of ranking is such that when two or more numbers are equal they all receive the same rank, which is the next ranking number, whatever it may be. One or more rank-numbers are then omitted before the next number in the array is assigned a rank. Thus in the array 7.0, 6.5, 4.0, 6.5, 2.1 the descending ranks would be respectively 1, 2, 4, 2, 5. The two numbers 6.5 are each placed second, followed by a 4.0 in fourth place. This system of ranking is useful for many purposes, but it is not the one needed for some statistical purposes, where ties have to be scored differently (see p. 322).

The program begins with an explicit type-declaration for the two arrays which will be used to accommodate the ranks, R1 for the descending series and R2 for the ascending series. Both sets of ranks will be integers in the method followed here. R1, R2 and A (accommodating the numbers to be ranked) are all dimensioned at 1000 in the program though this could readily be increased if necessary.

After writing a title the problem specification card is read, giving the problem reference number (NPROB) and the number of observations in the array A. This number, N, is sometimes referred to as the length of the array A or the length of the vector A. The significance of the term 'vector' for a linear array of numbers will become apparent later (p. 317). The variable NPROB is also used, as in previous programs, to allow a sentinel card (NPROB having a zero value) to stop execution when all the data has been dealt with. For each program the array A is read in by an implied DO-loop under a FORMAT of 7F10.0. This F-specification should accommodate any number likely to be encountered in practice since it will be remembered (p. 49) that a decimal point punched on the data card over-rides the exact form of the F-specification. F10.0 will therefore accommodate any number which, with its decimal point, can be punched in a field 10 columns wide.

The essential feature of the program then follows in the two nested DO-loops beginning DO 3 J = 1, N which rank the numbers in descending sequence. To understand this procedure, consider the first passage through the outer loop. This sets a counter NX = 1 and then, by means of the inner loop, compares the first number in the array—A(1)—with each number in the array in turn. Every time a number is encountered that exceeds A(1) the counter NX is incremented by 1 and when the inner loop is satisfied—i.e. when A(1) has been compared with all other numbers—the current value of NX gives the rank of A(1). This rank is then placed in location R1(1) and the outer loop is entered for a second time, performing the same series of operations for A(2), and so on.

An exactly similar procedure is then used to rank the numbers in ascending sequence, except that here the logical IF statement increments NX whenever A(I) is *less* than A(J).

The output lists each number in the array A with its position in the array and its ascending and descending ranks. Note how this is done by means of an

Flowchart for Program 5

Ranking an array of numbers in ascending and descending order.

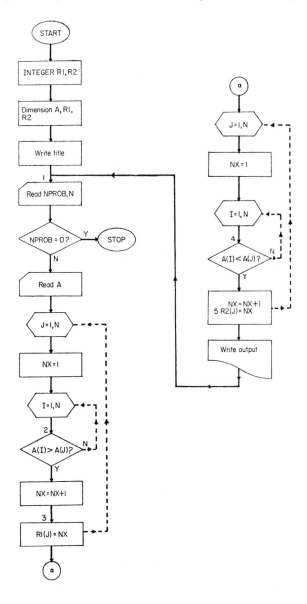

implied DO-loop in the list of the WRITE statement. Control then returns to statement 1 when the next problem specification card is read. If this is the sentinel card (NPROB set at zero) execution stops, otherwise the next problem is solved.

Program 5

Ranking an array of numbers in ascending and descending order; with sample data and output.

```
C
C............................................................
C
C  THIS PROGRAM RANKS A SET OF NUMBERS IN ASCENDING AND DESCENDING
C  ORDER AND PRINTS RANKS AND NUMBERS. INPUT PARAMETERS ARE AS FOLLOWS –
C     NPROB = NUMBER OF PROBLEM
C     N     = NUMBER OF OBSERVATIONS IN SAMPLE
C     A     = VECTOR OF OBSERVATIONS TO BE RANKED
C
C............................................................
C
      INTEGER R1, R2
      DIMENSION A(1000), R1(1000), R2(1000)
C
C  WRITE TITLE.
      WRITE (6,100)
  100 FORMAT (1H1, 8X, 23HRANKED ARRAY OF NUMBERS)
C
C  READ PROBLEM SPECIFICATION.
    1 READ (5,101) NPROB, N
  101 FORMAT (2I5)
      IF (NPROB .EQ. 0) STOP
C
C  READ ARRAY A.
      READ (5,102) (A(I), I = 1,N)
  102 FORMAT (7F10.0)
C
C  RANK IN DESCENDING ORDER (LARGEST = 1).
      DO 3 J = 1,N
      NX = 1
      DO 2 I = 1,N
    2 IF (A(I) .GT. A(J)) NX = NX + 1
    3 R1(J) = NX
C
C  RANK IN ASCENDING ORDER (SMALLEST = 1).
      DO 5 J = 1,N
      NX = 1
      DO 4 I = 1,N
    4 IF (A(I) .LT. A(J)) NX = NX + 1
    5 R2(J) = NX
C
C  OUTPUT RESULTS.
      WRITE (6,103) NPROB, (I, A(I), R1(I), R2(I), I = 1,N)
  103 FORMAT (////8H PROBLEM, I4/ 12H ************/// IX, 11HPOSITION IN,
     A4X, 6HNUMBER, 4X, 10HDESCENDING, 3X, 9HASCENDING/
     B4X, 7HPROBLEM, 17X, 5HRANKS, 8X, 5HRANKS//(I10, F12.2, 2I12))
C
C  READ NEXT SAMPLE.
      GO TO 1
      END
```

Program 5 (continued)

DATA

```
   1   9
       5.7    3.9    1.1    2.5    2.5    8.9    8.9
       6.3    3.1
   0
```

OUTPUT

RANKED ARRAY OF NUMBERS

PROBLEM   1
****

POSITION IN PROBLEM	NUMBER	DESCENDING RANKS	ASCENDING RANKS
1	5.70	4	6
2	3.90	5	5
3	1.10	9	1
4	2.50	7	2
5	2.50	7	2
6	8.90	1	8
7	8.90	1	8
8	6.30	3	7
9	3.10	6	4

Program 6: Tabulation of frequency tables

This program is intended to read a large number of observations, punched so many to a card, and to tabulate them as a frequency distribution. There are so many ways in which such a program can be constructed, depending on the initial assumptions made, that one needs briefly to set these out before describing the program. It is assumed first that the variable may be continuous or discrete and that we do not know how many observations have been made altogether, so that they will need to be counted in the course of the program. It is further assumed that the user will know the value of the largest and smallest observations so that he can choose suitable class boundaries and a class interval and so specify them and the number of classes in which the distribution is to be tabulated. The essential information to be provided is the lower limit of the lowest class, the number of classes and the class interval. Different authorities define the class limits in different ways and it is essential to be clear on the way these are defined in the program. Suppose, for example, that we wished to tabulate the distribution of ages (in years) in a human population. This cannot be less than zero and, we may suppose, it cannot be greater than 125. We could therefore decide to tabulate the distribution in 25 classes, starting with age 0 and going up with a class interval of 5 years to 125. Thus the limits of the lowest class would be from zero up to, but not including, 5 years. The second class would contain people whose ages ran from 5 years up to but not including 10 years, and so on up to, but not including, 125 years. If we were measuring height, instead of age, the position would be a little more complicated. We might record all heights to the nearest inch and find they ranged from 49 in. to 82 in. However, since these figures are accurate only to 1 in. we must fix the lower limit of the lowest class as not more than 48.5 in. and the upper limit of the uppermost class as not less than 82.5 in. The exact limits we choose may then be round numbers which fit a suitable class interval and a suitable number of classes. We could, for example, choose 48 in. as the lower limit of the lowest class, 84 in. as the upper limit of the uppermost class and 3 in. as the class interval, so giving us 12 classes. The term 'class mark' is used in this program to refer to the mid-point of each class. Thus, in the height distribution just discussed the class mark of the first class would be 49.5, of the second 52.5 and so on at intervals of 3 in. up to 82.5 for the 12th class.

After a few comments listing the names of the input variables the program sets up dimensions for the various arrays to be used. These are: X, which will accommodate up to 10 observations punched on each card; INFMT which will hold the variable input FORMAT specification; NFREQ which will give the numerical frequency in each class; PFREQ, giving the corresponding proportional frequency; NCUM gives the cumulative frequency; PCUM gives the proportional cumulative frequencies and CLMK the class marks. The program as dimensioned will provide for 50 classes. Note that if percentage frequencies are required, they can readily be obtained by multiplying the corresponding proportional frequencies by 100.

The program continues by reading in the specification of the problem, requiring a data card on which are punched values for NPROB (the problem reference number), NCLASS (the number of classes), XMIN (the lower limit of the lowest class) and CLINT (the class interval). A test on NPROB enables a sentinel card to stop execution after all the data of all the problems have been read, and the remainder of the first major section of the program simply reads the variable format specification and writes headings for the frequency distribution table that will eventually be printed out.

We are now ready to begin the basic operation of the program, i.e. to tabulate the values of X in the array NFREQ, for which purpose we set all locations of the array to zero so that we can later accumulate the appropriate frequencies in each location. At the same time we also initialise at zero a counter, N, which will be used to accumulate the total number of observations (required later to compute proportional frequencies). Beginning with statement 3 of the program we then read the contents of a single data card on which are punched a set of up to 10 values of X, preceded by ISTOP and M.

ISTOP is used as in previous programs to know when the last data card of a problem is being read—it is left unpunched except on the last card where it is given a non-zero value (e.g. 9). M simply indicates the number of observations on the card and must, of course, precede the implied DO-loop by which these observations are read. Having placed the M observations in core storage we then execute the loop beginning DO 4 I = 1, M. It is this loop which actually accomplishes the tabulation. For each value of X it computes a value J using the arithmetic statement

$$J = (X(I) - XMIN)/CLINT + 1.0$$

If we imagine the value of X(I) in the age distribution example given above to be 73.0, with XMIN equal to 0.0 and CLINT to 5.0, then clearly J will be equal to $73/5 + 1.0 = 15.6$, truncated to give 15. This is actually a way of determining that an age of 73 is placed in the 15th class and the last statement of the loop would therefore increment NFREQ (15) by 1. The arithmetic statement for J thus represents a tabulating algorithm that forms the heart of the whole program. When the loop containing it has been satisfied, each of the M values on the card will have been placed in its correct class. The program then counts the observations just tabulated by incrementing N and uses ISTOP to test whether further cards are to be read. Assuming this is the case control returns to statement 3 and the whole process continues until all the observations making up the problem have been tabulated in NFREQ and their number recorded in N.

The program continues by using the values stored in NFREQ to compute the corresponding proportional frequencies. Each value in NFREQ is divided by the total number of observations to give the proportional frequencies placed in the array PFREQ. To ensure that we do not write mixed-mode expressions it is necessary to express N in real number form (as AN) and to do the same for each value of NFREQ by means of the Fortran function FLOAT which carries out conversion from integer to real numbers. There are, of course, alternative ways of doing this, but they would require an additional real array and the use of FLOAT is a convenient solution to the problem. Having computed NFREQ and PFREQ we can now proceed to evaluate the corresponding cumulative frequencies and the class-marks (CLMK) in a single DO-loop. We do this by first fixing the initial values— i.e. NCUM (1), PCUM (1) and CLMK (1). Once we have these it is then possible to derive all the other values in the arrays by the DO-loop beginning DO 6 I = 2, NCLASS. For example, the values of the second class of the cumulative frequency distribution would be obtained by adding to NCUM (1) the value NFREQ (2) and so on for all the other classes. The reader may care to follow out the workings of this DO-loop from the tabulated output given on p. 120. Having computed the values for all members of all five arrays, it

Flowchart for Program 6
Construction of frequency tables.

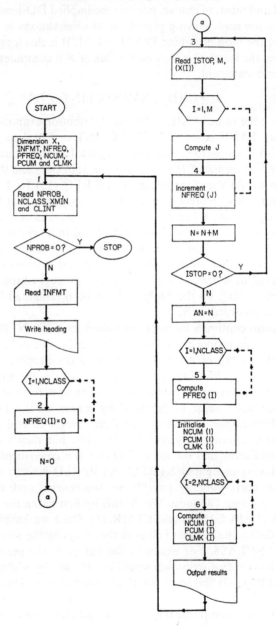

only remains to print them out under the headings previously written. The DO-loop DO 7 I = 1, NCLASS accomplishes this by writing out a line every time the loop is run through. A minor difficulty encountered by beginners in programming output of this kind is to arrange for the figures to fall in line with the headings! The best solution is to plan the desired lay-out on graph paper with $\frac{1}{10}$ in. squares *before* writing the formats for headings and numerical output. Spaces can easily be counted in this way and figures aligned appropriately.

Program 6

Construction of frequency tables; with sample data and output.

```
C
C..........................................................
C
C   THIS PROGRAM TABULATES A FREQUENCY DISTRIBUTION, PROPORTIONAL
C   FREQUENCY DISTRIBUTION AND CUMULATIVE DISTRIBUTIONS.
C   INPUT PARAMETERS ARE AS FOLLOWS –
C       NPROB  = NUMBER OF PROBLEM (ZERO ON LAST DATA CARD)
C       NCLASS = NUMBER OF CLASSES IN TABLE (MAXIMUM = 50)
C       XMIN   = LOWER LIMIT OF LOWEST CLASS
C       CLINT  = CLASS INTERVAL
C       INFMT  = VECTOR HOLDING INPUT FORMAT
C       ISTOP  = 9 ON LAST CARD OF PROBLEM, OTHERWISE DO NOT PUNCH
C       M      = NUMBER OF OBSERVATIONS ON CARD
C       X      = VECTOR OF OBSERVATIONS ON CARD
C
C..........................................................
C
        DIMENSION X(10), INFMT(20), NFREQ(50), PFREQ(50), NCUM(50),
       A PCUM(50), CLMK(50)
C
C   READ PROBLEM SPECIFICATION AND WRITE HEADING.
      1 READ (5,100) NPROB, NCLASS, XMIN, CLINT
    100 FORMAT (2I5, 2F10.0)
        IF (NPROB .EQ. 0) STOP
        READ (5,101) (INFMT(I), I = 1,20)
    101 FORMAT (20A4)
        WRITE (6,102) NPROB
    102 FORMAT (30H1FREQUENCY TABULATION. PROBLEM, I4///6H CLASS, 5X,
       A  5HCLASS, 5X, 9HFREQUENCY, 3X, 12HPROPORTIONAL, 3X,
       B  10HCUMULATIVE, 3X, 12HPROPORTIONAL/11X, 4HMARK, 19X,
       C  9HFREQUENCY, 5X, 9HFREQUENCY, 5X, 10HCUMULATIVE/
       D  62X, 9HFREQUENCY/)
C
C   INITIALIZE.
        DO 2 I = 1,NCLASS
      2 NFREQ(I) = 0
        N = 0
C
C   READ DATA AND TABULATE FREQUENCIES.
      3 READ (5,INFMT) ISTOP, M, (X(I), I = 1,M)
        DO 4 I = 1,M
        J = (X(I)–XMIN)/CLINT + 1.0
      4 NFREQ(J) = NFREQ(J) + 1
        N = N + M
        IF (ISTOP .EQ. 0) GO TO 3
```

Program 6 (*continued*)

```
C
C   COMPUTE CLASS MARKS AND PROPORTIONAL AND CUMULATIVE FREQUENCIES.
      AN = N
      DO 5 I = 1,NCLASS
    5 PFREQ(I) = FLOAT(NFREQ(I))/AN
      NCUM(1) = NFREQ(1)
      PCUM(1) = PFREQ(1)
      CLMK(1) = XMIN + 0.5*CLINT
      DO 6 I = 2,NCLASS
      NCUM(I) = NCUM(I - 1) + NFREQ(I)
      PCUM(I) = PCUM(I - 1) + PFREQ(I)
    6 CLMK(I) = CLMK(I - 1) + CLINT
C
C   OUTPUT RESULTS.
      DO 7 I = 1,NCLASS
    7 WRITE (6,103) I, CLMK(I), NFREQ(I), PFREQ(I), NCUM(I), PCUM(I)
  103 FORMAT (I5, F12.3, I11, F13.3, I14, F13.3)
C
C   START NEW PROBLEM.
      GO TO 1
      END

DATA
    1    25         0.0         5.0
(I1, I4, 10F6.2)
    10   3.21  12.70   7.31  31.25  51.37  40.99  68.70  72.15  90.08   8.79
    10   0.01  88.12  17.31  21.18  37.18  49.31  60.20  77.43  99.89  11.38
 9    7107.10  58.32  47.18  99.18   5.22   3.71   7.88
 0
```

OUTPUT

FREQUENCY TABULATION. PROBLEM 1

CLASS	CLASS MARK	FREQUENCY	PROPORTIONAL FREQUENCY	CUMULATIVE FREQUENCY	PROPORTIONAL CUMULATIVE FREQUENCY
1	2.500	3	0.111	3	0.111
2	7.500	4	0.148	7	0.259
3	12.500	2	0.074	9	0.333
4	17.500	1	0.037	10	0.370
5	22.500	1	0.037	11	0.407
6	27.500	0	0.	11	0.407
7	32.500	1	0.037	12	0.444
8	37.500	1	0.037	13	0.481
9	42.500	1	0.037	14	0.519
10	47.500	2	0.074	16	0.593
11	52.500	1	0.037	17	0.630
12	57.500	1	0.037	18	0.667
13	62.500	1	0.037	19	0.704
14	67.500	1	0.037	20	0.741
15	72.500	1	0.037	21	0.778
16	77.500	1	0.037	22	0.815
17	82.500	0	0.	22	0.815
18	87.500	1	0.037	23	0.852
19	92.500	1	0.037	24	0.889
20	97.500	2	0.074	26	0.963
21	102.500	0	0.	26	0.963
22	107.500	1	0.037	27	1.000
23	112.500	0	0.	27	1.000
24	117.500	0	0.	27	1.000
25	122.500	0	0.	27	1.000

Once the output has been printed control returns to statement 1 and begins the next problem (or stops execution if NPROB is set at zero by the sentinel card).

The main points of interest in this program are the use of an ingenious algorithm for placing each observation in its appropriate class and the method used to compute cumulative quantities. Both involve attention to subscripts. In the first case we see how a computed subscript can solve a tabulation problem and in the second we see how a little simple arithmetic with subscripts helps to derive one element of an array from another. The flow-chart on p. 118 enables the logic of the program to be examined at leisure.

Program 7: Construction of two-way contingency tables

In biological work it is not uncommon to collect data in which a number of variables have been measured on each individual. For example, one might have measured the dimensions of ten different bones in the skeletons of 25 rats or the numbers of six different species of arthropods present in a hundred samples of soil, and so on. It is often of interest to know whether there is an association between any two of the variables measured. Such a possibility may be most appropriately assessed by calculating correlation or regression coefficients, but association can also be detected by tabulating the observations in a contingency table, which is often also a convenient way of presenting summarised data on two variables at a time. Each table consists of two or more rows and columns, so that it is divided into a number of cells. The rows represent a set of classes into which the measurements of one variable fall, while the columns form a similar set of classes for the other variable. The numbers entered in each cell are then the frequencies with which the observations fall into the relevant classes of the two variables considered simultaneously. For example, in the 2×2 contingency table shown below the first row represents measurements of height below average and the second those measurements above average. Similarly the two columns represent weights below and above average. From the frequencies entered in the cells it can be

		Height	
		Below average	Above average
Weight	Below average	38	20
	Above average	17	52

seen that there is a tendency for individuals to be above average in both respects or below in both respects. In other words, there is some association between height and weight.

Since any one contingency table compares only two variables, a set of data in which each individual is measured in respect of n variables should result in a sufficient number of tables to compare all combinations of two from the total n. This is given by the formula

$$\binom{n}{2} = \frac{n(n-1)}{2}. \tag{1}$$

The object of this program, therefore, is to read data consisting of n measurements on any number of individuals and then to construct the resulting set of $n(n-1)/2$ contingency tables.

There are several ways in which to carry out this task, depending on the initial information or the assumptions one cares to make about the data. Here we shall assume as little as possible—we need to know only the number of variables we have measured and the number of rows and columns we would like in the contingency tables. These tables will consist of the same numbers of rows and columns, e.g. a 4×4 table, with 16 cells. The division of the range of each variable into equal classes will be carried out automatically and this requires that the program first finds the minimum and maximum values of each variable in the data. Having done this it is possible to use an extension of the tabulating algorithm of Program 6 to allocate observations to their cells in a particular table. This means that we must scan all the data twice—once to determine the minima and maxima and a second time in which we use this information to assign each observation to its cell. This double scan is best done by storing the data on magnetic tape or disk after it has been read originally from cards. This enables large bodies of data to be processed (far larger than could be accommodated at any one time in the central storage unit of the computer). Further, since we shall have $n(n-1)/2$ contingency tables, each with its set of rows and columns, we shall need to accumulate our frequencies in a 3-dimensional array, each element of which is denoted in the program by NF (K, L, M) where K represents the row-number, L the column-number and M the number of the table. We can in fact think of the contingency tables as a sequence of 'layers' in the three-dimensional array, each layer consisting of a number of cells. We must also arrange for the computer to record which pair of the original variables is compared in each contingency table identified by the subscript M. Since the number of tables rapidly increases as the number of variables increases, our 3-dimensional array will soon reach a considerable size. For 15 variables, for example, we would require $15 \times 14/2 = 105$ contingency tables to compare all possible pairs of variables. If, furthermore, we suppose that each table could contain up to a hundred cells (10 rows and 10 columns), the array NF would require $10 \times 10 \times 105 = 10,500$ storage locations. These are the limits we shall set in our program; they are greater than are likely to be needed in practice.

The program begins, then, by dimensioning the arrays we require, all 1-dimensional except for the triply subscripted array NF. NAME will hold the names of the variables, abbreviated if necessary to 6 characters each, so that a single variable name can be contained in one 36-bit computer word. If one were dealing with a computer of different word-structure then one would have to modify this part of the program slightly in accordance with the discussion of A-formats on p. 55. X is an array holding the values of up to 15 variables, measured on each individual. These values are punched on a single card (plus a continuation card if needed). The set of measurements for each individual must therefore always begin on a new data card. XMIN and XMAX are the arrays which will hold the minimum and maximum values for each variable while RANGE will hold the range. CLINT accommodates the class-interval sizes for each variable and NF is the final 3-dimensional frequency array from which the tables are eventually produced by writing it out one layer at a time.

Input begins by reading in the number of variables (NV), the logical number of the tape or disk unit to be used (NTAPE) and the number of rows and columns (NRC) needed in the tables. The logical tape-unit number must be one of those allotted to binary scratch tapes by the computer installation at which the program is being used. It is assumed not to be 5 or 6 as these numbers are conventionally allotted to input and output BCD tapes and are so used in all programs in this book. It is further assumed that scratch disk areas may be used in place of tapes without affecting the Fortran programming method. The reader who is uncertain how magnetic tapes are handled should refer to p. 449 for a brief introduction to the subject; only a minimum of explanation is given below when discussing the simple tape manipulation statements of this program.

A test on NV makes it possible to stop execution by a card punched with a zero value as the last data card. Input then continues with the names of the variables read under the format specification 12A6, i.e. each variable name punched in a field of six columns. Again, remember that this is adapted to the requirements of a computer with a 36-bit word structure. The program is then almost ready to begin reading in the major part of the data, one card at a time. Before starting this, however, two things are necessary. First, we need to initialize the arrays XMAX and XMIN so that they can be used to determine the maxima and minima. We choose initial settings which, for XMAX are all very much *smaller* than any data ever to be encountered, and for XMIN those which are all very much *larger*. The values chosen are actually -10^{10}

and $+10^{10}$ respectively. The way in which this initialisation operates will soon become clear. Secondly, we rewind the magnetic tape to its 'load point' (effectively the beginning of the tape).

Starting with statement 3, then, the program causes the first main data card to be read, using an implied DO-loop for the values of the different variables that make up the array X. NCHK is a device used for detecting the last of these data cards, where it is given a non-zero value (e.g. 9); on all other cards it is left unpunched. Note, by the way, the relevant FORMAT statement—103 FORMAT (I2, (10 F7.0)). The space for NCHK occupies the first two columns followed by up to 10 values of X, punched with a decimal point in fields 7 columns wide. The inner parentheses around 10 F7.0 denote that if there are more than 10 values of X(I)—and the program dimensions allow up to 15—the others can appear on a second (continuation) card, starting with the eleventh value of X(I) in columns 1–7. That is, the format repeats only on the parenthesised 10 F7.0 part of the whole specification and space is allowed for NCHK only on the first of the two cards. Immediately after the information punched on a card has been read into the locations reserved for X(I), it is also written (in binary form) on to the magnetic tape unit NTAPE. One should notice that this WRITE statement, which produces on tape what is called a single logical record, has no associated FORMAT statement. This is because it is written in binary form and binary records are non-formatted.

While it is still stored in core as the array X, the data will be used to help find minimum and maximum values for each variable. This is done by the loop beginning DO 5 I = 1, NV. The method used is to compare the current value of the variable X(I) with the values previously stored in arrays XMAX and XMIN. If X(I) is greater than XMAX(I) or smaller than XMIN(I) it will replace the latter and becomes in turn the stored value of XMAX(I) or XMIN(I) against which future values are compared. Since the arrays XMAX and XMIN were initialised to extremely small and extremely large numbers ($\pm 10^{10}$) the effect after the DO-loop is satisfied for the first time is to replace these initial values by the values of X(I) on the first data card. Subsequent data cards may have values which will increase XMAX(I) or decrease XMIN(I), and these cards will continue to be read until the last one—bearing the non-zero value of NCHK—is encountered. This marks the end of that set of data and the file of records on magnetic tape is closed by the instruction END FILE NTAPE.

It now remains to compute the range for each variable, which is easily done by subtracting XMIN(I) from XMAX(I), but the computation of the class-interval is slightly more troublesome. If we merely divide the range as it stands by the number of classes specified in NRC we shall find that the one or more observations with the maximum value will not be correctly assigned by the tabulating algorithm. These maximum values would, in fact, be allocated to the next, non-existent 'class', NRC + 1. We therefore expand the range by 1% and then find class-intervals which will now extend just sufficiently beyond the maxima to ensure that these are allocated to the uppermost (NRC-th) class. This is simply a rather formal version of what we would do intuitively in forming classes into which all values would fit.

We now have all the information needed to carry out the tabulation and we therefore first initialise our 3-dimensional frequency array NF to zero throughout, so that it can be used to accumulate frequencies. This initialisation involves the NRC rows, the NRC columns and the full number of contingency tables, which we calculate as NTAB from the number of variables involved. The program then prepares to read the data stored on tape by first rewinding the tape to its load point and then reading successive logical records into the core-locations X(I). After each record is read we calculate for the observation on each variable the three required subscripts of NF so that we can allocate that observation to the appropriate cell by incrementing NF(K, L, M) by 1. Two of the subscripts (K and L, referring respectively to rows and columns) are calculated by applying the tabulating algorithm to the values of those variables that will form the rows and columns of a particular table. The third subscript, M, which provides the number of the table— i.e. the 'layer' of NF—is computed by successive increments of 1 on each pass through the main tabulating loop that starts on the statement DO 11 I = 1, NVM. The loop has a rather complicated logical structure which is best understood by the reader following it through a few times with simple numerical data rather than by wrestling with a verbal description. Note especially that the indexing parameters are intended to combine systematically all possible pairs of variables. The effect of the loop is to increase by 1 the frequency record in a cell of the array NF every time an observation falls within the limits of that cell. The complete set of logical records is read by means of the outer loop formed between statement 9 and the test on NCHK (which has a non-zero value in the last record only).

Once all the frequencies have been recorded in NF, they can be printed out as contingency tables. The loop that accomplishes this starts on the statement DO 14 I = 1, NVM and has essentially the same logical structure as was used in the tabulating loop outlined in the previous paragraph. Its effect is to produce a set of contingency tables, each on a new page and headed with a number, the names of the variables involved and the numbered rows and columns of the contingency table with the frequencies occurring in each cell (Fig. 15). Note that the first-named variable in each heading refers to the successive rows of the table, while the second refers to successive columns. The output gives no information on the class limits defining the cells, but simply indicates that the range for each variable (strictly speaking the range plus 1%) has been divided equally into the numbered classes. It would be

Flowchart for Program 7
Construction of two-way contingency tables.

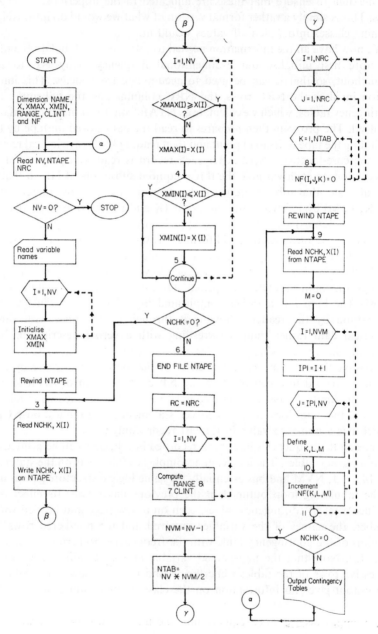

possible, however, to add the class limits without difficulty from the values of XMIN(I) and CLINT(I) that are still in core at the end of the program.

A flowchart for this tabulation of contingency tables is indicated on p. 126. The program has a relatively complicated logical structure and the reader who is still rather slow in grasping the significance of some of the procedures should not be unduly discouraged at this stage. The combination of subscript manipulation, tape operations and somewhat involved output are not easy to follow at first. The various segments of the program should be analysed in detail one at a time before trying to picture the over-all strategy. Ultimately, of course, it is the latter which is most interesting, but to evolve it requires some knowledge of what elementary operations can easily be performed in Fortran. Perhaps the main lesson to be learnt from this program is that there is a considerable gulf between Fortran coding and Fortran programming! Fortunately the remaining programs in this chapter are logically much simpler.

Program 7

Construction of two-way contingency tables; with sample data and output (the latter given for first contingency table only). This program uses one scratch tape or scratch area on disk.

```
C
C.................................................................
C
C  THIS PROGRAM CONSTRUCTS A SET OF TWO-WAY CONTINGENCY TABLES FROM
C  DATA ON ANY NUMBER OF CASES AND UP TO 15 VARIABLES. INPUT
C  PARAMETERS ARE AS FOLLOWS –
C      NV    = NUMBER OF VARIABLES (MAXIMUM = 15)
C      NTAPE = LOGICAL NUMBER OF BINARY I/O UNIT
C      NRC   = NUMBER OF ROWS AND COLUMNS NEEDED IN TABLE (MAXIMUM = 10)
C      NAME  = VECTOR HOLDING NAMES OF VARIABLES
C      NCHK  = 9 ON LAST CARD OF EACH SET OF DATA, OTHERWISE LEAVE BLANK
C      X     = VECTOR OF OBSERVATIONS (LENGTH NV)
C
C.................................................................
C
      DIMENSION NAME(15), X(15), XMAX(15), XMIN(15), RANGE(15),
     A  CLINT(15), NF(10,10,105)
C
C  READ PROBLEM SPECIFICATION.
    1 READ (5,100) NV, NTAPE, NRC
  100 FORMAT (3I5)
      IF (NV .EQ. 0) STOP
C
C  READ NAMES OF VARIABLES (NOT MORE THAN 6 CHARACTERS PER NAME).
      READ (5,101) (NAME(I), I = 1,NV)
  101 FORMAT (12A6)
C
C  READ DATA ONE CASE AT A TIME AND STORE ON TAPE, COMPUTING
C  MINIMUM AND MAXIMUM FOR EACH VARIABLE.
      DO 2 I = 1,NV
      XMAX(I) = – 1.0E10
    2 XMIN(I) = + 1.0E10
```

Program 7 (*continued*)

```
        REWIND NTAPE
     3  READ (5,102) NCHK, (X(I), I = 1,NV)
   102  FORMAT (I2, (10F7.0))
        WRITE (NTAPE) NCHK, (X(I), I = 1,NV)
        DO 5 I = 1,NV
        IF (XMAX(I) .GE. X(I)) GO TO 4
        XMAX(I) = X(I)
     4  IF (XMIN(I) .LE. X(I)) GO TO 5
        XMIN(I) = X(I)
     5  CONTINUE
        IF(NCHK) 6,3,6
     6  END FILE NTAPE
C
C    COMPUTE RANGE AND CLASS INTERVAL FOR EACH VARIABLE.
        RC = NRC
        DO 7 I = 1,NV
        RANGE(I) = XMAX(I) - XMIN(I)
        RANGE(I) = RANGE(I) + 0.01 * RANGE(I)
     7  CLINT(I) = RANGE(I) / RC
C
        SET FREQUENCY ARRAY TO ZERO.
        NVM = NV - 1
        NTAB = NV * NVM / 2
        DO 8 I = 1,NRC
        DO 8 J = 1,NRC
        DO 8 K = 1,NTAB
     8  NF(I,J,K) = 0
C
C    READ DATA FROM TAPE AND INCREMENT APPROPRIATE ELEMENTS
C    OF FREQUENCY ARRAY.
        REWIND NTAPE
     9  READ (NTAPE) NCHK, (X(I), I = 1,NV)
        M = 0
        DO 11 I = 1,NVM
        IP1 = I + 1
        DO 10 J = IP1, NV
        M = M + 1
        K = (X(I) - XMIN(I)) / CLINT(I) + 1.0
        L = (X(J) - XMIN(J)) / CLINT(J) + 1.0
    10  NF(K,L,M) = NF(K,L,M) + 1
    11  CONTINUE
        IF (NCHK) 12,9,12
C
C    OUTPUT CONTINGENCY TABLES.
    12  M = 0
        DO 14 I = 1,NVM
        IP1 = I + 1
        DO 13 J = IP1, NV
        M = M + 1
        WRITE (6,103) M, NAME(I), NAME(J), (IH, IH = 1,NRC)
   103  FORMAT (18H1CONTINGENCY TABLE, I4//11X, 9HVARIABLES, 1X, A6,
       A4H AND, 1X, A6////8X, 7HCLASSES, 6X, 10I10)
        DO 13 K = 1,NRC
    13  WRITE (6,104) K, (NF(K,L,M), L = 1,NRC)
   104  FORMAT (/I11, 10X, 10I10)
    14  CONTINUE
        GO TO 1
        END

DATA

        4    4    5
```

Program 7 (*continued*)

```
LENGTH  WIDTH  HEIGHT  WEIGHT
        2.0     4.0     6.0     8.0
        1.0     3.0     5.0     7.2
        3.2     2.3     4.1     9.3
99      4.7     4.8     5.8     6.8
 0
```

OUTPUT

CONTINGENCY TABLE 1

                  VARIABLES LENGTH AND WIDTH

CLASSES	1	2	3	4	5
1	0	1	0	0	0
2	0	0	0	1	0
3	1	0	0	0	0
4	0	0	0	0	0
5	0	0	0	0	1

Program 8: Data-summarising Program 1

This program is essentially similar to Program 1 in that it calculates a few simple statistics for any number of consecutive samples. It is, however, intended to deal with relatively large numbers of samples and therefore provides a more economical form of output. Each sample is summarised in a single line of output under a set of column headings. In order to simplify the reader's task these column headings are repeated at the top of each page and a new page of printed output is started after 50 samples have been summarised. Essentially the program comprises two large loops. One of these, which is brought about through a test on LINE (the number of lines of printed output) includes the column headings while the other does not and is used for each of the remaining 49 lines on a page. The essential statistical computations have already been dealt with in Programs 1 and 7 (minima and maxima). They will therefore only be indicated briefly in the flowchart and in the following description.

The array of observations, X, is dimensioned at 1000 and the program prints a set of column headings at the top of a new page after first initialising NSAMP at zero. This variable is a counter which will be incremented after each sample has been dealt with and therefore provides automatically a consecutive set of sample numbers. There is no provision for the user to attach his own reference number to each sample. Immediately after writing the headings we initialise LINE to zero. Like NSAMP this is also a counter in which we shall record the number of lines of output printed and thereby provide a means of moving to a new page after 50 lines have been printed. The program then reads N, the number of observations in the first sample and

Flowchart for Program 8
Data-summarising Program 1

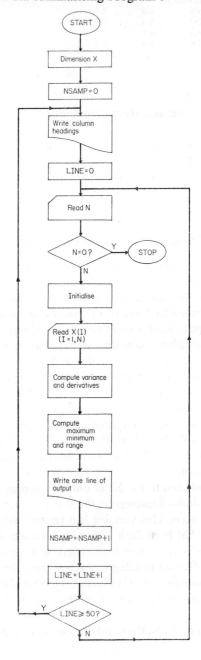

tests for a non-zero value so as to provide a mechanism for stopping the program (a zero punched for N on the last data card). The variables SX and SSQX are initialised to zero so that they can eventually be used to accumulate the sums and sums of squares for the sample, and N is given representation as a real number AN. The whole sample is then read into core-storage by the implied DO-loop.

Program 8

Data-summarising Program 1.

```
C
C............................................................
C
C   THIS PROGRAM SUMMARISES DATA FOR ANY NUMBER OF SAMPLES, EACH
C   CONTAINING UP TO 1000 OBSERVATIONS. IT COMPUTES MEAN, STANDARD
C   ERROR, STANDARD DEVIATION, VARIANCE, MINIMUM, MAXIMUM AND RANGE.
C   INPUT PARAMETERS ARE AS FOLLOWS –
C      N      = NUMBER OF OBSERVATIONS IN SAMPLE (ZERO ON LAST DATA CARD)
C      X      = VECTOR OF OBSERVATIONS (LENGTH N)
C
C............................................................
C
      DIMENSION X(1000)
      NSAMP = 0
    1 WRITE (6,100)
  100 FORMAT (1H1, 3X, 6HSAMPLE, 4X, 7HSIZE OF, 6X, 4HMEAN, 5X,
     A8HSTANDARD, 3X, 8HSTANDARD, 2X, 8HVARIANCE, 3X, 9HCOEFF. OF,
     B3X, 7HMAXIMUM, 4X, 7HMINIMUM. 4X, 5HRANGE/ 15X, 6HSAMPLE,
     C17X, 5HERROR, 3X, 9HDEVIATION, 13X, 9HVARIATION//)
      LINE = 0
C
C   READ NUMBER IN SAMPLE AND TEST FOR STOP.
    2 READ (5,101) N
  101 FORMAT (I4)
      IF (N .EQ. 0) STOP
C
C   INITIALISE.
      SX = 0.0
      SSQX = 0.0
      AN = N
C
C   READ SAMPLE.
      READ (5,102) (X(I), I = 1,N)
  102 FORMAT (10F7.0)
C
C   SUM X AND X–SQUARED AND COMPUTE MEAN.
      DO 3 I = 1,N
      SX = SX + X(I)
    3 SSQX = SSQX + X(I) * X(I)
      XBAR = SX / AN
C
C   COMPUTE VARIANCE AND DERIVATIVES.
      VAR = SSQX – SX * SX / AN
      VAR = VAR / (AN – 1.0)
      SDEV = SQRT(VAR)
      SERR = SDEV / SQRT(AN)
      COEFF = SDEV / XBAR * 100.0
```

132 SORTING, TABULATING AND SUMMARISING DATA

Program 8 (*continued*)

```
C
C   COMPUTE MAXIMUM, MINIMUM AND RANGE.
      XMAX = X(1)
      XMIN = X(1)
      DO 5 I = 1,N
      IF (XMAX .GE. X(I)) GO TO 4
      XMAX = X(I)
    4 IF (XMIN .LE. X(I)) GO TO 5
      XMIN = X(I)
    5 CONTINUE
      RANGE = XMAX - XMIN
      NSAMP = NSAMP + 1
      LINE = LINE + 1
C
C   OUTPUT.
      WRITE (6,103) NSAMP, N, XBAR, SERR, SDEV ,VAR, COEFF XMAX,
     AXMIN, RANGE
  103 FORMAT (4X, I6, 4X, I6, 2X, 8F11.4)
C
      IF (LINE .GE. 50) GO TO 1
      GO TO 2
      END
```

Further stages in the program involve summing the observations and their squares and computing the mean and the variance with its other derivatives. These are indicated in the program by the obviously mnemonic variables VAR, SDEV, SERR and COEFF. The computation of maxima and minima is carried out by essentially the same method as in Program 7, except that it is here possible to initialise XMAX and XMIN by placing them both equal to X(1), the first observation in the sample. RANGE is computed by taking the difference of XMAX and XMIN and the counters NSAMP and LINE are then both incremented by 1. After printing the required line of output a test on LINE or the unconditional GO TO 2 returns control to the appropriate starting point. The only novel feature of this program is the simple device of using the variable LINE to control the point at which new page headings are printed out.

Program 9: Data summarising Program 2

This program carries out similar statistical calculations to Program 8 but performs them on data presented in the form of a grouped frequency table. It therefore supplements Program 6, which tabulates a frequency distribution from a series of individual observations.

The DIMENSION statement refers to two arrays—NFREQ accommodates the frequencies (integer numbers) for each class of the frequency table and FREQ is a real array which will hold the same frequencies when these are required for real mode arithmetic later in the program. It would in fact have been possible to do without the array NFREQ if the original frequencies were read in as real numbers, but the integer array has been retained here because it agrees better with the output of frequency tabulating programs and makes it easier for anyone who wishes to link such a program with this one.

After writing a heading the program reads a sample specification card which provides values for the following variables:

NSAMP, the reference number of the sample being summarised.

XL, the lower limit of the lowest class.

CLINT, the class interval

and NCL, the number of classes in the tabulated frequency distribution.

A test on the first of these variables is used to stop execution by the user providing a sentinel card which is placed after the data cards for the last sample and bears a value of NSAMP equal to zero. Once again, it is necessary to be quite clear how the tabulated frequency distribution that forms the input to this program has been arranged, the method adopted being that of **Program 6**. The first class will contain those observations falling between the lower limit of the lowest class and the lower limit of the second class. It will contain any observations exactly equal to the lower limit of the lowest class, but will not contain observations exactly equal to the lower limit of the second class—these fall into the second class. Thus if the lower limit of the lowest class is 1.0 and the class interval is 0.2, the first class would contain all observations from 1.0 up to but not including 1.2; the second class would contain those from 1.2 up to but not including 1.4, and so on. The central value of each class—the so-called class-mark—is the lower limit of the class plus one-half of the class interval. Frequency tables which are not in this form must therefore first be adjusted before using them as input to this program, otherwise a systematic error will be introduced into the calculations.

Before reading in the frequencies for each class the program carries out an initialisation. SUMX and SSQX are both set at zero with the object of accumulating the sum and sum of squares of the central values for each class. The central value (class mark) of the lowest class (X) is established as $XL + 0.5 * CLINT$ and a counter N, initialised at zero, is intended to accumulate the total number of observations in the frequency table.

We now begin dealing with the frequency table by reading the frequencies (punched in accordance with FORMAT statement number 102) using an implied DO-loop. The frequencies are read into the linear array NFREQ, previously dimensioned at 50, a number large enough to accommodate most frequency tables. The frequencies are summed in N and both N and the integers in NFREQ are converted to real numbers for use in the ensuing computation.

Flowchart for Program 9
Data-summarising Program 2

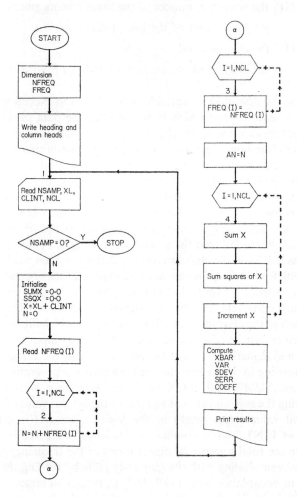

The program then begins to compute the statistics we require, namely the mean, the variance and the functions of the variance. The first step is taken by the loop beginning

$$DO\ 4\ I = 1, NCL$$

This accumulates the sum of the original variables (SUMX) by multiplying the class mark of each class by the class frequency and summing. It is the usual statistical convention to assume that within each class the original variables are distributed symmetrically about the class-mark. By a similar line of reasoning the sum of squares of the original variables (SSQX) is obtained by multiplying the square of each class mark by the class frequency and summing. At each cycle through the loop the class mark, X, is incremented by the class-interval CLINT. Knowing SUMX and the total number of observations (AN is the real equivalent of N), we calculate the mean XBAR, while from these and SSQX we can calculate the variance VAR. The expression used to compute VAR makes use of Sheppard's correction. This consists of reducing the variance as calculated in the usual way (formula 3 of Chapter IV, p. 83) by a quantity $h^2/12$ where h is the class interval. This correction is often recommended when variances are computed from grouped data.

The remainder of the program computes the standard deviation (SDEV), the standard error (SERR) and the coefficient of variation (COEFF) in the usual way and prints out the results in one line under the headings previously written. Control is then transferred to statement 1 which needs the specification card for the next frequency table. Should this be the sentinel, on which NSAMP has a zero value, the program stops execution.

The main novelty in this program is the statistical procedures necessary to compute the mean and variance from grouped data. In other respects the reader should have no difficulty in understanding it. A flow chart is given on p. 134.

Program 9

Data summarising Program 2

```
C
C.................................................................
C
C  THIS PROGRAM COMPUTES THE MEAN, VARIANCE AND RELATED STATISTICS
C  FOR ANY NUMBER OF SAMPLES OF DATA ARRANGED IN GROUPED FREQUENCY
C  TABLES.  INPUT PARAMETERS AS FOLLOWS—
C      NSAMP = NUMBER OF SAMPLE (ZERO ON LAST DATA CARD)
C      XL    = LOWER LIMIT OF LOWEST CLASS
C      CLINT = CLASS INTERVAL
C      NFREQ = FREQUENCY OF EACH CLASS (LENGTH NCL)
C      NCL   = NUMBER OF CLASSES
C
C.................................................................
```

Program 9 (continued)

```
C
      DIMENSION NFREQ(50), FREQ(50)
C
C WRITE HEADING.
      WRITE (6,100)
  100 FORMAT (1H1, 3X, 6HSAMPLE, 6X, 6HNO. IN, 9X, 4HMEAN, 6X,
     A8HSTANDARD, 4X, 8HVARIANCE, 4X, 8HSTANDARD, 4X, 9HCOEFF. OF/
     B4X, 6HNUMBER, 6X, 6HSAMPLE, 21X, 5HERROR, 17X, 9HDEVIATION,
     C3X, 9HVARIATION//)
C
    1 READ (5,101) NSAMP, XL, CLINT, NCL
  101 FORMAT (I10, 2F10.3, I10)
      IF (NSAMP .EQ. 0) STOP
C
C INITIALISATION.
      SUMX = 0.0
      SSQX = 0.0
      X = XL + 0.5 * CLINT
      N = 0
C
C READ CLASS FREQUENCIES, SUM AND CONVERT TO REAL NUMBERS.
      READ (5,102) (NFREQ(I), I = 1,NCL)
  102 FORMAT (10I7)
      DO 2 I = 1,NCL
    2 N = N + NFREQ(I)
      DO 3 I = 1,NCL
    3 FREQ(I) = NFREQ(I)
      AN = N
C
C COMPUTE STATISTICS.
      DO 4 I = 1,NCL
      SUMX = SUMX + X * FREQ(I)
      SSQX = SSQX + X * X * FREQ(I)
    4 X = X + CLINT
      XBAR = SUMX / AN
      VAR = (SSQX - XBAR * SUMX) / (AN - 1.0) - CLINT * CLINT /12.0
      SDEV = SQRT(VAR)
      SERR = SDEV / SQRT(AN)
      COEFF = SDEV * 100.0 / XBAR
C
C OUTPUT RESULTS.
      WRITE (6,103) NSAMP, N, XBAR, SERR, VAR, SDEV, COEFF
  103 FORMAT (1H0, 2(I8,4X), 5F12.4)
      GO TO 1
      END
```

Program 10: Data summarising Program 3

This program is intended to summarise data arranged in a somewhat more elaborate manner than in the two previous programs. Consider, for example, the table on p. 137.

Here we have 5 variables taking various values indicated in the table. These relate to two groups, the first having been sampled 4 times (4 sets of 5 variables) while the second group has been sampled 3 times (3 sets of 5 variables). Data of this kind frequently arise in biological work. For example the variables might be 5 different body dimensions, each set having been

			Variable			
Group	Set	1	2	3	4	5
	1	2.1	10.7	8.3	3.1	1.2
1	2	2.2	10.2	7.9	3.8	1.6
	3	2.5	9.8	8.1	2.9	1.4
	4	1.9	11.1	8.0	3.4	1.8
	1	4.0	15.7	10.1	6.0	3.0
2	2	3.6	12.6	10.8	6.1	3.2
	3	3.4	13.8	10.7	6.2	3.6

recorded from a different individual, those of group 1 reared on one diet, those of group 2 on another. The purpose of this program is to summarise, group by group, the data relating to each variable in turn. As before, we shall compute for each variable its mean, variance, standard deviation, standard error, coefficient of variation, minimum, maximum and range. The results will be tabulated group by group, the statistics for each variable occupying a separate line of printed output. We require a fairly flexible program which will deal not only with different numbers of groups, sets and variables, but will also cope with different formats for measurement in different groups. We might, for instance, wish to summarise data already punched in, say, 10F7.2 for one group and in 5F8.3 for another. Further, we have in mind a situation where data are collected and punched set by set, so that we shall in fact process the above table row by row, though the quantities we require to compute are based on the figures in each *column* of each group.

The program begins by establishing maximum dimensions for the subscripted arrays that will be used. One of these, INFMT, will hold the variable input FORMAT specification for each group. This works in the way described on p. 56 and already used in Programs 4 and 6. The remaining dimensioned arrays refer to the fact that we shall deal with up to 100 variables in a set—X(100)—and that for each of these we shall be calculating quantities like their sum SUMX, variance VAR, coefficient of variation COEFF and so on. These variable names are the same as those used in Programs 8 and 9.

The program then reads the number of groups in which the data is arranged (NGROUP) and begins the main loop for each of these with the statement

$$DO\ 4\ K = 1, NGROUP$$

Each passage through this loop will execute all the operations required to process a group and print out the results. Within this loop are other smaller ones, so that the greater part of the program forms a typical set of nested

DO-loops. The first set within the main loop is to read the variable input format and the number of sets and variables in the group (NSET and NVAR). knowing the number of variables we can then initialize at zero the two arrays SUMX and SSQX which will be used to assumulate a sum and sum of squares for each variable.

The program now contains a loop beginning

$$DO\ 3\ J = 1, NSET$$

which is intended to deal with one set of data after another (i.e. one row of the table after another for a particular group). The events occurring within this loop can be explained a little more fully. First, a set of observations is read in by the implied DO-loop under the format specification stored in INFMT. The program then begins (or continues) the assumulation of sums and sums of squares for each variable and also stores in XMAX and XMIN the largest and smallest value for each variable. Note, incidentally, that we have here used a rather different method of computing XMAX and XMIN from that used in previous programs. In this case we utilise one of the 'built-in' Fortran functions AMAX1 or AMIN1 which will give us the maximum or minimum of any number of variables set as arguments of the function. In this loop the functions always have two arguments, the value already stored in XMAX or XMIN and the current value of X. This means that we must, on the first run through the loop beginning DO 3 J = 1, NSET, arrange for XMAX and XMIN to be set equal to X. It also caters for the possibility that a particular group might contain only a single set. The reader will see that other methods of obtaining maximum and minimum values could equally well have been used here—there are usually several ways of writing any segment of a program.

Having satisfied the loop ending on statement 3 we have now obtained all the quantities needed to complete the calculation. The program therefore writes the heading for the group and using the values stored in SUMX and SSQX it proceeds to calculate the mean, variance, standard deviation and so on for each variable. All these calculations are carried out in a single loop, the last statement of which (statement 4 in the program) prints out for each variable the quantities needed. This statement is the last one in the range of the main loop for groups, so that once all the values for a group have been printed out looping continues with the next group until the main loop has been satisfied.

The main point of interest in this program lies in the fact that a very large table of data can be processed one row at a time, without the need to store the whole table. This is made possible by the use of formula 3 of Chapter IV (p. 83) for calculating the variances. Although the comments in Program 10 indicate limits for the size of each group and the number of groups, these can

Flowchart for Program 10

Data-summarising Program 3

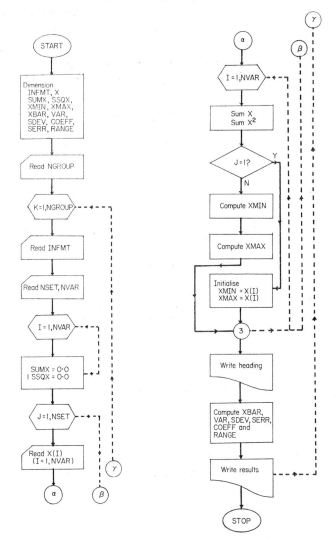

readily be increased almost without limit if the formats are changed to allow larger values of NGROUP and NSET to be read in and printed out. The storage capacity of the computer only places a limit on the number of *variables* and even this could easily be increased beyond any likely figures (e.g. a computer with a storage capacity of 32K words could run through this program with the dimensions increased to accommodate 2,000 variables! Apart from its general use in data summarising the program is particularly useful for reducing data to be used in multivariate analysis.

The flow chart given on p. 139 will enable the logic of the program to be followed readily.

Program 10

Data-summarising Program 3.

```
C
C................................................................
C
C   THIS PROGRAM SUMMARISES DATA IN UP TO 999 GROUPS, EACH GROUP
C   COMPRISING UP TO 9999 SETS OF OBSERVATIONS, EACH SET COMPRISING
C   UP TO 100 VARIABLES. INPUT PARAMETERS ARE AS FOLLOWS –
C       NGROUP= NUMBER OF GROUPS
C       INFMT  = VECTOR HOLDING VARIABLE INPUT FORMAT
C       NSET   = NUMBER OF SETS IN GROUP
C       NVAR   = NUMBER OF VARIABLES IN GROUP
C       X      = VECTOR OF OBSERVATIONS ON CARD (LENGTH NVAR)
C
C................................................................
C
      DIMENSION INFMT(20), X(100), SUMX(100), SSQX(100), XMIN(100),
     AXMAX(100), XBAR(100), VAR(100), SDEV(100), COEFF(100),
     BSERR(100), RANGE(100)
C
C   READ NUMBER OF GROUPS (NGROUP).
      READ (5,100) NGROUP
  100 FORMAT (I3)
C
C   BEGIN MAIN LOOP, READING NUMBER OF SETS IN GROUP (NSET) AND
C   NUMBER OF VARIABLES IN SET (NVAR).
      DO 4 K = 1,NGROUP
      READ (5,101) (INFMT(I), I = 1,20)
  101 FORMAT (20A4)
      READ (5,102) NSET, NVAR
  102 FORMAT (2I5)
C
C   INITIALISE FOR SUBSEQUENT ACCUMULATION OF SUMS.
      DO 1 I = 1,NVAR
      SUMX(I) = 0.0
    1 SSQX(I) = 0.0
C
C   LOOPS FOR EACH SET.
      DO 3 J = 1,NSET
      READ (5,INFMT) (X(I), I = 1,NVAR)
      DO 3 I = 1,NVAR
      SUMX(I) = SUMX(I) + X(I)
      SSQX(I) = SSQX(I) + X(I) * X(I)
      IF (J .EQ. 1) GO TO 2
      XMIN(I) = AMIN1(X(I), XMIN(I))
```

Program 10 (*continued*)

```
       XMAX(I) = AMAX1(X(I), XMAX(I))
       GO TO 3
    2  XMIN(I) = X(I)
       XMAX(I) = X(I)
    3  CONTINUE
C
C  WRITE HEADING AND COMPUTE STATISTICS.
C
       WRITE (6,103) K, NSET
   103 FORMAT (6H1GROUP, I4, 13H  (CONTAINING, I5,
      X 23H  SETS OF OBSERVATIONS)
      A///3X, 8HVARIABLE, 6X, 4HMEAN, 6X, 8HSTANDARD, 4X, 8HVARIANCE,
      B4X, 8HSTANDARD, 4X, 9HCOEFF. OF, 4X, 7HMINIMUM, 5X, 7HMAXIMUM,
      C6X, 5HRANGE/ 29X, 5HERROR, 17X, 9HDEVIATION 3X, 9HVARIATION/)
C
       AN = NSET
       DO 4 I = 1,NVAR
       XBAR(I) = SUMX(I) / AN
       VAR(I) = (SSQX(I) - SUMX(I) * XBAR(I)) / (AN - 1.0)
       SDEV(I) = SQRT(VAR(I))
       COEFF(I) = SDEV(I) * 100.0 / XBAR(I)
       SERR(I) = SDEV(I) / SQRT(AN)
       RANGE(I) = XMAX(I) - XMIN(I)
    4  WRITE (6,104) I, XBAR(I), SERR(I), VAR(I), SDEV(I), COEFF(I),
      AXMIN(I), XMAX(I), RANGE(I)
   104 FORMAT (I8, 3X, 8F12.4)
       STOP
       END
```

Program 11: Calculation of running means

Biological data frequently take the form of a time-series, where observations are made at successive points in time. Examples include the numbers of small mammals trapped annually in a certain area or the number of insects collected each night in a light-trap. These data often show quite wide and apparently random fluctuations and it is often desirable to smooth them by computing a running mean. Thus if the observations form the sequence

$$x_1, x_2, x_3, x_4, x_5, \ldots x_n$$

the running means of order 2 are computed by averaging successive pairs of observations so as to yield the new sequence

$$\frac{x_1 + x_2}{2}, \frac{x_2 + x_3}{2}, \frac{x_3 + x_4}{2}, \ldots \frac{x_{n-1} + x_n}{2}$$

Similarly running means of order 3 are formed by averaging successive groups of three observations and so on.

The program dimensions an array to hold the original observations (X), another to hold the running means (XRUN) and a third to contain a variable

Flowchart for Program 11
Calculation of running means.

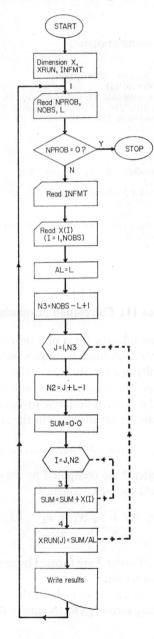

input format as already used in Programs 4, 6 and 10. The problem specification card is then read, followed by the usual device for stopping execution by a final data card on which NPROB is punched as zero. NOBS denotes the number of observations in the original series and L the order of the required running mean—say 3 or 4 if one is interested only in producing a somewhat smoother series. The variable input format is read from the next card and then the primary observations in the form of the singly subscripted array X, punched in the specified format and read in by the implied DO-loop.

The computation of the running means follows in the next program segment. The first step is to establish the upper indexing parameter N3 of the main loop. Obviously the series of running means will contain a smaller number of terms than the original series. It will in fact contain NOBS − L + 1 terms and this number forms N3, so that the elements of the array XRUN will eventually bear subscripts from 1 to N3. We now form the indexing parameters for the inner DO-loop, which will be used to accumulate the sum of the primary observations used to give the running mean. The lower parameter (denoted by J in the program) will be set by the value of the index J of the outer loop. The upper parameter N2 is calculated as J + L − 1. Thus, if one were to run through these loops at the beginning of a series with L = 3, the outer loop would set J = 1 and hence N2 = 1 + 3 − 1 = 3, which would cause the inner loop to sum X(1), X(2) and X(3) as required, the final division to give the running mean being accomplished through statement 4. On the next run through these loops J = 2, N2 = 4 and the terms summed would be X(2), X(3) and X(4). The outer loop would be satisfied when J reached the value N3, in this case two less than the value NOBS, so that the last three members of X are summed to form the last running mean. Whereas many of our data-sorting and tabulating programs make use of computed subscripts, the present program shows how the computation of indexing parameters within an outer DO-loop can also be usefully employed in programming.

Program 11

Calculation of running means; with sample data and output.

```
C
C.........................................................
C
C   THIS PROGRAM COMPUTES THE RUNNING MEANS (OF ANY DESIRED ORDER)
C   FOR ANY NUMBER OF SETS OF OBSERVATIONS. INPUT PARAMETERS ARE
C   AS FOLLOWS –
C       NPROB = PROBLEM REFERENCE NUMBER (ZERO ON LAST DATA CARD ONLY)
C       NOBS  = NUMBER OF OBSERVATIONS
C       L     = ORDER OF RUNNING MEAN
C       X     = VECTOR OF OBSERVATIONS (LENGTH NOBS)
C
C.........................................................
C
```

Program 11 (*continued*)

```
      DIMENSION X(1000), XRUN(1000), INFMT(20)
C
C  READ SAMPLE SPECIFICATION, VARIABLE INPUT FORMAT, AND DATA.
    1 READ (5,100) NPROB, NOBS, L
  100 FORMAT (3I5)
      IF (NPROB .EQ. 0) STOP
      READ (5,101) (INFMT(I), I = 1,20)
  101 FORMAT (20A4)
      READ (5,INFMT) (X(I), I = 1,NOBS)
C
C  COMPUTE RUNNING MEANS AND STORE IN XRUN(J).
      AL = L
      N3 = NOBS - L + 1
      DO 4 J = 1, N3
      N2 = J + L - 1
      SUM = 0.0
      DO 3 I = J,N2
    3 SUM = SUM + X(I)
    4 XRUN(J) = SUM / AL
C
C  OUTPUT RESULTS.
      WRITE (6,102) NPROB, NOBS, L, (XRUN(I), I = 1,N3)
  102 FORMAT (///14H SAMPLE NUMBER, I5, 5X, I5, 13H OBSERVATIONS//
     A23H RUNNING MEANS OF ORDER, I4//(10F10.3))
      GO TO 1
      END
```

DATA

```
    1   11    5
(10F5.0)
 50.0 36.5 43.0 44.5 38.9 38.1 32.6 38.7 41.7 41.1
 33.8
    0
```

OUTPUT

SAMPLE NUMBER   1   11 OBSERVATIONS

RUNNING MEANS OF ORDER   5
    42.580   40.200   39.420   38.560   38.000   38.440   37.580

Chapter 6

Analysis of Variance

In the discussion of Program 3 (p. 97) we saw that the difference between the means of two samples could be tested for significance by a t-test or a normal deviate test. The question immediately arises as to how one would test the significance of differences between the means of several samples. To conduct all the possible t-tests between pairs of samples would not be an efficient procedure, and the usual method employed is to carry out an *analysis of variance*. This statistical technique, which originated in the work of R. A. Fisher and is widely used in a variety of forms, is discussed in all modern textbooks of statistics (see especially Li, 1964) and in special works by Huitson (1966) and Scheffé (1959), the last-mentioned book being an advanced mathematical treatment of the subject.

The analysis of variance takes its name from the fact that when two or more factors influence a variable, then provided they act independently the total variance of the observations is equal to the sum of the variances associated with each of the factors. It is therefore possible to analyse or partition this total into several components, the relative importance of which may then be assessed. For example, suppose that one wished to compare the yields of four different varieties of wheat. We could arrange a trial, from the results of which we could separate that part of the total variance due to differences between varieties from that due to the replicates within the varieties. Alternatively, if we wished to assess the effect of, say, two fertilisers on crop yield, we might design a field experiment in which each fertiliser (i.e. each *factor*) was tested at several rates of application (i.e. at different *levels*). The effects of all possible combinations of the levels of the two factors would be measured and, if the experiment were suitably conducted, one might hope to disentangle by an analysis of variance the *main effects*, exerted by each factor independently, as well as the effects of an *interaction* between the factors. This capacity for assessing the influence of factors separately and in combination by means of a single experiment is one reason why the analysis of variance is a particularly valuable statistical technique.

Biologists sometimes find it difficult to understand how an analysis carried out on variances can be used to assess the significance of differences between the mean values for the different levels of a single factor. The explanation is that in computing the different parts of the total variance we begin by assuming that there are no differences between these means and thereby obtain two separate estimates of a given variance. We can then compare these two estimates by a variance-ratio test (F-test, p. 98). If they differ significantly we reject our hypothesis of equal means, i.e. within the probability limit involved in the F test we have demonstrated a significant difference between the means.

This explanation will perhaps become clearer if we return to the example of 4 varieties of wheat, each grown in, say, 10 suitably arranged plots of ground. This may be generalised to p varieties grown in n replicates, giving a total of $np = N$ samples on which to conduct an analysis. The yield of each sample may be expressed as x_{ij} (i.e. the yield of the j-th replicate of the i-th variety) and we can set up the model

$$x_{ij} = \mu + F_i + \varepsilon_{ij}. \tag{1}$$

That is, the value of each individual sample can be represented as the sum of three quantities: μ is the over-all mean, F_i is the effect due to the i-th variety and ε_{ij} is the *residual*, a term taking into account all the unknown or uncontrolled factors which might affect yield. It is assumed that for each variety the observations are normally distributed about a mean $(\mu + F_i)$ with a common variance σ^2.

We can now draw up an analysis of variance table that takes the following traditional form:

Source of variation	Sum of Squares	Degrees of Freedom	Mean Square
Between varieties	S_1	$(p-1)$	$M_1 = \dfrac{S_1}{(p-1)}$
Within varieties	S_2	$p(n-1) = N - p$	$M_2 = \dfrac{S_2}{(N-p)}$
Total	S	$N-1$	—

The calculation of the numbers of degrees of freedom associated with each source of variation is easily performed, but how do we calculate the sums of squares in column 2 of the table? These take the form:

$$\left.\begin{aligned} S_1 &= n \, \Sigma \, (x_{i.} - x_{..})^2 \\ S_2 &= \Sigma \, (x_{ij} - x_{i.})^2 \\ \text{and } S &= \Sigma \, (x_{ij} - x_{..})^2 \end{aligned}\right\} \quad (2)$$

where Σ is the usual notation for summing and the dot suffix indicates a mean, i.e. $x_{i.}$ denotes each of the means for varieties averaged over the n replicates and $x_{..}$ denotes the over-all mean, averaged over all varieties and replicates. A computationally simpler version of (2) is as follows:

$$\left.\begin{aligned} S_1 &= n \, \Sigma \, x_{i.}^2 - (\Sigma \, x_{ij})^2 / N \\ S &= \Sigma \, x_{ij}^2 - (\Sigma \, x_{ij})^2 / N \\ S_2 &= S - S_1 \end{aligned}\right\} \quad (3)$$

In this form both S and S_1 involve the term $(\Sigma \, x_{ij})^2 / N$, known as the correction term and comparable to that given in equation (3) on p. 83, where the calculation of sample variances was outlined. Division of these sums of squares by the corresponding degrees of freedom yields the mean squares of the fourth column. When these have been calculated we have a situation in which, adopting the null hypothesis that the F_i are equal, M_1 and M_2 will both be unbiassed estimates of σ^2, so that a variance ratio test carried out on them is in effect a test of the equality of the variety means.

An important and not always fully appreciated fact is that in many analysis of variance designs the particular mean squares to be compared in a variance-ratio test depend on (a) whether the factors exert fixed, systematic effects, or (b) whether their effects are random, or (c) whether some factors are systematic and others are random. There is a good discussion of fixed and random factors from a biological standpoint in Simpson, Roe and Lewontin (1960) and a convenient statistical introduction in Huitson (1966). In general it may be said that factors such as deliberately planned physical or chemical treatments, seasons, clearly defined environmental variables, altitude or latitude, taxonomic categories, age, sex and sampling methods are usually fixed factors, since one is interested in the particular levels chosen or determined by the nature of the problem. On the other hand, localities chosen merely as convenient representatives of a range of possible environments, or observers selected from a multitude of possible ones are factors with random effects. A program that catered for a variety of options as to which mean squares were compared in the variance ratio test would not be difficult to write, but it seems less confusing at this stage to present one set of variance-ratio tests (those for the fixed-factor model) in each program and to indicate where others might be more appropriate. Fortunately in several of the simple cases considered here it makes no difference which model is adopted.

This very brief introduction to some of the principles and assumptions underlying the analysis of variance cannot do justice to the great variety of experimental designs in which it may be used. From the standpoint of the computer programmer, these present considerable difficulty in that a single program sufficiently general to cope with all or most of them must inevitably be a very elaborate affair. Hartley (1960) and Yates and Anderson (1966) have devised suitable methods of programming a general analysis, but these involve sophisticated statistical techniques and have the drawback of requiring the data and results in the form of a 'fully randomised' design, from which the user subsequently constructs the analysis of the particular experimental design he had employed. Those interested will find a Fortran IV program based on these principles among the sample programs that accompany the Scientific Subroutines Package issued by IBM (Manual 260A–CM–03X Version III (1968)). Another similar program is available in the BMD package programs and is described in the accompanying manual (Dixon, 1970). Further general details of these collections of programs are given on p. 472. The alternative approach adopted here is to discuss separately a few of the simpler and more widely used experimental designs, for each of which a program has been provided. The designs we shall deal with are:

Program 12: Two factors without replication

Program 13: Two factors with replication

Program 14: Three factors without replication

Program 15: Latin Square design

Program 16: Balanced incomplete block design

While most textbooks tend to discuss analysis of variance and factorial designs in terms of agricultural field experiments (for the analysis of which they were first developed), the same methods can be used in the planning and analysis of laboratory experiments and of data relating to a wide range of other situations. The assumptions involved in an analysis, the methods used to partition the total variance, and the techniques employed in comparing individual levels of a factor, all require careful consideration. The programs given here may relieve the user of routine computations, but they do not absolve him from careful thought in the planning of even the simplest factorial experiment. Brownlee (1951) can be recommended as a simple, non-mathematical exposition of many procedures with large numbers of numerical examples; more advanced aspects of experimental design are discussed by Cochran and Cox (1957) and by Cox (1958).

Program 12: Two-factor analysis of variance without replication

In this type of design, each of the two factors is studied at two or more levels but the effects of any combination of levels is assessed by a single unreplicated observation. Thus, if one factor were studied at 2 levels and the other at 3, there would be a total of 6 observations represented by a two-way table of observations, x_{ij}. The model for this analysis is then

$$x_{ij} = \mu + F_i + G_j + \varepsilon_{ij} \tag{4}$$

in which it is assumed that there is no interaction between the two factors, whose main effects are denoted by F_i and G_j.

The basis of the computer program is the calculation of four 'intermediate variables' denoted by A, B, C and D, from which the sums of squares are subsequently calculated. They are defined as follows:

A = Sum of squares of the individual observations in the two-way table.

B = Sum of the squares of the totals for each row of the table, divided by the number of columns.

C = Sum of the squares of the totals for each column of the table, divided by the number of rows.

D = Grand total of all observations, squared and divided by the total number of observations. This is the 'correction term' referred to above.

The between-row variation is then measured by a sum of squares equal to (B − D), the between-column variation by (C − D) and the total variation by (A − D). The difference between the total and the sum of the other two sums of squares—i.e. the quantity (A + D − B − C)—represents the 'residual sum of squares' or 'error sum of squares'. It is the ratios of the mean squares for rows or columns to the residual mean square which will provide the variance ratios we require.

The program begins by establishing the values of DIMENSION for the variable X. This is set at 50 × 50 which would allow for up to 50 levels of each of the factors, far more than are likely to be needed in most analyses. The DIMENSION statement also includes values for the sums of the rows—RSM (50)—and columns—CSM (50). The title and column headings for the analysis of variance table are then printed out and the first data card is read.

This is a problem specification card on which are punched three quantities:

NPROB = Reference number of problem.
NR = Number of rows in table.
NC = Number of columns in table.

A decision box at this point enables one to stop the program when all problems have been dealt with by using a sentinel card in which NPROB has been punched at zero and which is placed at the end of the data cards. The various locations in which sums are to be accumulated are initialized at zero, using DO-loops to do this for the singly subscripted arrays RSM and CSM and simple assignment statements for the variables A, B, C and D.

The array of observations X(I, J) is then read in using implied DO-loops. The observations—as indicated in the 102 FORMAT statement—are punched 10 to a card (10F7.2), and are arranged to run on continuously row by row from the first observation in the first row to the last in the last row. It is most important that the data are punched in this way so that they correspond with the arrangement of DO-loops in the READ statement. Users uncertain of the various ways in which arrays can be read in should consult the account on p. 219.

The major computational part of the program then follows, using nested DO-loops to compute the sum of squared observations A, the row sums RSM(I) and summed squares B, the column sums RCM(I) and summed squares C, as well as the total of observations which will eventually give D. The computation of B, C and D is then completed after converting the integers NR and NC to corresponding real variables ANR and ANC.

It is next possible to compute the various quantities which will appear in the final analysis of variance table:

BR
BC
RES
TOT
} These are respectively the sums of squares for rows, columns, residual and total, all derived from the intermediate variables A to D.

NDFR
NDFC
NRES
NTOT
} The degrees of freedom for rows, columns, residual and total, all derived from the values given to NR and NC.

BRM
BMC
RESM
} The between-rows mean square, between-columns mean square and residual (=error) mean square.

FR
FC
} The variance ratios (F-values) for rows and columns.

Flowchart for Program 12

Analysis of variance: two factors without replication.

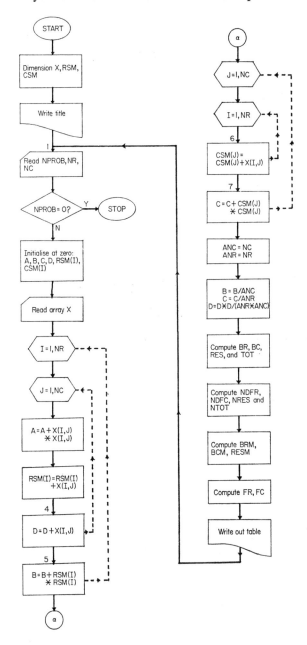

The program then prints out these values in tabular form. Finally a test on NPROB from the next data card decides whether another problem is dealt with or whether (in the event of NPROB = 0 on the sentinel card) the program stops.

Program 12

Analysis of variance: two factors without replication; with sample data and output.

```
C
C.................................................................
C
C   THIS PROGRAM CARRIES OUT A TWO-FACTOR ANALYSIS OF VARIANCE.
C   INPUT PARAMETERS ARE AS FOLLOWS –
C       NPROB = NUMBER OF PROBLEM
C       NR    = NUMBER OF ROWS IN DATA ARRAY
C       NC    = NUMBER OF COLUMNS IN DATA ARRAY
C       X     = DATA ARRAY
C
C.................................................................
C
        DIMENSION X(50,50), RSM(50), CSM(50)
C
C   WRITE HEADINGS.
        WRITE (6,100)
    100 FORMAT (21H1ANALYSIS OF VARIANCE////1X, 7HPROBLEM, 11X,
       A9HSOURCE OF, 9X, 6HSUM OF, 12X, 10HDEGREES OF, 8X, 4HMEAN, 14X,
       B8HVARIANCE/1X, 6HNUMBER, 12X, 8HVARIANCE, 10X, 7HSQUARES, 11X,
       C7HFREEDOM, 11X, 6HSQUARE, 12X, 5HRATIO //)
C
C   READ PROBLEM SPECIFICATIONS AND INITIALISE.
      1 READ (5,101) NPROB, NR, NC
    101 FORMAT (I6, I3, I3)
        IF (NPROB .EQ. 0) STOP
        A = 0.0
        B = 0.0
        C = 0.0
        D = 0.0
        DO 2 I = 1,NR
      2 RSM(I) = 0.0
        DO 3 J = 1,NC
      3 CSM(J) = 0.0
C
C   READ DATA AND COMPUTE INTERMEDIATE VARIABLES.
        READ (5,102) ((X(I,J), J = 1,NC), I = 1,NR)
    102 FORMAT (10F7.2)
        DO 5 I = 1,NR
        DO 4 J = 1,NC
        A = A + X(I,J) * X(I,J)
        RSM(I) = RSM(I) + X(I,J)
      4 D = D + X(I,J)
      5 B = B + RSM(I)*RSM(I)
        DO 7 J = 1,NC
        DO 6 I = 1,NR
      6 CSM(J) = CSM(J) + X(I,J)
      7 C = C + CSM(J) * CSM(J)
        ANC = NC
        ANR = NR
```

Program 12 (continued)

```
      B = B/ANC
      C = C/ANR
      D = D *D /(ANR * ANC)
C
C  COMPUTE MEAN SQUARES AND VARIANCE RATIOS.
      BR = B - D
      BC = C - D
      RES = A + D - B - C
      TOT = A - D
      NDFR = NR - 1
      NDFC = NC - 1
      NRES = NDFR * NDFC
      NTOT = NR * NC - 1
      BRM  = BR / (ANR - 1.0)
      BCM  = BC / (ANC - 1.0)
      RESM = RES / ((ANR - 1.0) * (ANC - 1.0))
      FR = BRM /RESM
      FC = BCM /RESM
C
C  OUTPUT RESULTS.
      WRITE (6,103) NPROB, BR, NDFR, BRM, FR, BC, NDFC, BCM, FC, RES,
     A    NRES, RESM, TOT, NTOT
  103 FORMAT (1X,I6,12X,4HROWS,10X,E16.8,6X,I6,8X,E16.8,
     A6X,F6.2 /19X,7HCOLUMNS,7X,E16.8,6X,I6,8X,E16.8,6X,
     BF6.2/19X,8HRESIDUAL,6X,E16.8,6X,I6,8X,E16.8//
     C19X,5HTOTAL,9X,E16.8,6X,I6///)
      GO TO 1
      END
```

DATA

```
  1  6  4
  4.4    5.9    6.0    4.1    3.3    1.9    4.9    7.1    4.4    4.0
  4.5    3.1    6.8    6.6    7.0    6.4    6.3    4.9    5.9    7.1
  6.4    7.3    7.7    6.7
  2  4  6
  4.4    3.3    4.4    6.8    6.3    6.4    5.9    1.9    4.0    6.6
  4.9    7.3    6.0    4.9    4.5    7.0    5.9    7.7    4.1    7.1
  3.1    6.4    7.1    6.7
  0  0  0
```

OUTPUT

ANALYSIS OF VARIANCE

PROBLEM NUMBER	SOURCE OF VARIANCE	SUM OF SQUARES	DEGREES OF FREEDOM	MEAN SQUARE	VARIANCE RATIO
1	ROWS	0.31652138E 02	5	0.63304275E 01	4.82
	COLUMNS	0.31412964E 01	3	0.10470987E 01	0.80
	RESIDUAL	0.19716194E 02	15	0.13144129E 01	
	TOTAL	0.54509620E 02	23		
2	ROWS	0.31412964E 01	3	0.10470987E 01	0.80
	COLUMNS	0.31652138E 02	5	0.63304275E 01	4.82
	RESIDUAL	0.19716194E 02	15	0.13144129E 01	
	TOTAL	0.54509620E 02	23		

A typical example of output from Program 12 is shown after the listing. The sums of squares and mean squares are given in an E-specification so as to allow for a very wide variation in the magnitudes of these quantities. The variance ratios should be tested for significance in ithe usual way, using tabulated values of F as given, for example, by Fisher and Yates (1963).

Program 13: Incomplete three-factor analysis of variance (two factors with replication)

This program undertakes a somewhat more elaborate analysis, again involving two factors (each at two or more levels) but this time allowing an estimate of the row-column interaction. This is possible because each combination of the different levels of the two factors is now assessed on two or more replicates. The model therefore takes the form:

$$x_{ijk} = \mu + F_i + G_j + I_{ij} + \varepsilon_{ijk} \tag{5}$$

where I_{ij} is the interaction term and the additional subscript k indicates the replicate. The analysis of variance table for a design based on this model will take the following form, there being a total of $N = pqr$ observations where r is the number of replicates (which must be the same for all combinations of rows and columns).

Source of Variation	Sum of Squares	Degrees of freedom	Mean Square
Rows	S_1	$(p-1)$	$M_1 = S_1/(p-1)$
columns	S_2	$(q-1)$	$M_2 = S_2/(q-1)$
rows × columns	S_3	$(p-1)(q-1)$	$M_3 = S_3/(p-1)(q-1)$
residual	S_4	$(N-pq)$	$M_4 = S_4/N-pq)$
Total	S	$(N-1)$	—

The variance-ratios computed by the program are those required when the effects of the factors are systematic, comparing M_1, M_2 and M_3 in turn with M_4. It is sometimes suggested that if the variance-ratio for the rows × columns interaction is not significant, one should obtain a new estimate of the residual mean square by pooling S_3 and S_4 in the above table. This procedure is not to be recommended (Huitson, 1966). If the effects of one or both factors are random, the user must ignore the printed variance-ratios and work out the appropriate ones by hand from the computed mean squares.

There are several ways of formulating a program to carry out an analysis of variance on these lines. The one given below involves a three-dimensional array, the data being arranged in NR rows, NC columns and NL 'layers', the latter representing the replicates while the rows correspond to the various levels of one factor and the columns to those of the other factor. The contents of the array are therefore represented by a triply subscripted variable X(I,J,K).

The DIMENSION declaration sets the numbers of rows, columns and layers at 10 each, but these could be increased if necessary according to the storage space available. The remaining dimensioned variables are:

RSM = sums for each row (summed over columns and layers)
CSM = sums for each column
RCSM = sums for each cell of the 2-way table derived by summing over layers
RMEAN = means for each row
CMEAN = means for each column

After printing a heading, the program reads a specification card on which are punched the following variables

NPROB = reference number of problem
NR   = number of rows
NC   = number of columns
NL   = number of layers

A decision-box then enables the execution of the program to be stopped by a card in which NPROB = 0. This card must be placed at the end of all other data cards. The arrays RSM, CSM, RCSM and the intermediate variables A to E are then initialised at zero so that sums may subsequently be accumulated in these locations. The intermediate variables, A, B, C and D are similar to those in the previous program, bearing in mind that summation now occurs over layers as well as over rows or columns. Intermediate variable E is defined as follows: square the value in each cell of the two-way table obtained by summing over layers, add the squares and divide by the number of layers.

Having initialised, the main computational part of the program then begins by reading in the three-dimensional array X(I,J,K). This is done by an implied DO-loop which, in effect, causes the rightmost subscript to vary most rapidly and the leftmost least rapidly. The data must therefore be punched on cards so that all replicates of the first row–column combination are followed by all replicates for the second and so on as indicated in Fig. 22.

Having read in the complete array the program accumulates various sums in a series of DO-loops and the calculation of the intermediate variables is completed by division (after converting the integers NR, NC and NL to their real equivalents ANR, ANC and ANL). The row- and column-means and the sums of squares are then computed, followed by the mean squares and variance ratios, after which a fairly long output sequence arranges the results of the analysis in tabular form. Finally control is again transferred to statement 1 so that the data for the next problem can be read. Should there be no further problems the sentinel card (NPROB = 0) ensures that execution stops.

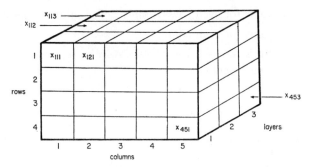

FIG. 22. Arrangement of data in rows, columns and layers.

The kind of design analysed in this program is sometimes known as a two-way crossed classification since every level of the first factor is represented in combination with every level of the second. It often happens in biological work that a different model—that of a two-way nested classification (or hierarchical model)—is more appropriate. Here, the various levels of one of the factors are different within each level of the other factor, so that the second factor is said to be nested within the first. For example, suppose that the rows in the original data matrix correspond to a number of species whose geographical distributions do not overlap and that the columns correspond to several localities. These cannot be the same for any two species and the column factor 'locality' is therefore nested within the row factor 'species'. Under these circumstances model (5) above is not appropriate and must be replaced by a model for the two-way nested classification.

$$x_{ijk} = \mu + F_i + I_{j(i)} + \varepsilon_{ijk} \tag{6}$$

where F_i corresponds to the effect of the main factor and $I_{j(i)}$ to the effect of the nested factor for a particular level i of the main factor. Fortunately the computation carried out by this program can easily be made to apply to a nested design. Let us suppose that the row factor is the main one and the

Flowchart for Program 13

Analysis of variation: two factors with replication.

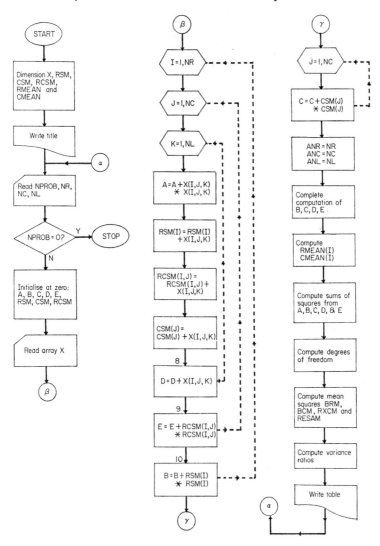

column factor nested within it. The appropriate analysis of variance table is then as follows, using the same symbols as in the table on p. 154.

Source of variation	Sum of squares	Degrees of freedom	Mean square
Rows	S_1	$(p-1)$	$M_1 = S_1/(p-1)$
Between cells (within rows)	$S_2 + S_3$	$p(q-1)$	$M_2 = (S_2 + S_3)/p(q-1)$
Within cells	S_4	$(N-pq)$	$M_4 = S_4/(N-pq)$
Total	S	$(N-1)$	—

The only difference is that the table for a nested classification has a second row that consists of the pooled sums of squares and degrees of freedom for the second and third rows of the crossed classification table. The calculation of the variance ratio is somewhat more complicated, however, and is discussed by Huitson (1966: p. 20). If the row factors have systematic effects, the ratios M_1/M_4 and M_2/M_4 will test the main effect of the row factor and the nested effect of the column factor.

Program 13

Analysis of variance: two factors with replication; with sample data and output.

```
C
C.................................................................
C
C   THIS PROGRAM EXECUTES AN INCOMPLETE 3-FACTOR ANALYSIS OF VARIANCE
C   (TWO FACTORS WITH EQUAL NUMBERS OF REPLICATES). INPUT PARAMETERS
C   ARE AS FOLLOWS -
C       NPROB = NUMBER OF PROBLEM
C       NR    = NUMBER OF ROWS IN DATA ARRAY
C       NC    = NUMBER OF COLUMNS IN DATA ARRAY
C       NL    = NUMBER OF LAYERS (REPLICATES) IN DATA ARRAY
C       X     = DATA ARRAY
C
C.................................................................
C
      DIMENSION X(10,10,10), RSM(10), CSM(10), RCSM(10,10),
     ARMEAN(10), CMEAN(10)
C
      WRITE (6,100)
  100 FORMAT (///1H1, 40X, 20HANALYSIS OF VARIANCE///)
    1 READ (5,101) NPROB, NR, NC, NL
  101 FORMAT (I6, 3I3)
      IF (NPROB .EQ. 0) STOP
```

Program 13 (*continued*)

```
C
C   INITIALISE FOR SUBSEQUENT ACCUMULATION OF SUMS.
        A = 0.0
        B = 0.0
        C = 0.0
        D = 0.0
        E = 0.0
        DO 3 I = 1,NR
        DO 2 J = 1,NC
      2 RCSM(I,J) = 0.0
      3 RSM(I) = 0.0
        DO 5 J = 1,NC
      5 CSM(J) = 0.0
C
C   READ AND STORE ARRAY
        READ (5,102) (((X(I,J,K), K = 1,NL), J = 1,NC), I = 1,NR)
    102 FORMAT (10F7.2)
C
C   LOOPS TO BEGIN ACCUMULATING SUMS FOR ROWS, COLUMNS, LAYERS AND
C   INTERMEDIATE VARIABLES A TO E.
        DO 10 I = 1,NR
        DO 9 J = 1,NC
        DO 8 K = 1,NL
        A = A + X(I,J,K) * X(I,J,K)
        RSM(I) = RSM(I) + X(I,J,K)
        RCSM(I,J) = RCSM(I,J) + X(I,J,K)
        CSM(J) = CSM(J) + X(I,J,K)
      8 D = D + X(I,J,K)
      9 E = E + RCSM(I,J) * RCSM(I,J)
     10 B = B + RSM(I) * RSM(I)
        DO 12 J = 1,NC
     12 C = C + CSM(J) * CSM(J)
C
C   COMPLETE COMPUTATION OF INTERMEDIATE VARIABLES A TO E.
        ANR = NR
        ANC = NC
        ANL = NL
        B = B / (ANC * ANL)
        C = C / (ANR * ANL)
        D = D * D / (ANR * ANC * ANL)
        E = E / ANL
C
C   COMPUTE MEANS OF ROWS AND COLUMNS.
        DO 15 I = 1,NR
     15 RMEAN(I) = RSM(I) / (ANC * ANL)
        DO 16 J = 1,NC
     16 CMEAN(J) = CSM(J) / (ANR * ANL)
C
C   COMPUTE SUMS OF SQUARES FROM INTERMEDIATE VARIABLES A TO E.
        BR = B - D
        BC = C - D
        RXC = D + E - B - C
        RESA = A - E
        TOT = A - D
C
C   COMPUTE DEGREES OF FREEDOM.
        NDFR = NR - 1
        NDFC = NC - 1
        NRXC = (NR - 1) * (NC - 1)
        NRESA = NR * NC * (NL - 1)
        NTOT = NR * NC * NL - 1
```

Program 13 (*continued*)

```
C
C   COMPUTE MEAN SQUARES.
        BRM = BR/(ANR - 1.0)
        BCM = BC/(ANC - 1.0)
        RXCM = RXC / ((ANR - 1.0) * (ANC - 1.0))
        RESAM = RESA / (ANR * ANC * (ANL - 1.0))
C
C   COMPUTE VARIANCE RATIOS.
        FR = BRM / RESAM
        FC = BCM / RESAM
        FRXC = RXCM / RESAM
C
C   OUTPUT ANALYSIS TABLES.
        WRITE (6,103) NPROB, (I. RMEAN(I), I = 1,NR)
    103 FORMAT (12H PROBLEM NO., I8///11X, 9HROW MEANS/
       A(5(12X, I2, F7.2)))
        WRITE (6,105) (J, CMEAN(J), J = 1,NC)
    105 FORMAT (/11X, 12HCOLUMN MEANS/(5(12X, I2, F7.2)))
C
        WRITE (6,104) BR, NDFR, BRM, FR, BC, NDFC, BCM, FC,
       ARXC, NRXC, RXCM, FRXC, RESA, NRESA, RESAM, TOT, NTOT
    104 FORMAT (///11X,9HSOURCE OF, 9X,6HSUM OF,12X,10HDEGREES OF, 8X,
       A4HMEAN, 14X, 8HVARIANCE/11X, 8HVARIANCE, 10X,7HSQUARES, 11X,
       B7HFREEDOM, 11X, 6HSQUARE, 12X, 5HRATIO///
       C11X, 4HROWS, 10X,E16.8, 6X, I6, 8X, E16.8, 6X, F6.2/
       D11X, 7HCOLUMNS, 7X, E16.8, 6X, I6, 8X, E16.8, 6X, F6.2/
       F11X, 3HRXC, 11X, E16.8, 6X, I6, 8X, E16.8, 6X, F6.2/
       I11X, 8HRESIDUAL, 6X, E16.8, 6X, I6, 8X, E16.8//
       J11X, 5HTOTAL, 9X, E16.8, 6X, I6///)
C
        GO TO 1
        END
```

```
DATA
    1   2   3   2
   3.1  2.1  2.0  1.6  1.3  1.9  4.3  6.1  2.8  3.9
   4.1  2.6
    0   0   0   0
```

OUTPUT

ANALYSIS OF VARIANCE

PROBLEM NO. 1

ROW MEANS
  1 2.00   2 3.97

COLUMN MEANS
  1 3.90   2 2.57   3 2.47

SOURCE OF VARIANCE	SUM OF SQUARES	DEGREES OF FREEDOM	MEAN SQUARE	VARIANCE RATIO
ROWS	0.11603332E 02	1	0.11603332E 02	16.94
COLUMNS	0.50616683E 01	2	0.25308342E 01	3.69
RXC	0.62166691E 00	2	0.31083345E 00	0.45
RESIDUAL	0.41099996E 01	6	0.68499994E 00	
TOTAL	0.21396667E 02	11		

Program 14: Three factor analysis of variance without replication

This design is one in which the experimenter examines the effect of three factors, each operating at two or more levels, and in which every possible combination of levels for every factor is assessed by a single unreplicated observation. It is therefore the three-factor equivalent of Program 12, but the computations have a closer resemblance to Program 13 since we shall again operate with a triply subscripted data array. The model for the analysis may be represented as

$$x_{ijk} = \mu + F_i + G_j + H_k + I_{ij} + J_{ik} + K_{jk} + \varepsilon_{ijk} \qquad (7)$$

where F_i, G_j and H_k represent the main effects of the three factors and I_{ij}, K_{ik} and K_{jk} are the first-order interaction terms. The model assumes that there is no second-order interaction and the analysis of variance table will therefore have the following form, where p, q and r are the number of levels for the row-, column- and layer-factors respectively.

Source of variation	Sums of squares	Degrees of freedom	Mean square
Rows	S_1	$(p-1)$	$M_1 = S_1/(p-1)$
Columns	S_2	$(q-1)$	$M_2 = S_2/(q-1)$
Layers	S_3	$(r-1)$	$M_3 = S_3/(r-1)$
Rows × columns	S_4	$(p-1)(q-1)$	$M_4 = S_4/(p-1)(q-1)$
Rows × layers	S_5	$(p-1)(r-1)$	$M_5 = S_5/(p-1)(r-1)$
Columns × layers	S_6	$(q-1)(r-1)$	$M_6 = S_6/(q-1)(r-1)$
Residual	S_7	$(p-1)(q-1)(r-1)$	$M_7 = S_7/(p-1)(q-1)(r-1)$
Total	S_8	$pqr-1$	—

Since the program follows the pattern of the previous two very closely, the flow-chart and description are briefer than would otherwise be necessary and concentrate mainly on defining the variables.

The DIMENSION statement includes the following variables:
- X: the triply subscripted array of observations
- RSM: sums of rows
- CSM: sums of columns

ESM: sums of layers (the mnemonic LSM is not used as it implies an integer variable).

RCSM: Sum of observations in each row-column cell of the 2-way table obtained by summing over layers.

RLSM: sum of observations in each row-layer cell of the 2-way table obtained by summing over columns.

CLSM: sum of observations in each column-layer cell of the 2-way table obtained by summing over rows.

RMEAN: row means

CMEAN: column means

LMEAN: layer means

The initialisation for accumulation of sums includes the six sums listed above (RSM to CLSM) as well as the following intermediate variables:

A = sum of squares of each observation in the original data array.

B = sum of squared row totals divided by the number of individual observations used to form each row total.

C = the same for columns.

D = grand total of observations, squared and then divided by the total number of individual observations in the data array.

E = sum of squares of the RCSM (I,J) divided by the number of layers.

F = sum of squares of ESM (K) divided by the total number of individual observations in each layer sum.

G = sum of squares of the RLSM (I,K) divided by the number of columns.

H = sum of squares of CLSM (J,K) divided by the number of rows.

Note that A–E are the same as in the previous program.

Having read in the primary data array (bearing in mind that the arrangement of punched data must correspond with the structure of the implied DO-loop, so that observations are properly allocated to rows, columns and layers), the intermediate variables are then calculated in two stages. From them the sums of squares are determined in the usual way and hence the mean squares and variance ratios. By returning control to statement 1 after printing out the results it is possible to run a series of problems consecutively. As before, a sentinel card (NPROB punched as zero) results in the program stopping execution when all the data have been analysed. The accompanying flow-chart is abbreviated, emphasising only the outlines of the program. Details correspond to the chart for Program 13 with the difference that the

Outline flowchart for Program 14

Analysis of variance: three factors without replication.

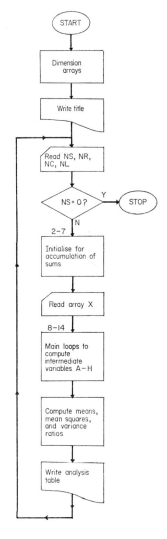

layers are here treated as levels of a main factor and the number of interactions is increased as indicated above.

A few words are necessary on the calculation of variance-ratios in this design. The program computes them as the ratios of each of the mean squares (M_1 to M_6 of the table on p. 161) to the residual mean square M_7. This is appropriate only if all the factors exert fixed, systematic effects. If they are such that the random factor or the mixed model applies, then the calculation of variance-ratios must depend on principles discussed in the references previously cited. The mean squares and degrees of freedom required for these calculations are, of course, those given in the printed output for the program.

Program 14

Analysis of variance: three factors without replication.

```
C
C.............................................................
C
C   THIS PROGRAM CARRIES OUT A THREE-FACTOR ANALYSIS OF VARIANCE.
C   INPUT PARAMETERS ARE AS FOLLOWS -
C       NPROB = NUMBER OF PROBLEM
C       NR    = NUMBER OF ROWS IN DATA ARRAY
C       NC    = NUMBER OF COLUMNS IN DATA ARRAY
C       NL    = NUMBER OF LAYERS IN DATA ARRAY
C       X     = DATA ARRAY (NR ROWS, NC COLUMNS, NL LAYERS)
C
C.............................................................
C
        DIMENSION X(10,10,10), RSM(10), CSM(10), ESM(10), RCSM(10,10),
       ARLSM(10,10), CLSM(10,10), RMEAN(10), CMEAN(10), EMEAN(10)
C
C   WRITE HEADINGS AND READ PROBLEM SPECIFICATION.
        WRITE (6,100)
  100   FORMAT (1H1, 40X, 20HANALYSIS OF VARIANCE///)
    1   READ (5,101) NPROB, NR, NC, NL
  101   FORMAT (I6, 3I3)
        IF (NPROB .EQ. 0) STOP
C
C   INITIALISE FOR SUBSEQUENT ACCUMULATION OF SUMS.
        A = 0.0
        B = 0.0
        C = 0.0
        D = 0.0
        E = 0.0
        F = 0.0
        G = 0.0
        H = 0.0
        DO 3 I = 1,NR
        DO 2 J = 1,NC
    2   RCSM(I, J) = 0.0
    3   RSM(I) = 0.0
        DO 5 J = 1,NC
        DO 4 K = 1,NL
    4   CLSM(J,K) = 0.0
    5   CSM(J) = 0.0
        DO 7 K = 1,NL
        DO 6 I = 1,NR
```

Program 14 (*continued*)

```
      6 RLSM(I,K) = 0.0
      7 ESM(K) = 0.0
C
C   READ AND STORE ARRAY.
        READ (5,102) (((X(I,J,K), K = 1,NL), J = 1,NC), I = 1,NR)
  102 FORMAT (10F7.2)
C
C   LOOPS TO BEGIN ACCUMULATING SUMS FOR ROWS, COLUMNS, LAYERS AND
C   INTERMEDIATE VARIABLES A TO H.
        DO 10 I = 1,NR
        DO 9 J = 1,NC
        DO 8 K = 1,NL
        A = A + X(I,J,K) * X(I,J,K)
        RSM(I) = RSM(I) + X(I,J,K)
        RCSM(I,J) = RCSM(I,J) + X(I,J,K)
        CSM(J) = CSM(J) + X(I,J,K)
        CLSM(J,K) = CLSM (J,K) + X(I,J,K)
        ESM(K) = ESM(K) + X(I,J,K)
        RLSM(I,K) = RLSM (I,K) + X(I,J,K)
      8 D = D + X(I,J,K)
      9 E = E + RCSM(I,J) * RCSM(I,J)
     10 B = B + RSM(I) * RSM(I)
        DO 12 J = 1,NC
        DO 11 K = 1,NL
     11 H = H + CLSM(J,K) * CLSM(J,K)
     12 C = C + CSM(J) * CSM(J)
        DO 14 K = 1,NL
        DO 13 I = 1,NR
     13 G = G + RLSM(I,K) * RLSM(I,K)
     14 F = F + ESM(K) * ESM(K)
C
C   COMPLETE COMPUTATION OF INTERMEDIATE VARIABLES A TO H.
        ANR = NR
        ANC = NC
        ANL = NL
        B = B / (ANC * ANL)
        C = C / (ANR * ANL)
        D = D * D / (ANR * ANC * ANL)
        E = E / ANL
        F = F / (ANR * ANC)
        G = G / ANC
        H = H / ANR
C
C   COMPUTE MEANS OF ROWS, COLUMNS AND LAYERS.
        DO 15 I = 1,NR
     15 RMEAN(I) = RSM(I) / (ANC * ANL)
        DO 16 J = 1,NC
     16 CMEAN(J) = CSM(J) / (ANR * ANL)
        DO 17 K = 1,NL
     17 EMEAN(K) = ESM(K) / (ANR * ANC)
C
C   COMPUTE SUMS OF SQUARES FROM INTERMEDIATE VARIABLES A TO H.
        BR = B - D
        BC = C - D
        BL = F - D
        RXC = D + E - B - C
        RXL = D + G - B - F
        CXL = D + H - C - F
        RESB = A + B + C - D - E + F - G - H
        TOT = A - D
```

Program 14 (*continued*)

```
C
C   COMPUTE DEGREES OF FREEDOM.
        NDFR = NR - 1
        NDFC = NC - 1
        NDFL = NL - 1
        NRXC = (NR - 1) * (NC - 1)
        NRXL = (NR - 1) * (NL - 1)
        NCXL = (NC - 1) * (NL - 1)
        NRESB = (NR - 1) * (NC - 1) * (NL - 1)
        NTOT = NR * NC * NL - 1
C
C   COMPUTE MEAN SQUARES.
        BRM = BR/(ANR - 1.0)
        BCM = BC/(ANC - 1.0)
        BLM = BL/(ANL - 1.0)
        RXCM = RXC / ((ANR - 1.0) * (ANC - 1.0))
        RXLM = RXL / ((ANR - 1.0) * (ANL - 1.0))
        CXLM = CXL / ((ANC - 1.0) * (ANL - 1.0))
        RESBM = RESB / ((ANR - 1.0) * (ANC - 1.0) * (ANL - 1.0))
C
C   COMPUTE VARIANCE RATIOS.
        FR = BRM / RESBM
        FC = BCM / RESBM
        FL = BLM / RESBM
        FRXC = RXCM / RESBM
        FRXL = RXLM / RESBM
        FCXL = CXLM / RESBM
C
C   OUTPUT ANALYSIS TABLES.
            WRITE (6,103) NPROB, (I, RMEAN(I), I = 1,NR)
        103 FORMAT (15H PROBLEM NUMBER, I5///11X, 9HROW MEANS/
       A(5(12X,I2,F7.2)))
            WRITE (6,105) (J, CMEAN(J), J = 1,NC)
        105 FORMAT (/11X, 12HCOLUMN MEANS/(5(12X,I2,F7.2)))
            WRITE (6,106) (K, EMEAN(K), K = 1,NL)
        106 FORMAT (/11X, 11HLAYER MEANS/(5(12X, I2, F7.2)))
C
            WRITE (6,104) BR, NDFR, BRM, FR, BC, NDFC, BCM, FC, BL, NDFL,
       ABLM, FL, RXC, NRXC, RXCM, FRXC, RXL, NRXL, RXLM, FRXL, CXL, NCXL,
       BCXLM, FCXL, RESB, NRESB, RESBM, TOT, NTOT
        104 FORMAT (///11X, 9HSOURCE OF, 9X, 6HSUM OF, 12X,10HDEGREES OF, 8X,
       A4HMEAN, 14X, 8HVARIANCE/11X, 8HVARIANCE, 10X,7HSQUARES, 11X,
       B7HFREEDOM, 11X, 6HSQUARE, 12X, 5HRATIO///
       C11X, 4HROWS, 10X, E16.8, 6X, I6, 8X, E16.8, 6X, F6.2/
       D11X, 7HCOLUMNS, 7X, E16.8, 6X, I6, 8X, E16.8, 6X, F6.2/
       E11X, 6HLAYERS, 8X, E16.8, 6X, I6, 8X, E16.8, 6X, F6.2/
       F11X, 3HRXC, 11X, E16.8, 6X, I6, 8X, E16.8, 6X, F6.2/
       G11X, 3HRXL, 11X, E16.8, 6X, I6, 8X, E16.8, 6X, F6.2/
       H11X, 3HCXL, 11X, E16.8, 6X, I6, 8X, E16.8, 6X, F6.2//
       I11X, 8HRESIDUAL, 6X, E16.8, 6X, I6, 8X, E16.8//
       J11X, 5HTOTAL, 9X, E16.8, 6X, I6///)
C
        GO TO 1
        END
```

Program 15: Analysis of variance: Latin square design

In the previous three programs the design has been fully factorial, in the sense that each level of every factor is investigated in combination with each level of every other factor. The result is that large numbers of observations may be needed. For example three factors, each at six levels, would require (even without replication) $6 \times 6 \times 6 = 216$ observations. Experiments have therefore been devised for carrying out only part of the fully factorial design, but to do this in a balanced way so that the effects may be correctly estimated. One such design is the so-called Latin square, in which three factors, each at the same number of levels, say n, may be investigated using n^2 observations instead of the n^3 required in a fully factorial design. The principles involved can be appreciated from Fig. 23 in which we have a 5×5 Latin square. The five levels of the first factor are represented by the rows and of the second factor by the columns. The third factor, at five levels A–E is then combined with the rows and columns in such a way that each level (often referred to as a treatment or a variety) is represented once in each row and once in each column. There are clearly many different ways in which the letters of Fig. 23 can be arranged so as to satisfy this constraint. Fisher and Yates (1963) tabulate the possibilities systematically for various sizes of Latin square, thus enabling designs to be selected at random.

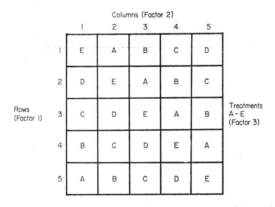

FIG. 23. Latin Square design with five treatments (A to E).

Typically a Latin square forms a convenient layout for an agricultural experiment in which Fig. 23 would be a plan of the field plots in which five different crop varieties A to E or five different fertilizer or insecticide treatments A to E are tested. The design then has the obvious advantage of allowing for the effects of soil heterogeneity in two directions at right angles to each other. The statistical principles are, however, equally applicable to other laboratory or field experiments involving three factors. For example Fig. 23 might represent the behavioural responses to five different stimuli (A to E) applied on five different occasions to five different strains of a species.

The model underlying a Latin square design may be expressed as

$$x_{ij} = \mu + F_i + G_j + H_k + \varepsilon_{ij} \tag{8}$$

where F_i, G_j and H_k are the effects of the i-th row, j-th column and k-th 'treatment'. A major assumption of the design is that there is no interaction between any of the factors, so that one is in effect carrying out a 3-factor experiment like that of Program 14 but using only n^2 instead of n^3 observations. The importance of the assumption of no interaction must be remembered, since a significant interaction between any two factors would appear as a spurious 'effect' of the third factor. The analysis of variance table for a $n \times n$ Latin square is:

Source of variation	Sums of squares	Degrees of freedom	Mean squares
Rows	S_1	$(n-1)$	$M_1 = S_1/(n-1)$
Columns	S_2	$(n-1)$	$M_2 = S_2/(n-1)$
Treatments	S_3	$(n-1)$	$M_3 = S_3/(n-1)$
Residual	S_4	$(n-1)(n-2)$	$M_4 = S_4/(n-1)(n-2)$
Total	S	$n^2 - 1$	—

Variance-ratios are calculated by comparing M_1, M_2 and M_3 in turn with M_4, irrespective of whether the factors have fixed or random effects.

The program bears sufficient general resemblance to the previous ones not to require detailed discussion. The input for each problem consists of (a) a problem specification card on which are punched values of NPROB (problem reference number) and N (number of rows, columns and treatments); (b) a card on which are punched the totals for the N individual observations which make up each of the N treatments. These treatment totals constitute the singly subscripted variable TX and must be worked out by hand before the data cards are punched; (c) the observations in each cell of the data array X, such as is shown in Fig. 23. These observations are read in row by row, using the DO-loops ending on statement 2; the data must accordingly be punched so that each row begins on a new card.

When all the data are in core-storage the program computes the grand total of observations (GRAND) and the totals for each row and column (SR and SC). These computations are accomplished by the DO-loops ending on statement 4 after previous initialisation of GRAND, SR and SC to zero. Again, we begin the computation of sums of squares for rows, columns, treatments,

Outline flowchart for Program 15

Analysis of variance: Latin square design.

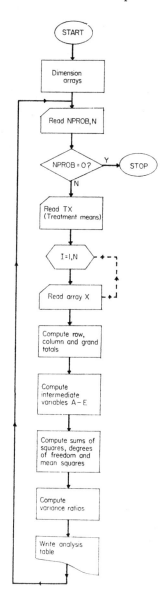

residual and the total by evaluating five intermediate variables A to E, defined this time as follows:

A = sum of squares of each observation in the N × N data array
B = sum of squared row totals, divided by N
C = sum of squared column totals, divided by N
D = sum of squared treatment totals, divided by N
E = grand total of all observations, divided by number of observations in the array (the correction term).

The similarity of these to some of the intermediate variables in previous programs is clear. The computation of the following sums of squares then follows directly:

SSQR   = sum of squares for rows
SSQC   = sum of squares for columns
SSQTR   = sum of squares for treatments
SSQRES = residual sum of squares
SSQTOT = total sum of squares.

The corresponding mean squares (VR, VC, VTR and VRES) and the variance ratios FR, FC and FTR for rows, columns and treatments respectively are computed as usual and the results written out.

Program 15

Analysis of variance: Latin square design; with sample data and output.

```
C
C.................................................................
C
C  THIS PROGRAM EXECUTES AN ANALYSIS OF VARIANCE BASED ON A LATIN SQUARE
C  DESIGN. FOR COMPUTATIONAL DETAILS SEE BROWNLEE (1949) INDUSTRIAL
C  EXPERIMENTATION, PAGES 144–145. INPUT PARAMETERS ARE AS FOLLOWS –
C     NPROB = NUMBER OF PROBLEM
C     N     = NUMBER OF TREATMENTS
C     TX    = VECTOR OF TREATMENT TOTALS (LENGTH N)
C     X     = DATA ARRAY (ORDER N BY N)
C
C.................................................................
C
      DIMENSION X(50,50), TX(50), SR(50), SC(50)
C
C  READ PROBLEM SPECIFICATION AND DATA.
    1 READ (5,100) NPROB, N
  100 FORMAT (2I5)
      IF (NPROB .EQ. 0) STOP
      READ (5,101) (TX(I), I = 1,N)
      DO 2 I = 1,N
```

Program 15 (*continued*)

```
    2 READ (5,101) (X(I,J), J = 1,N)
  101 FORMAT (10F7.0)
C
C   COMPUTE ROW, COLUMN AND GRAND TOTALS.
      GRAND = 0.0
      DO 3 I = 1,N
      SR(I) = 0.0
    3 SC(I) = 0.0
      DO 4 I = 1,N
      DO 4 J = 1,N
      SR(I) = SR(I) + X(I,J)
      SC(J) = SC(J) + X(I,J)
    4 GRAND = GRAND + X(I,J)
C
C   COMPUTE INTERMEDIATE VARIABLES A TO E.
      AN = N
      A = 0.0
      B = 0.0
      C = 0.0
      D = 0.0
      E = GRAND * GRAND / (AN*AN)
      DO 5 I = 1,N
      B = B + SR(I) * SR(I)
      C = C + SC(I) * SC(I)
      D = D + TX(I) * TX(I)
      DO 5 J = 1,N
    5 A = A + X(I,J) * X(I,J)
      B = B / AN
      C = C / AN
      D = D / AN
C
C   COMPUTE SUMS OF SQUARES, DEGREES OF FREEDOM AND VARIANCES.
      SSQR = B - E
      SSQC = C - E
      SSQTR = D - E
      SSQTOT = A - E
      SSQRES = A - B - C - D + 2.0*E
      NDF = N - 1
      NDFTOT = N * N - 1
      NDFRES = (N - 1) * (N - 2)
      DF = NDF
      DFRES = NDFRES
      VR = SSQR / DF
      VC = SSQC / DF
      VTR = SSQTR / DF
      VRES = SSQRES / DFRES
C
C   COMPUTE VARIANCE RATIOS.
      FR = VR / VRES
      FC = VC / VRES
      FTR = VTR / VRES
C
C   OUTPUT RESULTS.
      WRITE (6,102) NPROB, SSQR, NDF, VR, FR, SSQC, NDF, VC, FC,
     A  SSQTR, NDF, VTR, FTR, SSQRES, NDFRES, VRES, SSQTOT, NDFTOT
  102 FORMAT (28H1LATIN SQUARE DESIGN PROBLEM, I4//// 7X,
     A  9HSOURCE OF, 8X, 7HSUMS OF, 8X, 10HDEGREES OF, 9X, 4HMEAN,
     B  10X, 8HVARIANCE/ 7X, 8HVARIANCE, 8X, 7HSQUARES, 9X, 7HFREEDOM,
     C  10X, 6HSQUARE, 16X, 5HRATIO// 9X, 4HROWS, 8X, 1PE12.4, I11, 10X,
     D  E12.4, 10X, E12.4/ 8X, 7HCOLUMNS, 6X, E12.4, I11, 10X, E12.4,
```

Program 15 (continued)

```
F    10X, E12.4/ 7X, 10HTREATMENTS, 4X, E12.4, I11, 10X, E12.4,
G    10X, E12.4/ 8X, 8HRESIDUAL, 5X, E12.4, I11, 10X, E12.4//
H    9X, 5HTOTAL, 7X, E12.4,I11)
     GO TO 1
     END
```

DATA

```
  1     5
325.0  410.0  305.0  393.0  321.0
 17.0   39.0   65.0   19.0   12.0
 32.0   33.0   61.0   71.0   94.0
 56.0   49.0   84.0   90.0  100.0
 76.0   81.0   97.0   98.0  100.0
 93.0   90.0   97.0  100.0  100.0
  0
```

OUTPUT

LATIN SQUARE DESIGN PROBLEM   1

SOURCE OF VARIANCE	SUMS OF SQUARES	DEGREES OF FREEDOM	MEAN SQUARE	VARIANCE RATIO
ROWS	1.4165E 04	4	3.5413E 03	1.8937E 01
COLUMNS	3.1946E 03	4	7.9864E 02	4.2706E 00
TREATMENTS	1.7874E 03	4	4.4684E 02	2.3894E 00
RESIDUAL	2.2441E 03	12	1.8701E 02	
TOTAL	2.1391E 04	24		

Program 16: Analysis of variance: balanced incomplete block design

Like the Latin square, a balanced incomplete block design is intended to use only some of the combinations of levels represented in a fully factorial design, doing so in a balanced way that permits an analysis of variance. It may be understood by reference first to a randomised block design in which the observations are arranged in homogeneous 'blocks', each block containing every treatment represented once. This is a case of the two-factor analysis of variance without replication (Program 13) and it requires bt observations, where b is the number of blocks and t the number of treatments. A situation could arise, however, where each block is not large enough to contain all treatments, though it could hold more than one treatment. Under these circumstances we can arrange for each block to be *incomplete* in the sense that it contains less than the full number of treatments, but for the design to be *balanced*, in the sense that every pair of treatments occurs together in a block the same number of times. An example of such a design is shown graphically in Fig. 24, where the treatments are denoted by letters A–E, the blocks by I–V and the replicates by 1–4. Thus, the first block would be divided into four units to which treatments A, C, D and E would be applied as indicated, the second block would receive treatments A, B, D and E and so on. Thus, whereas the randomised block design for 5 treatments in 5 blocks would require a total of 25 observations (assuming no replication) the balanced incomplete block design would need only 20. Further, for similar numbers of observations a balanced incomplete block design may be statistically more efficient than a randomised block lay-out. As in the Latin-square designs, the original uses of balanced incomplete blocks in agricultural field experiments can be extended to a wide variety of other situations in which one is concerned with the effects of two factors.

		Blocks			
	I	II	III	IV	V
A	1	2	3	4	–
B	–	1	2	3	4
C	4	–	1	2	3
D	3	4	–	1	2
E	2	3	4	–	1

Treatments

FIG. 24. Balanced incomplete block design with 5 treatments (A to E) and 5 blocks (I to V). The dashes denote that the treatment concerned does not appear in that block.

The experimental lay-out indicated in Fig. 24 is only one of a number containing 5 treatments and 5 blocks with 4 replicates per treatment and 4 units per block. Other arrangements of the replicates 1–4 may be constructed from tabulated information on the various combinatorial solutions in Fisher and Yates (1963). It is, however, necessary to observe two constraints in making up these designs. If we denote the numbers of elements in the experiment as:

b = number of blocks

k = number of units per block

r = number of replicates per treatment

t = number of treatments

then the arrangements must be such that:

(i) $bk = rt$

and (ii) $\dfrac{r(k-1)}{(t-1)}$ must be an exact integer (usually denoted by λ and

representing the number of blocks in which any given pair of treatments will occur). It is easy to verify that the lay-out of Fig. 24 conforms to these requirements and that in it $\lambda = 3$, as can easily be seen in the diagram. Other possible values of b, t, k and r are tabulated in detail by Fisher and Yates (1963); the important thing to realise is that only certain values are permitted.

The model representing a balanced incomplete block design is:

$$x_{ij} = \mu + T_i + B_j + e_{ij} \qquad (9)$$

where T_i denotes the effect of treatments and B_j that of blocks. It must be remembered that not all possible combinations of i and j are allowed and that the model assumes no interaction between treatments and blocks. The analysis of variance table is therefore as follows:

Source of variation	Sums of squares	Degrees of freedom	Mean squares
Between treatments	S_1	$(t-1)$	$M_1 = S_1/(t-1)$
Between blocks	S_2	$(b-1)$	$M_2 = S_2/(b-1)$
Residual	S_3	$(rt-b-t+1)$	$M_3 = S_3/(rt-b-t+1)$
Total	S	$(rt-1)$	—

The method by which the sums of squares are computed and the analysis carried out is based on the work of Yates (1936) as illustrated by Brownlee (1951). In order to simplify programming it is assumed here that the recorded observations never have zero values. This is not a serious limitation, since if zero values do occur the whole set of recorded observations can be coded, increasing them all by a constant chosen to eliminate the zeros. This will not affect the subsequent significance tests, though the means printed in the output from the program will require adjustment by subtracting from each mean whatever constant was originally added. The complete $b \times t$ data array (corresponding to that shown in Fig. 24) will therefore consist of non-zero observations with zeros inserted wherever a block does not contain a particular treatment (i.e. zeros replace the dashes in Fig. 24). This allows summation and summation of squares to be carried out by DO-loops systematically over the whole array, since the zeros contribute nothing to the totals. A more cumbersome alternative to this procedure consists of arranging for all unrepresented combinations to be denoted by some very large positive or negative number which is then excluded from summation or squaring by an IF statement in the DO-loop. The reader might experiment with various methods of programming this frequently occurring situation in which arrays contain some elements that have to be incorporated in a calculation and others that must be omitted from it.

The program, then, is dimensioned in respect of the treatment × block array X and five singly-subscripted variables B, T, Q, V and CB with meanings which will become clear later. The problem specification card is punched to give values of NPROB (the usual problem reference number, used also to stop execution by a sentinel card on which NPROB = 0), NT, NK, NR and NB, which are respectively the numbers of treatments, units per block, replicates per treatment, and blocks. The data array X is then read, punched so that each row (treatment) begins on a new card and with zeros punched in those treatment-block combinations for which no observation exists. The next segment of the program initialises the variables T, B and G to zero and the DO-loops ending on statement 5 accumulate sums in them as follows:

T(I) = sum of replicates in *i*-th treatment

B(J) = sum of units in *j*-th block

G   = grand total of all observations.

The conversion of the integers N, NT and NK to real equivalents AN, AT and AK also occurs in this segment of the program though they are not needed until later.

The computation of the sums of squares for treatments takes up the next segment of the program. The nested DO-loops ending on statement 6 begin this process by accumulating in array Q the sums of the totals for every block containing each treatment in turn. Thus in passing along the first row (treatment A) of Fig. 24 to obtain Q(1), we add the totals for blocks I, II, III and IV, omitting block V since it does not contain an observation for treatment A. Similarly for each of the other treatments. Note here that we again make use of the zeros in array X, this time to exclude the block totals when a particular treatment is omitted from the block. The DO-loop ending on 7 then completes the calculations of the elements of array Q by subtracting the original contents of Q(I) from the NK * T(I). Finally we use the next four statements of this program segment to evaluate the sum of squares for treatments (SSQT), using the formula

$$S_1 = \frac{(t-1)}{tr(k-1)} \sum_{i=1}^{t} Q_i^2 \qquad (10)$$

with which is associated a number of degrees of freedom (NDFT) equal to (NT − 1).

Having computed the sum of squares for treatments we next calculate the sums of squares for blocks (SSQB) and the total sum of squares (SSQTOT). SSQB is obtained by squaring the block totals, summing the squares and subtracting the usual correction term (G*G/AN). Similarly, the total sum of squares is obtained by squaring each observation, summing the squares and subtracting the same correction term as before. The sum of squares for error (SSQERR) is then found by difference:

$$SSQERR = SSQTOT - SSQB - SSQT$$

The degrees of freedom for all these sums of squares are calculated in the usual way (NDFB, NDFTOT and NDFERR).

The next segment of the program is quite simple: from the sums of squares and degrees of freedom (converted to their real equivalents DFB, DFT and DFERR) one computes the mean squares for blocks, treatments and error (VB, VT and VERR) and then from these the variance-ratios FB and FT for blocks and treatments respectively. The analysis of variance table corresponding to that on p. 174 can then be printed out in the usual way.

To complete the computations, however, we need to calculate the treatment means corrected for block differences and the block means corrected for treatment differences. The first of these is computed by the DO-loop ending on statement 11, which forms the t quantities v_i from the relationship

$$v_i = \frac{(t-1)}{tr(k-1)} Q_i \tag{11}$$

where v_i is the deviation of the i-th treatment from the grand mean. The same loop adds this grand mean (GN) to v_i so that the one set of locations V will finally contain the corrected means. The standard error of the difference between two treatment means (SE) is also computed from the expression

$$\sqrt{\frac{2\sigma_k^2 . k(t-1)}{tr(k-1)}} \tag{12}$$

where σ_k^2 is the error mean square (VERR). The corrected treatment means and the standard error are printed and the last segment of the program completes the analysis by computing the corrected block means and their standard error. These corrected block means are stored in the array CB and are computed by forming for each block

(i) the block total, less the sum of the corrected treatment means for those treatments represented in the block;

(ii) dividing (i) by the number of units per block;

(iii) adding the grand mean.

All these operations are performed by the nested DO-loops ending on

Flowchart for Program 16
Analysis of variance: balanced incomplete block design.

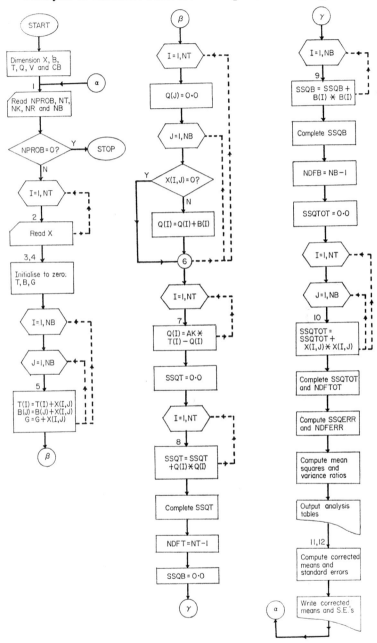

statement 13. The standard error of the difference between any two corrected block means is given by the expression

$$\sqrt{\frac{2\sigma_k^2}{k}\left(1 + \frac{(t-1)(t-k)}{t(k-1)(b-1)}\right)} \tag{13}$$

The accompanying flow-chart will help the reader to understand the program.

Program 16

Analysis of variance: balanced incomplete block design; with sample data and output.

```
C
C...........................................................
C
C   THIS PROGRAM EXECUTES AN ANALYSIS OF VARIANCE BASED ON A BALANCED
C   INCOMPLETE BLOCK DESIGN. FOR COMPUTATIONAL METHOD AND NOTATION
C   SEE YATES (1936), ANNALS OF EUGENICS, VOL. 7., P. 121, AND BROWNLEE
C   (1949) INDUSTRIAL EXPERIMENTATION, PP. 147-149. INPUT PARAMETERS ARE
C   AS FOLLOWS -
C       NT = NUMBER OF TREATMENTS
C       NK = NUMBER OF UNITS PER BLOCK
C       NR = NUMBER OF REPLICATES PER TREATMENT
C       NB = NUMBER OF BLOCKS
C       X  = MATRIX OF OBSERVED VALUES OF ORDER NT BY NB (X = ZERO WHEN
C            TREATMENT IS NOT REPRESENTED IN A PARTICULAR BLOCK)
C   NOTE. NB * NK MUST EQUAL NT * NR. ALL TREATMENTS ARE ASSUMED TO HAVE
C   NON-ZERO VALUES OF X.
C
C...........................................................
C
      DIMENSION X(20,20), B(20), T(20), Q(20), V(20), CB(20)
C
C   READ PROBLEM SPECIFICATION AND DATA.
    1 READ (5,100) NPROB, NT, NK, NR, NB
  100 FORMAT (5I5)
      IF (NPROB .EQ. 0) STOP
      DO 2 I = 1,NT
    2 READ (5,101) (X(I,J), J = 1,NB)
  101 FORMAT (10F7.0)
C
C   COMPUTE SUMS OF OBSERVATIONS FOR BLOCKS AND TREATMENTS.
      DO 3 I = 1,NT
    3 T(I) = 0.0
      DO 4 I = 1,NB
    4 B(I) = 0.0
      G = 0.0
      N = NT * NR
      AK = NK
      AT = NT
      AN = N
      DO 5 I = 1,NB
      DO 5 J = 1,NT
      T(I) = T(I) + X(I,J)
      B(J) = B(J) + X(I,J)
    5 G = G + X(I,J)
```

Program 16 (*continued*)

```
C
C   COMPUTE SUMS OF SQUARES FOR TREATMENTS.
        DO 6 I = 1,NT
        Q(I) = 0.0
        DO 6 J = 1,NB
        IF (X(I,J) .EQ. 0.0) GO TO 6
        Q(I) = Q(I) + B(J)
      6 CONTINUE
        DO 7 I = 1,NT
      7 Q(I) = AK * T(I) - Q(I)
        SSQT = 0.0
        DO 8 I = 1,NT
      8 SSQT = SSQT + Q(I) * Q(I)
        SSQT = SSQT * (AT-1.0) / (AN*AK*(AK - 1.0))
        NDFT = NT - 1
C
C   COMPUTE SUMS OF SQUARES FOR BLOCKS, TOTAL AND ERROR.
        SSQB = 0.0
        DO 9 I = 1,NB
      9 SSQB = SSQB + B(I) * B(I)
        SSQB = SSQB / AK - G*G/AN
        NDFB = NB - 1
        SSQTOT = 0.0
        DO 10 I = 1,NT
        DO 10 J = 1,NB
     10 SSQTOT = SSQTOT + X(I,J) * X(I,J)
        SSQTOT = SSQTOT - G*G/AN
        NDFTOT = N - 1
        SSQERR = SSQTOT - SSQB - SSQT
        NDFERR = N - NB - NT + 1
C
C   COMPUTE VARIANCES AND VARIANCE RATIOS.
        DFB = NDFB
        DFT = NDFT
        DFERR = NDFERR
        VB = SSQB / DFB
        VT = SSQT / DFT
        VERR = SSQERR / DFERR
        FB = VB / VERR
        FT = VT / VERR
C
C   OUTPUT ANALYSIS OF VARIANCE.
        WRITE (6,102) NPROB, SSQB, NDFB, VB, FB, SSQT, NDFT, VT, FT,
      A  SSQERR, NDFERR, VERR, SSQTOT, NDFTOT
    102 FORMAT (41H1BALANCED INCOMPLETE BLOCK DESIGN PROBLEM, I4////
      A  7X, 9HSOURCE OF, 8X, 7HSUMS OF, 8X, 10HDEGREES OF, 9X, 4HMEAN,
      B  16X, 8HVARIANCE/ 7X, 8HVARIANCE, 9X, 7HSQUARES, 9X, 7HFREEDOM,
      C  10X, 6HSQUARE, 16X, 5HRATIO// 8X, 6HBLOCKS, 7X, 1PE12.4, I11,
      D  10X, E12.4, 10X, E12.4/ 6X, 10HTREATMENTS, 5X, E12.4, I11,
      E  10X, E12.4, 10X, E12.4/ 8X, 5HERROR, 8X, E12.4, I11, 10X,
      F  E12.4// 8X, 5HTOTAL, 8X, E12.4, I11)
C
C   COMPUTE AND PRINT TREATMENT MEANS CORRECTED FOR BLOCK DIFFERENCES.
        GN = G / AN
        DO 11 I = 1,NT
     11 V(I) = (AT-1.0) / (AN*(AK-1.0)) * Q(I) + GN
        SE = 1.4142136 * SQRT(VERR*AK*(AT-1.0)/ (AN*(AK-1.0)))
        WRITE (6,103) (V(I), I = 1,NT)
    103 FORMAT (///26H CORRECTED TREATMENT MEANS // (1P10E12.4))
        WRITE (6,104) SE
```

Program 16 (*continued*)

```
  104 FORMAT (/   59H STANDARD ERROR OF DIFFERENCE BETWEEN TWO TREATMENT
     AMEANS = , 1PE12.4)
C
C  COMPUTE AND PRINT BLOCK MEANS CORRECTED FOR TREATMENT DIFFERENCES.
      AB = NB
      DO 13 J = 1,NB
      SUM = 0.0
      DO 12 I = 1,NT
      IF (X(I,J) .EQ. 0.0) GO TO 12
      SUM = SUM + V(I)
   12 CONTINUE
   13 CB(J) = (B(J) – SUM) / AK + GN
      SEB = SQRT(2.0 * VERR / AK * (1.0 + (AT–1.0)*(AT–AK) / AT/
     A  (AK–1.0) / (AB–1.0)))
      WRITE (6,105) (CB(I), I = 1,NB)
  105 FORMAT (/// 22H CORRECTED BLOCK MEANS // (1P10E12.4))
      WRITE (6,106) SEB
  106 FORMAT (/55H STANDARD ERROR OF DIFFERENCE BETWEEN TWO BLOCK MEANS
     A = , 1PE12.4)
      GO TO 1
      END
```

DATA

```
  1     7     3     3     7
 50    42    91     0     0     0     0
  0     0   118    94    94     0     0
 76     0     0    64     0    80     0
  0     0    72     0     0    53    31
 44     0     0     0    65     0    54
  0   102     0     0   119    92     0
  0    38     0    38     0     0    37
  0
```

OUTPUT

BALANCED INCOMPLETE BLOCK DESIGN PROBLEM 1

SOURCE OF VARIANCE	SUMS OF SQUARES	DEGREES OF FREEDOM	MEAN SQUARE	VARIANCE RATIO
BLOCKS	6.7258E 03	6	1.1210E 03	1.3463E 01
TREATMENTS	7.6659E 03	6	1.2777E 03	1.5345E 01
ERROR	6.6610E 02	8	8.3262E 01	
TOTAL	1.5058E 04	20		

CORRECTED TREATMENT MEANS
   5.7238E 01   9.2524E 01   7.9095E 01   4.6381E 01   5.7667E 01   1.0552E 02   4.6238E 01

STANDARD ERROR OF DIFFERENCE BETWEEN TWO TREATMENT MEANS = 8.4479E 00

CORRECTED BLOCK MEANS

   6.1238E 01   6.0238E 01   9.7524E 01   6.1952E 01   7.6667E 01   6.7238E 01   5.9810E 01

STANDARD ERROR OF DIFFERENCE BETWEEN TWO BLOCK MEANS = 8.4479E 00

Chapter 7

Correlation and Regression Analysis

Some of the elementary ideas underlying the statistical techniques of correlation and regression have already been discussed in connection with Program 2 (pp. 88). The two techniques are closely related and employ many common computational features so it is necessary to be clear how they differ. Correlation techniques assume that the variables are linearly related but they also suppose that the data are derived from a bivariate normal distribution. This may be represented diagrammatically as in Fig. 25 and

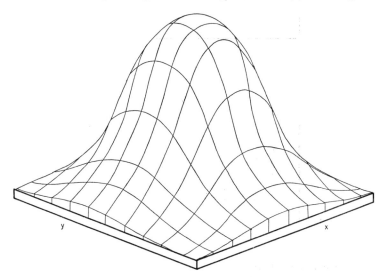

FIG. 25. Diagrammatic representation of a bivariate normal distribution. Frequencies are expressed on the vertical axis.

implies that for any chosen value of either x or y the distribution of the other variable is normal. A full mathematical test for bivariate normality of data is possible but is rarely if ever carried out by those embarking on the use of correlation coefficients. In practice it is usually considered enough to see whether the values of x and y, when plotted as a scatter diagram, fall approximately in the shape of a more or less elongate ellipse (Fig. 26) with the greatest density of points towards the centre of the ellipse and progressively fewer as one moves outwards. Data of this kind commonly arise when a random sample is taken and the variables x and y are ones which, when considered separately, have normal distributions. An example might be the height and length of leg in a random sample of adult males. If the data do not conform approximately to these assumptions, then although a correlation coefficient can always be calculated it may be difficult or impossible to attach much significance to it.

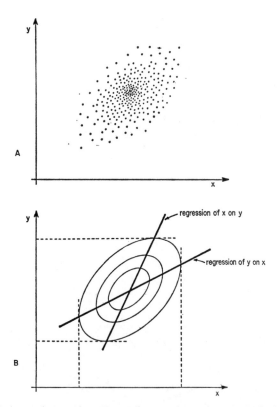

FIG. 26. Correlation and regression. Upper diagram shows the scatter obtained when x and y are positively correlated. Lower diagram expresses the relationship between the regression lines and the elliptical contours of a bivariate normal distribution.

Regression techniques are based on somewhat different assumptions, the basic model for a simple linear regression of y on x being:

$$y_i = \gamma + \beta (x_i - \bar{x}) + \varepsilon_i \qquad (1)$$

where the errors in y (denoted by ε) are normally distributed with zero mean and constant variance. For a given value of x, therefore, the values of y will also be normally distributed with constant variance. Nothing, however, is said in the model about the distribution of x, which may be irregular or arbitrary. Regression techniques are therefore applicable when correlation methods should not be used; for example when the various values of x are determined systematically by the experimenter or are selected in non-random fashion and then a value or values of y recorded for each value of x. It is also assumed that a distinction can be made between the independent variable (usually denoted as x) and the dependent variable y. Often this expresses a causal physical relationship—e.g. a change in light intensity (x) may cause a change in photosynthetic activity (y) while the reverse relationship is inconceivable. In other cases the essential feature is that one wishes to estimate or predict values of y from those of x, the technique being designed to minimise the squares of the deviations of y about the regression line. Regression techniques can be used when the data are distributed in bivariate normal fashion but in these cases there are two regressions involved, that of y on x and the other of x on y, Fig. 26. The two regression lines coincide only if there is a perfect correlation between the two variables, otherwise the two will have different slopes, b and b' related by the expression $r = \sqrt{b \cdot b'}$ where r is the correlation coefficient. Each line is intended to predict or estimate most accurately the value of the dependent variable for a given value of the independent variable; neither line expresses the intensity of the association in the way that the correlation coefficient does. When the problem before one is such that both variables are subject to error but one is primarily interested in the nature of the functional relationship between them, then another method of analysis is required, discussed in Program 22.

The programs to be described in this section cater for all the possibilities indicated above but it is for the user to decide how they should be applied in analysing the data of a particular problem. The fact that Program 17, for example, carries out both a correlation and regression analysis does not mean that both (or either!) are applicable to a given set of data. The programs involved are:

Program 17: Correlation and regression of data arranged in a grouped frequency table.

Program 18: Regression with tests of significance and linearity by analysis of variance.

184 CORRELATION AND REGRESSION ANALYSIS

Program 19: Comparison of two regression lines by analysis of variance.

Program 20: Autocorrelations in time series.

Program 21: Cross-correlations in time series.

Program 22: Best-fit linear relationship.

Program 17: Correlation and regression of data arranged in a grouped frequency table

A contingency table such as that produced by Program 7 indicates, by the way in which the frequencies are distributed among the various cells, the kind of association that exists between the two variables arranged in the rows and columns. It does not, however, measure this association or express it in a form suitable for prediction or estimation. The data in the table can, however, be used to measure the correlation coefficient between the two or to yield a regression equation.

The program is intended, then, to operate on a two-dimensional array of frequencies (F) in which the rows (NY in number) refer to equal sub-intervals of the variable y (assumed to be the dependent variable in the regression analysis) and the columns (NX in number) refer to sub-intervals of the variable x (the independent variable in the regression analysis). It is also necessary for class-marks for columns and rows to be available. These are the mid-points of the sub-intervals for each variable: the NX class-marks for x are denoted by the singly-subscripted Fortran variable X and the NY class-marks for y by the similar Fortran variable Y.

After establishing dimensions for X,Y and the main frequency array F, the program reads a specification card giving, as usual, the problem reference number (NPROB), NX and NY. The two sets of class-marks, X and Y, are then read in by implied DO-loops. Note that these arrays must be punched consecutively, the last observation in the X array being followed immediately by the first in the Y array. The main frequency array is then read in from data punched row by row, i.e. the first cell-frequency for each row begins on a new card.

When all the data have been read into core-storage the sums and uncorrected sums of squares and cross-products are computed in the usual way for grouped frequencies, multiplying each value by the appropriate frequency on principles similar to those already explained for Program 9. The variable names SX, SX2, SY, SY2, SXY for these sums and products are self-explanatory. From the uncorrected sums of squares and cross-products the ensuing program segment computes the corrected sums of squares (SSQX and SSQY) as well as the covariance and variances of X and Y (denoted by COVAR, VARX and VARY). The variances are adjusted by Sheppard's correction (p. 135) to allow for the fact that they are based on grouped data. The correlation coefficient (R) is then calculated from equation (9) (p. 89) and the slope and intercept of the regression line (B and A) are calculated from equations (13) and (14) (p. 91).

The remaining computations are concerned with significance testing. The first is a t-test for the significance of the correlation coefficient and involves the calculation of

$$t = \frac{r\sqrt{n-2}}{\sqrt{1-r^2}} \qquad (2)$$

which is distributed as Student's t with $(n-2)$ degrees of freedom. The second test involves the calculation of the quantity z, defined as:

$$z = \tfrac{1}{2} \log_e \frac{(1+r)}{(1-r)}. \qquad (3)$$

This transformation of r to z enables one to compare the computed value of r with some standard value ρ. To do this one transforms ρ in the same way to give

$$\zeta = \tfrac{1}{2} \log_e \frac{(1+\rho)}{(1-\rho)} \qquad (4)$$

then tests the difference $z - \zeta$ as a normal deviate with zero mean and variance $1/(n-3)$. The z-transformation also enables one to compare two correlation coefficients r_1 and r_2. For each of these one takes the z-values z_1 and z_2 as printed out by this program, then notes that $(z_1 - z_2)$ is normally distributed with zero mean and variance equal to $1/(n_1 - 3) + 1/(n_2 - 3)$. The reader who is uncertain of the application of the z-transformation in testing the significance of correlation coefficients will find an explanation in most modern textbooks of statistics (e.g. Bailey, 1959; Hoel, 1971).

The next significance test concerns the regression line. We first calculate the variance of the deviations of y from the regression line (VDY), using the formula:

$$s^2 = \frac{1}{(n-2)} \left\{ \Sigma(y-\bar{y})^2 - \frac{[\Sigma(x-\bar{x})(y-\bar{y})]^2}{\Sigma(x-\bar{x})^2} \right\} \qquad (5)$$

and then the standard error of the slope of the line given by

$$e = \frac{s}{\sqrt{\Sigma(x-\bar{x})^2}}. \qquad (6)$$

If the sample is a large one ($n > 30$), as would be expected if the data are arranged in a grouped frequency table, then the significance and confidence

Outline flowchart for Program 17

Correlation and regression of grouped data.

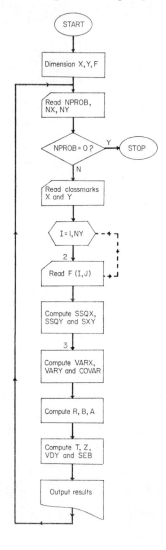

limits of the slope can be obtained as usual by a normal deviate test. The program finally prints out the results of the computation and returns to statement 1 to read the next data card.

From a programming point of view there is little in this program that has not already been covered in previous ones, though experience in summing squares and cross-products always helps to promote facility in writing statistical programs. The accompanying flow-chart outlines the main structure of the program rather than the details.

Program 17

Correlation and regression of grouped data; with sample data and output.

```
C
C......................................................................
C
C   THIS PROGRAM COMPUTES THE CORRELATION COEFFICIENT AND REGRESSION OF
C   Y ON X FROM DATA ARRANGED IN A GROUPED FREQUENCY TABLE OF NY ROWS
C   AND NX COLUMNS. FOR COMPUTATIONAL DETAILS SEE BAILEY (1959),
C   STATISTICAL METHODS IN BIOLOGY, P.81. INPUT PARAMETERS ARE
C   AS FOLLOWS −
C      NPROB = PROBLEM REFERENCE NUMBER (ZERO ON LAST DATA CARD)
C      NX    = NUMBER OF COLUMNS IN FREQUENCY MATRIX
C      NY    = NUMBER OF ROWS IN FREQUENCY MATRIX
C      X     = VECTOR OF COLUMN CLASS−MARKS (LENGTH NX)
C      Y     = VECTOR OF ROW CLASS−MARKS (LENGTH NY)
C      F     = FREQUENCY MATRIX (NY BY NX)
C
C......................................................................
C
      DIMENSION X(50), Y(50), F(50,50)
C
C   READ PROBLEM SPECIFICATION, CLASS−MARKS AND FREQUENCIES.
    1 READ (5,100) NPROB, NX, NY
  100 FORMAT (3I5)
      IF (NPROB .EQ. 0) STOP
      READ (5,101) (X(I), I = 1,NX), (Y(I), I = 1,NY)
  101 FORMAT (10F7.0)
      DO 2 I = 1,NY
    2 READ (5,102) (F(I,J), J = 1,NX)
  102 FORMAT (20F4.0)
C
C   COMPUTE SUMS OF SQUARES AND CROSS−PRODUCTS.
      SX = 0.0
      SX2 = 0.0
      SY = 0.0
      SY2 = 0.0
      SXY = 0.0
      AN = 0.0
      DO 3 I = 1,NY
      DO 3 J = 1,NX
      SX = SX + X(J) * F(I,J)
      SX2 = SX2 + X(J) * X(J) *F(I,J)
      SY = SY + Y(I) * F(I,J)
      SY2 = SY2 + Y(I) * Y(I) * F(I,J)
      SXY = SXY + X(J) * Y(I) * F(I,J)
    3 AN = AN + F(I,J)
```

Program 17 (continued)

```
C
C   COMPUTE COVARIANCE AND VARIANCES (WITH SHEPPARD'S CORRECTION).
        HX = ABS(X(2)-X(1))
        HY = ABS(Y(2)-Y(1))
        SSQX = SX2 - SX*SX/AN
        SSQY = SY2 - SY*SY/AN
        VARX = SSQX/(AN-1.0) - HX*HX/12.0
        VARY = SSQY/(AN-1.0) - HY*HY/12.0
        COVAR = (SXY-SX*SY/AN) / (AN-1.0)
C
C   COMPUTE CORRELATION COEFFICIENT AND REGRESSION EQUATION.
        R = COVAR / SQRT(VARX*VARY)
        B = (SXY-SX*SY/AN) / SSQX
        A = SY/AN - B*SX/AN
C
C   COMPUTE STATISTICS NEEDED FOR SIGNIFICANCE TESTS.
        N = AN
        T = ABS((R*SQRT(AN-2.0)) / SQRT(1.0-R*R))
        NDF = N - 2
        Z = 0.5 * ALOG((1.0+R)/(1.0-R))
        VDY = 1.0/(AN-2.0) * (SSQY-(SXY-SX*SY/AN)**2 / SSQX)
        SEB = SQRT(VDY) / SQRT(SSQX)
C
C   OUTPUT RESULTS.
        WRITE (6,103) NPROB, R, B, A, N, T, NDF, Z, VDY, SEB
    103 FORMAT (49H1CORRELATION AND REGRESSION EQUATION (Y = A + BX)//
       A    8H PROBLEM, I3///6X, 25HCORRELATION COEFFICIENT = , F8.4//
       B    6X, 11HB (SLOPE) = ,1PE15.7// 6X, 15HA (INTERCEPT) = , 1PE15.7//
       C    6X, 27HDATA FOR SIGNIFICANCE TESTS//
       D    11X,24HNUMBER OF OBSERVATIONS = ,I5//11X,11HT (FOR R) = ,0PF9.3,
       E    5H WITH, I5,19H DEGREES OF FREEDOM//11X, 3HZ = , F8.4//
       F    11X,50HVARIANCE OF DEVIATIONS OF Y FROM REGRESSION LINE = ,
       G    1PE15.7//11X, 21HSTANDARD ERROR OF B = , 1PE15.7)
        GO TO 1
        END
```

DATA

1	14	15							
61	62	63	64	65	66	67	68	69	70
71	72	73	74	61	62	63	64	65	66
67	68	69	70	71	72	73	74	75	
0	0	0	0	1	0	1	0	0	0	0	0	0
0	0	1	2	2	3	1	0	0	0	0	0	0
0	2	2	3	3	5	2	1	1	1	0	0	0
0	2	4	6	5	4	3	2	0	1	0	0	0
1	3	3	6	9	8	7	4	1	1	0	1	0
2	4	5	8	9	13	9	5	6	3	4	0	1	0
0	2	3	6	7	9	17	13	11	4	5	2	1	0
0	1	2	5	8	8	11	14	9	6	2	2	1	0
0	0	1	4	5	6	9	11	8	6	3	2	1	1
0	0	1	1	5	4	4	9	11	3	3	1	2	0
0	0	0	1	2	3	4	4	5	6	3	1	2	1
0	0	0	0	1	1	2	2	3	4	4	2	3	1
0	0	0	0	1	0	0	2	1	3	1	1	0	0
0	0	0	0	0	0	0	1	2	1	1	0	0
0	0	0	0	0	0	0	0	0	1	1	0	1
0												

OUTPUT

Program 17 (*continued*)

CORRELATION AND REGRESSION EQUATION (Y = A + BX)

PROBLEM 1

    CORRELATION COEFFICIENT = 0.4985

    B (SLOPE) = 4.9064614E-01

    A (INTERCEPT) = 3.4636548E 01

    DATA FOR SIGNIFICANCE TESTS

        NUMBER OF OBSERVATIONS = 493

        T (FOR R) =   12.742 WITH 491 DEGREES OF FREEDOM

        Z = 0.5473

        VARIANCE OF DEVIATIONS OF Y FROM REGRESSION LINE = 5.3496028E 00

        STANDARD ERROR OF B =   3.9116633E-02

Program 18: Regression with tests of significance and linearity by analysis of variance

The two things one usually wishes to know in a regression analysis is whether the slope of the line differs significantly from zero, (i.e. whether there is a significant regression) and whether the deviations from linearity are significant. Both these can be tested by an analysis of variance provided that for each value of x we have two or more replicated values of y. The primary data will therefore consist of a singly-subscripted array of x-values (denoted by the Fortran variable X and containing N elements) and a doubly-subscripted array of y-values (Y) consisting of N columns and NREP rows, where NREP denotes the number of replicates of each y-value.

Dimensions are provided for both X and Y as well as for YBAR, a singly subscripted array that will be used to hold the sums of the replicates for each column of Y. The input segment of the program has no special features but one should note that the Y array is read in row by row with the first observation of each row starting a new card. The next segment carries out the first major part of the computation, namely the partitioning of the sums of squares of Y into between-column and within-column components. To achieve this we compute the following intermediate variables:

    SY2   = the sum of the squares of each observation in the Y array.

    TWO   = the sum of the squares of each column total, divided by the number of replicates in each column.

    THREE = the sum of all observations in Y, squared and divided by the total number of observations.

Outline flowchart for Program 18

Linear regression with tests of significance by analysis of variance.

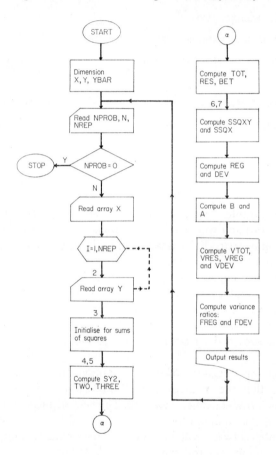

The corrected sums of squares are then derived very simply:

TOT = total sum of squares (SY2 − THREE)

RES = within-column sum of squares (SY2 − TWO), which will give us the residual sum of squares in the final analysis of variance table.

BET = between-column sum of squares (TWO − THREE).

The degrees of freedom for the total and residual variation (NDFT and NDFRES) are then calculated and the first stage of the computation is over.

The next stage consists of partitioning the between-column sum of squares derived above into a component due to linear regression and one due to deviations from it. Essentially this involves accumulating uncorrected sums of squares and cross-products involving X (accomplished by the DO-loop ending on statement 7), followed by correction to give:

SSQXY = corrected sum of cross-products

and SSQX = corrected sum of squares of X.

The regression sum of squares (REG) is then computed by a simple arithmetic statement and the deviation sum of squares (DEV) is obtained by difference. REG is associated with 1 degree of freedom (NDFR) and the number associated with DEV is obtained by difference (NDFD).

The last computational segment in the program provides the estimates of slope (B) and intercept (A) for the regression line, together with the conversion of sums of squares into variance estimates: VTOT, VRES, VREG, and VDEV, corresponding respectively to the total and residual components and those associated with linear regression and deviation. From the latter three components we also derive variance ratios for the final analysis of variance. All the calculations of this segment are based on simple arithmetic and assignment statements. The output is on lines similar to that used for forming analysis of variance tables in Programs 12–16. The significance tests based on the variance ratios are carried out by the user in the usual way.

Program 18

Linear regression with tests of significance by analysis of variance.

```
C
C................................................................
C
C   THIS PROGRAM COMPUTES THE LINEAR REGRESSION OF Y ON X USING SEVERAL
C   REPLICATES OF Y FOR EACH VALUE OF X AND TESTING SIGNIFICANCE AND
C   LINEARITY OF REGRESSION BY ANALYSIS OF VARIANCE. FOR COMPUTATIONAL
C   DETAILS SEE BROWNLEE (1949) INDUSTRIAL EXPERIMENTATION, PP. 69–71.
C   INPUT PARAMETERS ARE AS FOLLOWS –
C       NPROB = PROBLEM REFERENCE NUMBER (ZERO ON LAST DATA CARD)
```

Program 18 (*continued*)

```
C     N     = NUMBER OF VALUES OF X
C     NREP  = NUMBER OF REPLICATES OF Y FOR EACH VALUE OF X
C     X     = VECTOR OF X VALUES (LENGTH N)
C     Y     = MATRIX OF Y VALUES (NREP BY N)
C
C.............................................................
C
      DIMENSION X(50), Y(50,50), YBAR(50)
C
C     READ PROGRAM SPECIFICATION AND DATA.
    1 READ (5,100) NPROB, N, NREP
  100 FORMAT (3I5)
      IF (NPROB .EQ. 0) STOP
      READ (5,101) (X(I), I = 1,N)
  101 FORMAT (10F7.0)
      DO 2 I = 1,NREP
    2 READ (5,101) (Y(I,J), J = 1,N)
C
C     PARTITION SUMS OF SQUARES OF Y INTO BETWEEN- AND WITHIN-COLUMNS
C     COMPONENTS.
      SY = 0.0
      SY2 = 0.0
      AN = N
      REP = NREP
      AREP = AN * REP
      DO 3 I = 1,N
    3 YBAR(I) = 0.0
C
      DO 4 I = 1,NREP
      DO 4 J = 1,N
      SY = SY + Y(I,J)
      SY2 = SY2 + Y(I,J) * Y(I,J)
    4 YBAR(J) = YBAR(J) + Y(I,J)
      TWO = 0.0
      DO 5 I = 1,N
    5 TWO = TWO + YBAR(I) * YBAR(I)
      TWO = TWO / REP
      THREE = SY * SY / AREP
      TOT = SY2 - THREE
      RES = SY2 - TWO
      BET = TWO - THREE
      NDFT = AREP - 1.0
      NDFRES = NDFT - N + 1
C
C     PARTITION BETWEEN-COLUMNS SUMS OF SQUARES INTO REGRESSION
C     AND DEVIATION COMPONENTS.
      SX = 0.0
      SX2 = 0.0
      SXY = 0.0
      DO 6 I = 1,N
      SX = SX + X(I)
    6 SX2 = SX2 + X(I) * X(I)
      SX = SX * REP
      SX2 = SX2 * REP
      DO 7 I = 1,NREP
      DO 7 J = 1,N
    7 SXY = SXY + X(J) * Y(I,J)
      SSQXY = SXY - SX*SY/AREP
      SSQX = SX2 - SX*SX/AREP
      REG = SSQXY * SSQXY / SSQX
      DEV = BET - REG
```

Program 18 (*continued*)

```
        NDFR = 1
        NDFD = NDFT - NDFR - NDFRES
C
C   COMPUTE REGRESSION COEFFICIENTS, VARIANCES AND VARIANCE RATIOS.
        B = SSQXY / SSQX
        A = SY/AREP - B*SX/AN
        DFT = NDFT
        DFRES = NDFRES
        DFR = NDFR
        DFD = NDFD
        VTOT = TOT / DFT
        VRES = RES / DFRES
        VREG = REG / DFR
        VDEV = DEV / DFD
        FREG = VREG / VRES
        FDEV = VDEV / VRES
C
C   OUTPUT RESULTS.
        WRITE (6,102) NPROB, B, A, REG, NDFR, VREG FREG, DEV, NDFD,
       A    VDEV, FDEV, RES, NDFRES, VRES, TOT, NDFT
  102   FORMAT (26H1LINEAR REGRESSION PROBLEM, I4///12H B (SLOPE) = ,
       A    1PE15.7// 16H A (INTERCEPT) = , E15.7/// 7X, 9HSOURCE OF,
       B    8X, 7HSUMS OF, 8X, 10HDEGREES OF, 9X, 4HMEAN, 16X, 8HVARIANCE/
       C    7X, 8HVARIANCE, 9X, 7HSQUARES, 9X, 7HFREEDOM, 10X, 6HSQUARE,
       D    16X, 5HRATIO,// 6X, 10HREGRESSION, 5X, E12.4, I11,10X,
       E    E12.4, 10X, E12.4/ 6X, 9HDEVIATION, 6X, E12.4, I11, 10X, E12.4,
       F    10X, E12.4/ 6X, 8HRESIDUAL, 7X, E12.4, I11, 10X, E12.4//
       G    8X, 5HTOTAL, 8X, E12.4, I11)
        GO TO 1
        END
```

Program 19: Comparison of two regression lines by analysis of variance

A not infrequent situation in biological work requires the comparison of two regression lines and a decision as to whether they are distinct or whether the observations can be pooled to yield a common line. The present program deals with this problem by providing estimates of the parameters for the two lines separately and for a common line, with an analysis of variance that tests the significance of the observed differences in slope and position of the two lines. It is assumed that the values of the independent variable (X) are fixed as the same for the two sets of dependent variables (YA and YB). There is no replication of the observations of YA and YB.

The dimensions of X, YA and YB are established and the input of the three arrays (all with N observations) proceeds as usual. An initialisation segment begins the calculation of the sums of the variables and the uncorrected sums of squares and cross-products, and the process is completed in the DO-loop ending on statement 2. The following segment consists entirely of arithmetic statements which successively compute the sums of squares and regression parameters for the two lines separately. It is perhaps best to list these variables so that the reader can follow the computations which, in themselves, are very simple.

ST	= Total sum of squares for all the observations of Y (in both lines)
SSQX	= corrected sum of squares for X
STA	= corrected sum of squares for line A (i.e. the YA array)
SSQXYA	= corrected sum of cross products for line A
SA	= sum of squares accounted for by regression line A
BA	= slope of line A
AA	= intercept of line A

The last five of these quantities are then computed also for line B and stored as the correspondingly named variables STB, SSQXYB, SB, BB, AB. We finally compute the residual sum of squares (SRES = STA − SA + STB − SB) and divide it by the corresponding number of degrees of freedom to obtain the residual variance.

The next program segment carries out an analogous set of operations for the common line in which the Fortran variables being defined are:

SX	= sum of X's, multiplied by 2 to allow for the fact that we are now dealing with the data for both sets of Y-values.
SSQX	= corrected sum of squares for X, again twice the value previously used.
SSQXY	= corrected sum of cross-products for all observations.
SC	= sum of squares accounted for by common line.
SDIFF	= sum of squares due to improvement through use of two lines in place of a single common line.
SINT	= sum of squares due to difference in intercepts of lines A and B (found by difference).
BC	= slope of common line.
AC	= intercept of common line.

Finally, we obtain—in the last computational segment of the program—the variance ratios on which the significance tests depend. These are:

FC	= variance ratio associated with common regression line.
FDIFF	= variance ratio associated with difference between common line and separate lines.
FINT	= variance ratio associated with difference in intercepts of lines.

Outline Flowchart for Program 19

Comparison of regression lines by analysis of variance.

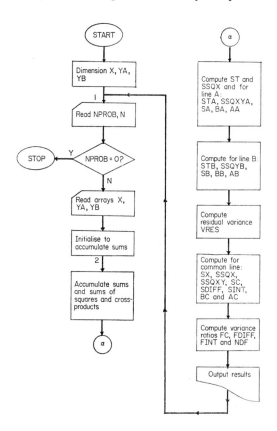

A single degree of freedom is associated with the common regression line and the two differences; the other degrees of freedom are associated with the residual sums of squares and their number is printed in the output (NDF). It will be recalled in carrying out the significance tests that when a variance ratio is less than unity one enters the table with its reciprocal and interchanges the numbers of degrees of freedom. The program segment concerned with output has no special features and control returns to statement 1 for further data to be read.

This program has few features that do not occur in others concerned with analysis of variance or regression; though the relatively large number of somewhat complicated arithmetic statements require care in writing they form a direct translation of the statistical operations involved.

Program 19

Comparison of regression lines by analysis of variance; with sample data and output.

```
C
C.................................................................
C
C   THIS PROGRAM COMPARES THE SLOPES AND INTERCEPTS OF TWO REGRESSION
C   LINES BY ANALYSIS OF VARIANCE. FOR COMPUTATIONAL DETAILS SEE
C   BROWNLEE (1949), INDUSTRIAL EXPERIMENTATION, PP.66-68. INPUT
C   PARAMETERS ARE AS FOLLOWS -
C       NPROB = PROBLEM REFERENCE NUMBER (ZERO ON LAST DATA CARD)
C       N     = NUMBER OF PAIRED OBSERVATIONS FOR EACH LINE
C       X     = VECTOR OF X VALUES (LENGTH N)
C       YA    = VECTOR OF Y VALUES IN LINE A (LENGTH N)
C       YB    = VECTOR OF Y VALUES IN LINE B (LENGTH N)
C
C.................................................................
C
        DIMENSION X(100), YA(100), YB(100)
C
C   READ PROBLEM SPECIFICATION AND DATA.
      1 READ (5,100) NPROB, N
    100 FORMAT (2I5)
        IF (NPROB .EQ. 0) STOP
        READ (5,101)(X(I), I = 1,N)
        READ (5,101) (YA(I), I = 1,N)
        READ (5,101)(YB(I), I = 1.N)
    101 FORMAT (10F7.0)
C
C   INITIALISE.
        SYA = 0.0
        SYB = 0.0
        SYA2 = 0.0
        SYB2 = 0.0
        SX = 0.0
        SX2 = 0.0
        SXYA = 0.0
        SXYB = 0.0
        AN = N
```

Program 19 (*continued*)

```
C
C   COMPUTE SUMS AND SUMS OF SQUARES AND CROSS-PRODUCTS.
       DO 2 I = 1,N
       SYA = SYA + YA(I)
       SYB = SYB + YB(I)
       SYA2 = SYA2 + YA(I) * YA(I)
       SYB2 = SYB2 + YB(I) * YB(I)
       SX = SX + X(I)
       SX2 = SX2 + X(I) * X(I)
       SXYA = SXYA + X(I) * YA(I)
     2 SXYB = SXYB + X(I) * YB(I)
C
C   COMPUTE REGRESSION PARAMETERS AND SUMS OF SQUARES FOR LINES A AND B.
       ST = SYA2 + SYB2 - (SYA+SYB) * (SYA+SYB) / (2.0*AN)
       SSQX = SX2 - SX*SX/AN
       STA = SYA2 - SYA*SYA/AN
       SSQXYA = SXYA - SX*SYA/AN
       SA = SSQXYA**2/SSQX
       BA = SSQXYA/SSQX
       AA = SYA/AN - BA*SX/AN
       STB = SYB2 - SYB*SYB/AN
       SSQXYB = SXYB - SX*SYB/AN
       SB = SSQXYB**2/SSQX
       BB = SSQXYB/SSQX
       AB = SYB/AN - BB*SX/AN
       SRES = STA - SA + STB - SB
       VRES = SRES / (2.0*AN-4.0)
C
C   COMPUTE REGRESSION PARAMETERS FOR COMMON LINE AND SUMS OF SQUARES
C   FOR COMMON LINE, DIFFERENCE, AND INTERCEPTS.
       SX = 2.0 * SX
       SSQX = 2.0 * SSQX
       SSQXY = SXYA + SXYB - SX *(SYA+SYB) / (2.0*AN)
       SC = SSQXY * SSQXY / SSQX
       SDIFF = SA + SB - SC
       SINT = ST - SRES - SDIFF - SC
       BC = SSQXY / SSQX
       AC = (SYA + SYB) / (2.0 * AN) - BC * SX / (2.0 * AN)
C
C   COMPUTE VARIANCE RATIOS.
       FC = SC / VRES
       FDIFF = SDIFF / VRES
       FINT = SINT / VRES
       NDF = 2.0 * AN - 4.0
C
C   OUTPUT RESULTS.
       WRITE (6,102) NPROB, BA, AA, BB, AB, BC, AC, FC, FDIFF, FINT, NDF
   102 FORMAT (40H1COMPARISON OF REGRESSION LINES. PROBLEM, I4////
      A  6X, 17HSLOPE OF LINE A = , 1PE12.4// 6X,
      B  21HINTERCEPT OF LINE A = , E12.4/// 6X, 17HSLOPE OF LINE B = ,
      C  E12.4// 6X, 21HINTERCEPT OF LINE B = , E12.4/// 6X,
      D  22HSLOPE OF COMMON LINE = , E12.4// 6X,
      E  26HINTERCEPT OF COMMON LINE = , E12.4/// 6X,
      F  43HVARIANCE RATIO FOR COMMON REGRESSION LINE = , E12.4// 6X,
      G  58HVARIANCE RATIO (DIFFERENCE OF COMMON AND SEPARATE LINES) = ,
      H  E12.4// 6X, 31HVARIANCE RATIO FOR INTERCEPTS = , E12.4//
      I  6X, 42HDEGREES OF FREEDOM FOR RESIDUAL VARIANCE = , I4)
C
       GO TO 1
       END
```

Program 19 (continued)

```
DATA
   1    4
   0.0    3.0    6.0    9.0
 100.0  230.0  301.0  506.0
 177.0  324.0  360.0  532.0
   0
```

OUTPUT

COMPARISON OF REGRESSION LINES. PROBLEM 1

    SLOPE OF LINE A = 4.2967E 01
    INTERCEPT OF LINE A = 9.0900E 01

    SLOPE OF LINE B = 3.6700E 01
    INTERCEPT OF LINE B = 1.8310E 02

    SLOPE OF COMMON LINE = 3.9833E 01
    INTERCEPT OF COMMON LINE = 1.3700E 02

    VARIANCE RATIO FOR COMMON REGRESSION LINE = 8.8212E 01
    VARIANCE RATIO (DIFFERENCE OF COMMON AND SEPARATE LINES) = 5.4582 E-01
    VARIANCE RATIO FOR INTERCEPTS = 5.0604E 00
    DEGREES OF FREEDOM FOR RESIDUAL VARIANCE = 4

Program 20: Autocorrelations in time-series

Many biological observations are repeated at regular intervals and therefore form some kind of time-series. Obvious examples which come to mind are periodic measurements of animal activity, population records over a more or less prolonged period, or light-trap data in entomology. The analysis of time series is now a major branch of statistics (e.g. Cox and Lewis (1966) and Quenouille (1968)) and no attempt will be made to pursue it beyond two very simple programs. The first of these undertakes the measurement of serial correlation or autocorrelation in a series, whereby in a set of n observations $x_1, x_2, x_3 \ldots x_n$ the i-th term of the series $(i = 1, 2, 3, \ldots (n-1))$ is compared with the $(i + 1)$-th term, as in the following example, where the *lag* $l = 1$. The data enclosed by the broken line in Fig. 27 form the two sets of $(n - 1)$

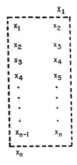

FIG. 27. Autocorrelation with lag = 1. The broken line encloses the terms to be correlated.

observations between which the correlation is computed, though since both are derived from the single series $x_1, x_2, x_3 \ldots x_n$ we need only operate on this one array.

To be of fairly wide application the program is designed to accommodate data in a variety of formats and to calculate the various correlation coefficients for lags of from 1 up to a maximum lag of L. The resulting set of coefficients are stored in the array R, so that the DIMENSION statement includes values for X (the array of primary observations), R and INFMT (the latter holding a variable input format). A problem specification card is then read to define:

NPROB = problem reference number, set at zero on last data card so as to terminate execution.
NOBS = number of observations in array X.
L = maximum lag.

After the problem specification card comes one bearing the variable input format under which array X is subsequently read in.

The main body of calculations is included in a large DO-loop that ends on statement 3 and enables the whole set of computations to be repeated L times for each successive value 1, 2, 3, ... L. If a first-order serial correlation only is required (i.e. x_i correlated with x_{i+1}, $i = 1, 2, 3, \ldots (n-1)$), L is read in as 1 and the DO-loop is satisfied after a single cycle. The first statement within the loop defines N1, the number of paired observations being correlated, which is obviously equal to (NOBS − J) where J is the current value of the lag. We then initialise for the sums of observations and their squares in the two sets of terms to be correlated, as well as for the sum of cross-products. These are denoted by the Fortran variables SUMX1, SSQX1, SUMX2, SSQX2 and SCROSS respectively. An inner DO-loop ending on statement 2 accumulates these sums, but in order to do so it is necessary to introduce a computed subscript IJ which will pick out, as it were, those observations of the original array X which form the right-hand set of terms in Fig. 27. It is not difficult to see that IJ will always equal the sum of the two indices J and I of the outer and inner DO-loops. The reader who is not convinced of this should follow through the loops once, as though computing a first-order serial correlation. Having defined IJ the accumulation of sums is straightforward and is followed by a set of arithmetic statements to form the corrected sums of squares and cross-products (DEV1, DEV2, CROSS). From these the correlation coefficient is computed and placed in the requisite location of array R by statement 3. It only remains to print out the results and return to statement 1 so that a new problem specification card can be read.

The most interesting feature of this program is, of course, the simple way in which an index IJ and an indexing parameter N1 are computed and then used

Flowchart for Program 20

Autocorrelation of time series.

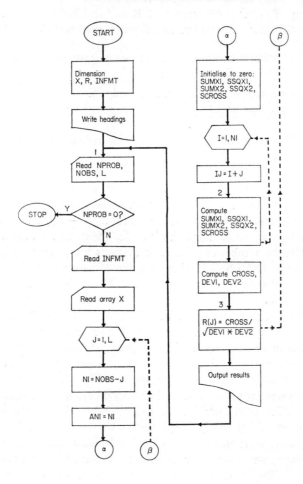

to select the relevant elements from the single primary array, thus conserving core-storage space. The correlation coefficients obtained in the program are, however, not easy to interpret. The significance tests usually available to assess a correlation coefficient assume that it is based on a set of random values from a joint distribution. Obviously this is not true of autocorrelation and the interpretation of such analyses goes beyond the scope of the present book. It is, however, worth mentioning that if the successive correlation coefficients are plotted against the corresponding values of the lag one obtains a graph known as the *correlogram*. This is not only a guide to some of the characteristics of the time series but also serves as the starting point for *power spectral analysis* of the series. Craddock (1968) gives an elementary outline of such methods which seem to deserve further attention by biologists. The complicated calculations involved demand the use of a computer and suitable programs are available in the BMD series (p. 475).

Program 20

Autocorrelation of time-series; with sample data and output.

```
C
C.................................................................
C
C  THIS PROGRAM COMPUTES AUTOCORRELATIONS FOR TIME-LAGS OF
C  1,2,3,... IN ANY NUMBER OF TIME SERIES. INPUT PARAMETERS ARE AS
C  FOLLOWS -
C     NPROB = PROBLEM REFERENCE NUMBER (ZERO ON LAST DATA CARD)
C     NOBS  = NUMBER OF OBSERVATIONS IN SERIES
C     L     = MAXIMUM TIME-LAG
C     INFMT = VECTOR HOLDING VARIABLE INPUT FORMAT
C     X     = VECTOR OF OBSERVATIONS
C  NOTE - L MUST NOT EXCEED (NOBS-2)
C
C.................................................................
C
      DIMENSION X(1000), R(1000), INFMT(20)
C
C  WRITE HEADING AND READ PROBLEM SPECIFICATION, VARIABLE INPUT FORMAT
C  AND SERIES DATA.
      WRITE (6,100)
  100 FORMAT (1H1, 30X, 30HAUTOCORRELATION OF TIME SERIES)
    1 READ (5,101) NPROB, NOBS, L
  101 FORMAT (3I5)
      IF (NPROB .EQ. 0) STOP
      READ (5,102) (INFMT(I), I = 1,20)
  102 FORMAT (20A4)
      READ (5,INFMT) (X(I), I = 1,NOBS)
C
C  BEGIN MAIN LOOP AND INITIALISE.
      DO 3 J = 1,L
      N1 = NOBS - J
      AN1 = N1
      SUMX1 = 0.0
      SSQX1 = 0.0
      SUMX2 = 0.0
      SSQX2 = 0.0
      SCROSS = 0.0
```

Program 20 (continued)

```
C
C   LOOP FOR EACH TIME-LAG.
        DO 2 I = 1,N1
        IJ = I + J
        SUMX1 = SUMX1 + X(I)
        SSQX1 = SSQX1 + X(I) * X(I)
        SUMX2 = SUMX2 + X(IJ)
        SSQX2 = SSQX2 + X(IJ) * X(IJ)
      2 SCROSS = SCROSS + X(I) * X(IJ)
        DEV1 = SSQX1 - SUMX1 * SUMX1 /AN1
        DEV2 = SSQX2 - SUMX2 * SUMX2 / AN1
        CROSS = SCROSS - SUMX1 * SUMX2 / AN1
      3 R(J) = CROSS / SQRT(DEV1 * DEV2)
C
C   OUTPUT RESULTS.
        WRITE (6,103) NPROB, NOBS, (J, R(J), J = 1,L)
    103 FORMAT (///14H SAMPLE NUMBER, I5, 12H (CONTAINING,
       AI4, 14H OBSERVATIONS)// (10(I5,F7.3)))
        GO TO 1
        END

DATA

    1    15    5
(10F5.0)
 1.0  1.5  2.0  3.0  3.5  4.0  5.0  5.5  6.0  7.0
 7.5  8.0  9.0  9.5 10.0
    2    15    5
(10F5.0)
 1.0  1.5  2.0  3.0  3.5  4.0  5.0  5.5  6.0  7.0
 7.5  8.0  9.0  9.5 10.0
    0

OUTPUT
            AUTOCORRELATION OF TIME SERIES


SAMPLE NUMBER      1 (CONTAINING 15 OBSERVATIONS)
   1  0.996    2  0.995    3  1.000    4  0.994    5  0.992


SAMPLE NUMBER      2 (CONTAINING 15 OBSERVATIONS)

   1  0.996    2  0.995    3  1.000    4  0.994    5  0.992
```

Program 21: Cross-correlations in time series

In this program we consider two separate time-series A and B and compute a set of correlation coefficients: (a) when each observation in A is paired with the corresponding observation in B; (b) when series A leads series B by 1, 2, 3, ... L terms; and (c) when A lags behind B by 1, 2, 3, ... L terms. The three possibilities can be illustrated as in Fig. 28 for L = 2, where the broken line encloses the terms to be correlated.

As in the previous program, the statistical interpretation of cross-correlations of this kind is difficult, but the problem has some interest for the programmer and is introduced here more for its manipulative aspects than anything else.

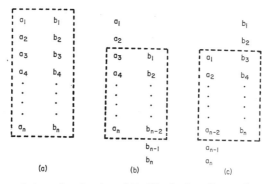

FIG. 28. Cross-correlation of series A and B. The broken line encloses the terms to be correlated. (a) A neither leads nor lags behind B; (b) A leads B by two terms; (c) A lags behind B by two terms.

The first segment of the program contains the usual input procedures and the following segment simply correlates the two series in exact correspondence without either leading or lagging behind the other. The Fortran variable names used here are very similar to those of Program 21, except that the last letters of each name (A and B) are used to denote the two series (e.g. SUMA is the sum of all NOBS terms in series A). The correlation coefficient resulting from this segment is denoted as R0.

It would now be possible to displace one of the series by the appropriate lead or lag and repeat the whole set of computations. A glance at Fig. 28 will show, however, that this would be a very wasteful procedure since almost all the computation involved in summing the observations and their squares is the same for cases (b) and (c) as for case (a). Clearly, the sum and uncorrected sum of squares for A in case (b) are equal to those in case (a) less the sum and sum of squares of a_1 and a_2. Similarly, for series B the values in case (b) are equal to those in case (a) less the sum and sum of squares of b_{n-1} and b_n. Only the uncorrected sum of cross-products needs to be worked out afresh each time. From these considerations we can devise a program segment which will compute a set of correlations for A leading B by 1, 2, 3, ... L.

Before starting such a segment, however, we preserve SUMA, SSQA, SUMB and SSQB, for reasons which will soon appear. This is done by simply duplicating their values under a parallel set of variable names PSUMA, PSSQA, etc. At the same time we also define INDEX = 1; this is to be used later in a branch instruction. We then begin the segment starting with statement 3. This is contained within an outer DO-loop ending on 5 that

Flowchart for Program 21

Cross-correlation for time series.

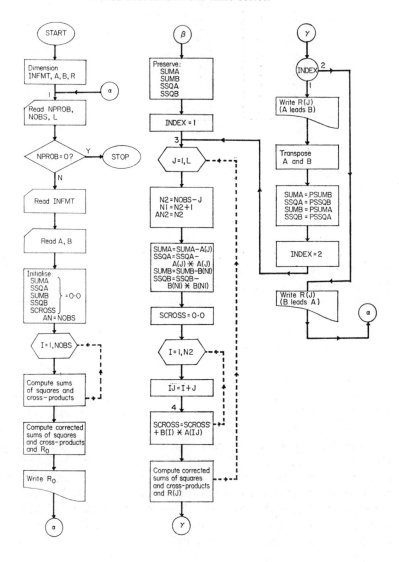

increments the lead of A over B successively from 1 to L. Within this DO-loop we first define the two integer variables N1 and N2. Of these, N2 is simply the number of paired observations to be correlated and is therefore equal to the original number NOBS less the current value of L. N1, which is simply N2 + 1, is the lowest excluded term of the lagging series. On the first pass through the outer loop, for example, it is used as a subscript to exclude observation b_n from the summation process. Using N1, therefore, we can progressively reduce SUMB and SSQB by the appropriate amount for each increasing value of the lead of A over B. Similarly, by using N2 as the upper indexing parameter in the inner DO-loop (ending on 4) we can sum cross-products over the appropriate number of pairs. This inner DO-loop deserves a little attention since each cross-product is computed between the i-th term of B and the $(i + 1)$-th term of A. The use of the computed subscript IJ achieves this. Once the new values for SUMA, SSQA, SUMB, SSQB and SCROSS have been calculated it is a simple matter to compute the correlation coefficient as the j-th element of the array R. When the outer DO-loop ending on 5 has been satisfied, control branches on the value of INDEX (defined earlier as 1) to print out the results for A leading B by 1, 2, 3, ... L.

It is now necessary to carry out a similar set of computations for the converse situation where B leads A (case (c) of Fig. 28). The obvious way to accomplish this is to interchange the two series and return to statement 3. This process begins with the DO-loop ending on 7 which exchanges A(I) and B(I) through the temporary 'holding' location TEMP. We then set up the new initial values of SUMA, SSQA, SUMB and SSQB (from the values preserved earlier, remembering to interchange them), we redefine INDEX and run once more through the calculations explained in the previous paragraph. When these are completed, control branches on the new value of INDEX (=2) to print out the results for A lagging behind B by 1, 2, 3, ... L. Control then returns to statement 1 and a new problem specification card is read.

Program 21

Cross-correlation of time-series; with sample data and output.

```
C
C.................................................................
C
C   THIS PROGRAM COMPUTES THE CROSS-CORRELATIONS BETWEEN TWO
C   TIME SERIES A AND B. A LEADS OR LAGS B BY 1,2,3,...L. INPUT
C   PARAMETERS ARE AS FOLLOWS -
C       NPROB = PROBLEM REFERENCE NUMBER (ZERO ON LAST DATA CARD)
C       NOBS  = NUMBER OF OBSERVATIONS IN SERIES
C       L     = MAXIMUM LEAD OR LAG
C       INFMT = VECTOR HOLDING VARIABLE INPUT FORMAT
C       A     = VECTOR OF OBSERVATIONS IN SERIES A (LENGTH NOBS)
C       B     = VECTOR OF OBSERVATIONS IN SERIES B (LENGTH NOBS)
C
C.................................................................
```

Program 21 (*continued*)

```
C
      DIMENSION INFMT (20), A(1000), B(1000), R(50)
C
C  READ PROBLEM SPECIFICATION, VARIABLE INPUT FORMAT AND SERIES DATA.
    1 READ (5,100) NPROB, NOBS, L
  100 FORMAT (3I5)
      IF (NPROB .EQ. 0) STOP
      READ (5,101) (INFMT(I), I = 1,20)
  101 FORMAT (20A4)
      READ (5,INFMT) (A(I), I = 1,NOBS), (B(I), I = 1,NOBS)
C
C  COMPUTE CROSS-CORRELATION WHEN LAG IS ZERO.
      SUMA = 0.0
      SUMB = 0.0
      SSQA = 0.0
      SSQB = 0.0
      SCROSS = 0.0
      AN = NOBS
      DO 2 I = 1,NOBS
      SUMA = SUMA + A(I)
      SUMB = SUMB + B(I)
      SSQA = SSQA + A(I) * A(I)
      SSQB = SSQB + B(I) * B(I)
    2 SCROSS = SCROSS + A(I) * B(I)
      DEVA = SSQA - SUMA * SUMA / AN
      DEVB = SSQB - SUMB * SUMB / AN
      CROSS = SCROSS - SUMA * SUMB / AN
      R0 = CROSS / SQRT(DEVA * DEVB)
      WRITE (6,102) NPROB, NOBS, R0
  102 FORMAT (8H1PROBLEM, I4, 13H (COMPRISING, I5,
     A23H PAIRS OF OBSERVATIONS)///5X,40HSERIES A NEITHER LEADS NOR LAGS
     B SERIES B//16X, 3HR = , F8.4)
C
C  PRESERVE SUMS AND SUMS OF SQUARES.
      PSUMA = SUMA
      PSUMB = SUMB
      PSSQA = SSQA
      PSSQB = SSQB
      INDEX = 1
C
C  COMPUTE CROSS CORRELATIONS WHEN A LEADS B BY 1,2,3...L.
    3 DO 5 J = 1,L
      N2 = NOBS - J
      N1 = N2 + 1
      AN2 = N2
      SUMA = SUMA - A(J)
      SSQA = SSQA - A(J) * A(J)
      SUMB = SUMB - B(N1)
      SSQB = SSQB - B(N1) * B(N1)
      SCROSS = 0.0
      DO 4 I = 1,N2
      IJ = I + J
    4 SCROSS = SCROSS + B(I) * A(IJ)
      DEVA = SSQA - SUMA * SUMA /AN2
      DEVB = SSQB - SUMB * SUMB /AN2
      CROSS = SCROSS - SUMA * SUMB / AN2
    5 R(J) = CROSS / SQRT(DEVA * DEVB)
      GO TO (6,8), INDEX
C
    6 WRITE (6,103) L, (I, R(I), I = 1,L)
  103 FORMAT (//5X, 35HSERIES A LEADS SERIES B BY 1,2,3..., I3//
     A(10X, 5(I7,F8.4)))
```

Program 21 (*continued*)

```
C
C   TRANSPOSE SERIES A AND B AND RECOMPUTE CORRELATIONS.
      DO 7 I = 1,NOBS
      TEMP = A(I)
      A(I) = B(I)
    7 B(I) = TEMP
      SUMA = PSUMB
      SUMB = PSUMA
      SSQA = PSSQB
      SSQB = PSSQA
      INDEX = 2
      GO TO 3
C
    8 WRITE (6,104) L, (I, R(I), I = 1,L)
  104 FORMAT (//5X, 35HSERIES B LEADS SERIES A BY 1,2,3...,I3//
     A(10X, 5(I7,F8.4)))
      GO TO 1
      END

DATA

    1   12   7
(10F7.0)
   1.0  1.5  2.0  3.0  3.5  4.0  5.0  5.5  6.0  7.0
   7.5  8.0  1.5  2.0  3.0  3.5  4.0  5.0  5.5  6.0
   7.0  7.5  8.0  9.0
    0

OUTPUT

PROBLEM  1 (COMPRISING  12 PAIRS OF OBSERVATIONS)

    SERIES A NEITHER LEADS NOR LAGS SERIES B
        R =  0.9950

    SERIES A LEADS SERIES B BY 1,2,3...   7
         1  1.0000    2  0.9930    3  0.9913    4  1.0000    5  0.9862
         6  0.9820    7  1.0000

    SERIES B LEADS SERIES A BY 1,2,3...   7
         1  0.9936    2  1.0000    3  0.9913    4  0.9880    5  1.0000
         6  0.9820    7  0.9705
```

There are three features of interest in this program. The first concerns the general formulation of the problem whereby we avoid repeated squaring and summation for each value 1, 2, 3, ... L. The second is related to the interchange of variables, enabling us to use the same program segment for two different calculations. The third concerns the use of computed indices and indexing parameters to accomplish the various arithmetic operations. This third feature is one which we have already noted in various forms (e.g. in Programs 6, 7, 11 and 20) and which we shall often meet again. It is difficult to lay down any general rules to guide the reader, but he should certainly try to develop facility in detecting manipulations that can be carried out neatly by the use of computed indices and indexing parameters.

Program 22: Best-fit linear functional relationship between two variables, each subject to error

In determining the regression of y (a dependent variable) on x (an independent variable) by the method of least squares we assume that x is measured without error whereas the errors in y are supposed to be normally distributed with zero mean and constant variance. The parameters a and b in the regression equation $y = a + bx$ are then calculated so as to allow the best prediction of y for given values of x. Even when we know that x is also subject to error, the same procedure may be used when we wish to use the regression equation for predicting values of y. Alternatively, if we wished to predict values of x from those of y we should compute the regression of x on y by the same method, giving a line whose equation may be written as $x = a_1 + b_1 y$. This and the previous regression line are the same only if there is a perfect linear correlation between x and y. Otherwise the two lines will be inclined at an angle to each other, intersecting at the point $\bar{x}$, $\bar{y}$ (Fig. 26).

It may be, however, that the two associated variables which we are studying are both subject to error and that we are not interested in predicting y from x or vice-versa, but in the nature of the true functional relationship between the variables. A good example of this occurs in the calculation of allometric growth parameters from the linear relation between logarithmically transformed measurements of two parts of the body. Under these circumstances we should not use the standard regression procedures; instead we require a method of computing a single best-fit linear relationship between x and y. This can be done by a technique due to Bartlett (1949) and discussed also by Simpson, Roe and Lewontin (1960). The procedure is outlined below in terms of the Fortran variables used in the program. By way of variation this is presented as a subroutine (SUBROUTINE LINE) with a short calling program that also controls input and output, but which we shall not discuss. The reader may like at this point to refer back to p. 69 for a discussion of subroutines in Fortran.

SUBROUTINE LINE begins with the normal statement of its name followed by the list of dummy arguments, indicating that these are being transmitted explicitly. Three of them, X, Y and N, are input parameters, defined by appearing in an input list of the calling program. The others are output parameters which are returned to the calling program when the subroutine has been executed. The only two dimensioned variables are X and Y, two arrays accommodating the values of the paired variables under analysis, each represented by N observations. The first short segment of the program then determines how many of the N observations will fall into each of three groups. These are as nearly equal in size as possible, and such that the first and third groups will contain equal numbers of observations while the middle group will contain either the same number (when N is exactly divisible by 3) or one or two less. The three statements ending on statement 1 therefore define K, the number of observations subsequently to be placed in the first and third groups.

We next arrange the stored values of X in ascending order and ensure that each associated value of Y forms the corresponding element of its own array. For example, if X(23) were the smallest value in array X we would transfer it to location X(1) and place the original Y(23) in location Y(1). This sorting procedure is relegated to a second subroutine (SUBROUTINE SORT) which is called by SUBROUTINE LINE. The sorting method is a slight extension of the ascending exchange sort used in Program 4 (p. 106). The difference lies in the fact that here the exchange of X's is immediately followed by an exchange of the corresponding Y's within the same DO-loop. The result of this is that on return to SUBROUTINE LINE the X's are now arranged in ascending order of size while the Y's are arranged in order of the size of their associated X's. The reader who finds difficulty in following SUBROUTINE SORT should study it in conjunction with the account of Program 4 on p. 106.

The next program segment computes the mean values of X and Y in the usual way, but the succeeding segment is more interesting and has two objects: first to assign the K lowest X's and their associated Y's to group 1, the K highest to group 3 and the remainder to group 2, then secondly to compute the means for each of these groups, storing them as XBAR1 and YBAR1, XBAR2 and YBAR2, and finally XBAR3 and XBAR3. To accomplish these two objectives we first compute three indexing parameters N1, N2 and N3. We then have a situation where group 1 consists of elements X(1) to X(K), group 2 contains X(N1) to X(N2) and group 3 contains X(N3) to X(N), all with the corresponding elements of Y. Once N1, N2 and N3 have been defined it is a simple matter to compute the group means for X and Y, utilising the three successive DO-loops that end on statements 4, 5 and 6 respectively. B and A, the required parameters in the equation $Y = A + BX$ are then calculated by simple arithmetic statements.

Having computed the equation to the line, the remainder of the program is concerned with calculating quantities required in significance tests. Our object is to calculate three t-variates, each with $(N-3)$ degrees of freedom, to test (i) the significance of the regression (i.e. whether B is significantly greater than zero); (ii) to test whether B differs significantly from 1 (a point of some importance in allometric growth studies, for instance), and (iii) to test whether the relationship between X and Y includes a significant non-linear component. These t-variates are denoted in the program as T0, T1 and TL respectively. To evaluate them we first compute the pooled sums of squares and cross-products about the group mean for each of the three groups, using three consecutive DO-loops (ending on statements 7, 8 and 9) in which the indexing parameters N1, N2 and N3 again appear. Each pooled sum of squares or cross-products is then divided by the corresponding degrees of freedom DF to give the pooled within-group variances of X and Y and the covariance of X and Y. Two further quantities XDIFF and SB2 are also computed: XDIFF is the difference between the means of the X's in the first and third group while SB2 is a variance estimate required in the significance tests. T0, T1 and TL are then computed directly by arithmetic statements, and control is returned to the main program which prints out the results as a line of answers for each set of input data. Since the three significance tests are probably not familiar to the student, it might be as well to summarise them algebraically below, along with some others that the reader can apply using various items in the printed output list. Full details of these tests are given by Bartlett (1949) and Simpson, Roe and Lewontin (1960).

(i) *Test for significance of regression (T0)*

$$t = \frac{(\bar{x}_3 - \bar{x}_1)\,b\sqrt{k/2}}{\sqrt{s_y^2}} \tag{7}$$

with $(n - 3)$ degrees of freedom.

(ii) *Test whether b differs significantly from a standard value β.*

$$t = \frac{(\bar{x}_3 - \bar{x}_1)(b - \beta)\sqrt{k/2}}{\sqrt{s_y^2 - 2\beta s_{xy} + \beta^2 s_x^2}} \tag{8}$$

with $(n - 3)$ degrees of freedom.

When $\beta = 1$ equation (8) reduces to the expression calculated in the program as T1, i.e.

$$t = \frac{(\bar{x}_3 - \bar{x}_1)(b - 1)\sqrt{k/2}}{\sqrt{s_y^2 - 2s_{xy} + s_x^2}}. \tag{9}$$

(iii) *Test for deviations from linearity (TL)*

$$t = \frac{\{(\bar{y}_1 + \bar{y}_3 - 2\bar{y}_2) - b(\bar{x}_1 + \bar{x}_3 - 2\bar{x}_2)\}\left\{\dfrac{2}{k} + \dfrac{4}{(n-k)}\right\}^{-\frac{1}{2}}}{s_B^2} \tag{10}$$

with $(n - 3)$ degrees of freedom.

(iv) The confidence limits for b can be obtained from the expression:

$$t^2 = \frac{(\bar{x}_3 - \bar{x}_1)^2\,(b - \beta)^2\,k/2}{s_y^2 - 2\beta s_{xy} + \beta^2 s_x^2} \tag{11}$$

by solving a quadratic equation for β after giving numerical values to all other variables and parameters. t is, of course, taken from tabulated values at $P = 0.95$ or 0.99 with $(n - 3)$ degrees of freedom. These limits can only be evaluated by the user if the output and subroutine arguments of the program are altered so as to print out the values of s_y^2, s_{xy} and s_x^2.

(v) The confidence limits for the parameter a are given approximately by:

$$\pm \frac{t\sqrt{s_B^2}}{\sqrt{n}} \tag{12}$$

where t is again the $P = 0.99$ or 0.95 tabulated value with $(n - 3)$ degrees of freedom.

(vi) The significance of a difference between the slopes of two regression lines is tested approximately by:

$$t = \frac{(\bar{x}_3 - \bar{x}_1 - \bar{x}_3' + \bar{x}_1')(b - b')}{\sqrt{\frac{2s_B^2}{k} + \frac{2s_{B'}^2}{k'}}} \tag{13}$$

with $(n + n' - 6)$ degrees of freedom.

(vii) The significance of a difference between the intercepts of two lines is tested approximately by:

$$t = \frac{a - a'}{\sqrt{\frac{s_B^2}{n} + \frac{s_{B'}^2}{n'}}} \tag{14}$$

In all the above tests the following notation applies:

a, a'	estimated intercept of line(s)
b, b'	estimated slope of line(s)
β	slope parameter being estimated
k, k'	number of observations in first and third group(s)
n, n'	total number of observations
s_x^2	pooled within-group sum of squares of x
s_{xy}	pooled within-group sum of cross products
s_y^2	pooled within-group sum of squares of y
$\bar{x}_1, \bar{x}_2, \bar{x}_3$	means of x in groups 1, 2 and 3
$\bar{y}_1, \bar{y}_2, \bar{y}_3$	means of y in groups 1, 2 and 3.

The accompanying flowchart is a summary of the operations involved in SUBROUTINE LINE.

Outline flowchart for Program 22

Best-fit linear relationship (Bartlett's method).

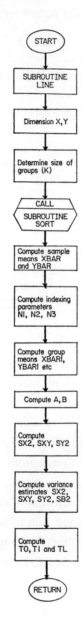

Program 22

Best-fit linear relationship (Bartlett's method). The listing comprises a short calling program and SUBROUTINE LINE, which itself calls SUBROUTINE SORT.

```
C
C   THIS PROGRAM CONTROLS INPUT/OUTPUT AND CALLS SUBROUTINES FOR
C   COMPUTING THE LINEAR FUNCTIONAL RELATION Y = A + BX.
C
      DIMENSION INFMT(20), X(200), Y(200)
C
      WRITE (6,99)
   99 FORMAT (8H1PROBLEM,8X,1HA,9X,1HB,8X,2HT0,8X,2HT1,8X,2HTL,7X,
     A5HXDIFF,8X,3HSB2,8X,1HK,8X,3HNDF)
    1 READ (5,100) NPROB, N
  100 FORMAT (2I5)
      IF (NPROB .EQ. 0) GO TO 2
      READ (5,101) (INFMT(I), I = 1,20)
  101 FORMAT (20A4)
      READ (5,INFMT) (X(I), Y(I), I = 1,N)
      CALL LINE (X,Y,N,A,B,T0,T1,TL,XDIFF,SB2,K,NDF)
      WRITE (6,102) NPROB, A, B, T0, T1, TL, XDIFF, SB2, K, NDF
  102 FORMAT (/I6, F14.3, 5F10.3, F12.3, I7, I10)
      GO TO 1
    2 STOP
      END


C
      SUBROUTINE LINE (X,Y,N,A,B,T0,T1,TL,XDIFF SB2,K,NDF)
C
C...............................................................
C
C   THIS SUBROUTINE COMPUTES THE BEST FIT LINEAR RELATION Y = A + BX
C   WHEN X AND Y ARE BOTH SUBJECT TO ERROR. SEE BARTLETT(1949) AND
C   SIMPSON, ROE AND LEWONTIN (1960). PARAMETERS ARE AS FOLLOWS -
C     X     = VECTOR OF X-VALUES (LENGTH N) (INPUT)
C     Y     = VECTOR OF CORRESPONDING Y-VALUES (LENGTH N) (INPUT)
C     N     = NUMBER OF PAIRED OBSERVATIONS (INPUT)
C     A     = INTERCEPT OF LINE (OUTPUT)
C     B     = SLOPE OF LINE (OUTPUT)
C     T0    = T FOR SIGNIFICANCE OF REGRESSION (OUTPUT)
C     T1    = T FOR DEVIATIONS FROM UNIT SLOPE (OUTPUT)
C     TL    = T FOR DEVIATIONS FROM LINEARITY (OUTPUT)
C     XDIFF = DIFFERENCE BETWEEN MEANS OF 1ST AND 3RD GROUPS (OUTPUT)
C     SB2   = VARIANCE OF ESTIMATED SLOPE (OUTPUT)
C     K     = NUMBER OF POINTS IN FIRST AND THIRD GROUPS (OUTPUT)
C     NDF   = DEGREES OF FREEDOM FOR T0, T1, AND TL (OUTPUT)
C   THIS SUBROUTINE CALLS SUBROUTINE SORT
C
C...............................................................
C
      DIMENSION X(200), Y(200)
C
C   DETERMINE SIZE OF GROUPS.
      K = N / 3
      IF (N - 3*K-1) 2,2,1
    1 K = K + 1
C
```

Program 22 (*continued*)

```
C     ARRANGE X'S IN ASCENDING ORDER, EACH WITH ASSOCIATED Y.
    2 CALL SORT (X,Y,N)
C
C     COMPUTE SAMPLE MEANS.
      AN = N
      XBAR = 0.0
      YBAR = 0.0
      DO 3 I = 1,N
      XBAR = XBAR + X(I)
    3 YBAR = YBAR + Y(I)
      XBAR = XBAR / AN
      YBAR = YBAR / AN
C
C     COMPUTE GROUP MEANS
      N1 = K + 1
      N3 = N − K + 1
      N2 = N3 − 1
      XBAR1 = 0.0
      YBAR1 = 0.0
      AK = K
      DO 4 I = 1,K
      XBAR1 = XBAR1 + X(I)
    4 YBAR1 = YBAR1 + Y(I)
      XBAR1 = XBAR1 / AK
      YBAR1 = YBAR1 / AK
C
      XBAR2 = 0.0
      YBAR2 = 0.0
      AM = N − 2 * K
      DO 5 I = N1,N2
      XBAR2 = XBAR2 + X(I)
    5 YBAR2 = YBAR2 + Y(I)
      XBAR2 = XBAR2 / AM
      YBAR2 = YBAR2 / AM
C
      XBAR3 = 0.0
      YBAR3 = 0.0
      DO 6 I = N3, N
      XBAR3 = XBAR3 + X(I)
    6 YBAR3 = YBAR3 + Y(I)
      XBAR3 = XBAR3 / AK
      YBAR3 = YBAR3 / AK
C
C     ESTIMATE PARAMETERS A AND B.
      B = (YBAR3 − YBAR1) / (XBAR3 − XBAR1)
      A = YBAR − B * XBAR
C
C     COMPUTE VARIANCES FOR SIGNIFICANCE TESTS
      SXY = 0.0
      SX2 = 0.0
      SY2 = 0.0
      DO 7 I = 1,K
      SX2 = SX2 + (X(I)−XBAR1) ** 2
      SXY = SXY + (X(I)−XBAR1)*(Y(I)−YBAR1)
    7 SY2 = SY2 + (Y(I)−YBAR1)**2
      DO 8 I = N1,N2
      SX2 = SX2 + (X(I)−XBAR2) ** 2
      SXY = SXY + (X(I)−XBAR2)*(Y(I)−YBAR2)
    8 SY2 = SY2 + (Y(I)−YBAR2)**2
      DO 9 I = N3,N
```

Program 22 (*continued*)

```
        SX2 = SX2 + (X(I)-XBAR3)**2
        SXY = SXY + (X(I)-XBAR3) * (Y(I)-YBAR3)
    9   SY2 = SY2 + (Y(I)-YBAR3) **2
        NDF = N - 3
        DF = NDF
        SX2 = SX2 / DF
        SY2 = SY2 / DF
        SXY = SXY / DF
        SB2 = SY2 - 2.0*B*SXY + B*B*SX2
        XDIFF = XBAR3 - XBAR1
        T0 = XDIFF * B * SQRT(0.5*AK) / SQRT(SY2)
        T1 = XDIFF * (B-1.0) * SQRT(0.5*AK) / SQRT(SY2-2.0*SXY+SX2)
        TL = ((YBAR1+YBAR3-2.0*YBAR2-B*(XBAR1+XBAR3-2.0*XBAR2)) /
       A SQRT(2.0/AK+4.0/(AN-AK))) / SQRT(SB2)
        RETURN
        END


        SUBROUTINE SORT (X,Y,N)
C
C   THIS SUBROUTINE ARRANGES THE ELEMENTS OF VECTOR X IN ASCENDING
C   ORDER OF SIZE WITH THE ASSOCIATED ELEMENTS OF Y IN
C   CORRESPONDING SEQUENCE.
C
        DIMENSION X(200), Y(200)
        LIMIT = N - 1
    1   INTER = 1
        DO 2 I = 1,LIMIT
        IF (X(I+1) .GE. X(I)) GO TO 2
        TEMP = X(I+1)
        X(I+1) = X(I)
        X(I) = TEMP
        TEMP = Y(I+1)
        Y(I+1) = Y(I)
        Y(I) = TEMP
        INTER = I
    2   CONTINUE
        IF (INTER .EQ. 1) GO TO 3
        LIMIT = INTER - 1
        GO TO 1
    3   RETURN
        END
```

Chapter 8

Matrix Methods

Definitions. As we have already seen, many of the operations of statistical analysis are carried out on sets of numbers arranged to form rectangular (two-dimensional) arrays. For example, a sequence of observations made on 4 variables on 3 successive days could be represented by a table of the form:

3.0	2.2	1.1	7.3
4.9	6.1	2.8	9.2
1.3	1.0	0.3	10.1

where each row consists of the four variables measured on a particular day and each column corresponds to the 3 successive observations of a particular variable.

Rectangular arrays of this kind are known as *matrices* and their mathematical properties form the subject matter of matrix algebra. Further accounts of this topic are available in mathematical works (e.g. Ferrar (1957) and more advanced accounts by Mirsky (1955), Bellman (1960) Fox (1964) and many others). There is also a very useful introductory account adapted to the needs of biologists by Searle (1966). The present chapter provides only an outline of those aspects of matrix algebra which are needed for the remainder of this book.

Matrices are usually denoted by an upper-case (capital) letter in heavy type, or by writing out the whole array enclosed in brackets, e.g.

$$\mathbf{A} = \begin{pmatrix} 2.1 & 1.3 \\ 4.0 & 0.0 \end{pmatrix}$$

The individual numbers, or *elements* of the matrix, may be represented more generally by subscripted lower case letters, the subscripts indicating the position of the element, e.g.

$$\mathbf{A} = \begin{pmatrix} a_{11} & a_{12} \\ a_{21} & a_{22} \end{pmatrix}$$

where a_{11} denotes the element in the first row and column, a_{12} that in the first row and second column and so on. Rows are numbered from above downwards and columns from left to right. The order of subscripts is important, the first figure always denoting the row while the second denotes the column. When it is necessary to indicate any element in general this can be done by referring to it as a_{ij}.

When there are m rows and n columns, the matrix is said to be of order $m \times n$ or is referred to as an $m \times n$ matrix. In some cases m and n are equal and one then has a square matrix (i.e. with equal numbers of rows and columns) of order m or n. e.g.

$$\mathbf{P} = \begin{pmatrix} p_{11} & p_{12} & p_{13} \\ p_{21} & p_{22} & p_{23} \\ p_{31} & p_{32} & p_{33} \end{pmatrix}$$

is a square matrix of order 3. The elements p_{11}, p_{22}, p_{33} (and so on in larger matrices) form the *main diagonal* of the matrix and the sum of these elements is referred to as the *trace* of the matrix:

$$\operatorname{tr} \mathbf{P} = p_{11} + p_{22} + p_{33}.$$

Not only may matrices be square or rectangular, they may also assume the special form of a single row or a single column. These, of course, are simply $1 \times n$ or $n \times 1$ matrices respectively, but they are usually referred to as *vectors* and are then generally denoted by lower-case heavy type letters or, of course, by the array of elements enclosed in brackets:

$$\mathbf{a}' = (a_1 \; a_2 \; a_3 \; a_4)$$

is the row-vector $\mathbf{a}'$, consisting of four terms $a_1 \; a_2 \; a_3$ and a_4, while

$$\mathbf{q} = \begin{pmatrix} q_1 \\ q_2 \\ q_3 \\ q_4 \end{pmatrix}$$

is the column vector $\mathbf{q}$, again composed of four elements q_1 to q_4. The 'prime' sign (′) is often used to distinguish the row vector **a**′ from the column vector **a** when both contain the same elements $a_1, a_2 \ldots a_n$. The above rather formal account may perhaps be made more concrete by a biological example. Suppose one sets out to measure skull-breadth, skull-length, humerus-length and femur-length in a sample of six individuals of a species of rodent. The results for each individual can be set out as a row of figures:

$$4.1 \quad 7.3 \quad 5.8 \quad 6.2$$

This may be spoken of as a vector of observations (a row-vector of four elements). Similar vectors can be obtained for the other five individuals and the whole set of data assembled to form a 6 × 4 matrix of observations. In statistical work it is often convenient to refer to such a matrix as the *primary data matrix*, to distinguish it from further matrices which may be obtained in the course of subsequent analysis. This same primary data matrix can also be regarded as made up of four column vectors, each representing the six measurements on a particular anatomical character.

As the reader knows from Chapter 3, Fortran provides a very straightforward method of referring to the elements of a matrix through its use of subscripted variables. Thus the matrix

$$\mathbf{A} = \begin{pmatrix} a_{11} & a_{12} & a_{13} & \ldots & a_{1m} \\ a_{21} & a_{22} & a_{23} & \ldots & a_{2m} \\ \vdots & & & & \vdots \\ a_{n1} & a_{n2} & \ldots & \ldots & a_{nm} \end{pmatrix}$$

is a rectangular matrix of n rows and m columns. Its elements are denoted in Fortran as A (I, J) where I can take any integer value from 1 to N and J from 1 to M. Similarly the singly subscripted Fortran variable P (I), where I varies from 1 to N, represents a vector of n elements or a 1 × n matrix. There is no way of distinguishing between a row vector and a column vector in Fortran coding with a single subscript. Which is intended will be apparent from the context of the computation.

Before reviewing matrix operations and the corresponding Fortran coding, two final points may be made regarding terminology. The matrices encountered in biological statistics almost always consist of real numbers (positive or negative and including zeros). They are spoken of as *real matrices* and are the only kind considered here. Frequently it happens that square matrices are *symmetric*, i.e. the triangular array of numbers to the left of the main diagonal is a mirror-image, so to speak, of the array to the right of the diagonal. For example, suppose one had five species A, B, C, D and E and on each one measured six characters. It would then be possible to calculate the correlation coefficients between any pair of species. These might be found to have the form:

	A	B	C	D	E
A	1.0	.7	.4	−.3	.8
B	.7	1.0	.5	.4	.1
C	.4	.5	1.0	.6	.9
D	−.3	.4	.6	1.0	.4
E	.8	.1	.9	.4	1.0

It is clear that the diagonal must consist of 1.0 1.0 1.0 1.0 1.0 since each diagonal element is the correlation coefficient of a species with itself! Similarly, the value .7 for the correlation between the first and second species (first row, second column) must be the same as that between the second and first species (second row, first column), and so on. A matrix of correlation coefficients or a variance–covariance matrix, where each element represents the correlation or covariance between pairs of species (the diagonals being unity for the correlation matrix and equal to the variances for the variance–covariance matrix) must necessarily be symmetric. This property, as we shall see later (pp. 445, 451), enables one to economise in computing time, print-out time and core-storage by operating on the main diagonal and one of the two identical off-diagonal parts of the matrix.

Input and Output of Matrices: Two of the basic operations on matrices consist of reading a data matrix into storage and causing the computer to print out matrices embodying the results of calculations. For convenience we shall consider mainly punched-card input and printed output and we shall assume in our coding that reading and printing are done by units 5 and 6. Despite their similarity it is simpler to deal separately with input and output.

Input: It is possible to read in a matrix element by element, but this is obviously an inefficient procedure. Three methods are used in practice, of which the second is more generally useful that the other two. The first method makes use of two implied DO-loops as in the following example:

READ (5, 100) ((X (I, J), J = 1, M), I = 1, N)
100 FORMAT (···

Here the matrix **X** is being read in, the elements being denoted by x_{ij} where i—the row number—varies from 1 to n and j—the column number—from 1 to m. In the READ statement there are, in effect, two nested DO-loops of which the innermost controls the column number and the outer one the row number. Operation of these DO-loops therefore causes the column number to vary more rapidly so that the elements are read into storage in the following sequence

$$X(1, 1) \ X(1, 2) \ X(1, 3) \ X(1, 4) \ ... \ X(1, M)$$
$$X(2, 1) \ X(2, 2) \ X(2, 3) \ X(2, 4) \ ... \ X(2, M)$$
$$\vdots$$
$$X(N, 1) \ X(N, 2) \ X(N, 3) \quad ... \quad X(N, M)$$

In other words the matrix is being read in row by row and it is absolutely imperature that the data cards are punched in the corresponding fashion, the elements being arranged one after the other in a continuous series as though one were following a page of written matter from left to right and top to bottom.

Thus, if a 4×3 matrix were being read in by this method, the 12 values might be arranged on cards according to the FORMAT 6F3.1:

   1st card 2.2 3.3 4.2 1.8 7.2 8.1
   2nd card 6.8 3.7 2.1 8.9 4.2 8.1

Starting with the first card the four values 2.2, 3.3, 4.2 and 1.8 would be read into the first 'row' of the storage space reserved for the matrix, then the remaining two values (7.2 and 8.1) and the first two of the second card (6.8 and 3.7) would form the second stored row, and so on.

This is probably the most straightforward way to read in a matrix, but it is obvious from the form of the READ statement above that the implied DO-loops could be nested in the opposite fashion, i.e.

  READ (5, 100) ((X(I, J), I = 1, N), J = 1, M)

Here it is the *row* subscript which is varying more rapidly so that the matrix would be read in column by column. And the data cards would then have to be punched in such a way as to correspond with this different version of the double implied DO-loop method of reading in a matrix.

The second main method of matrix input is to make the outer DO-loop explicit so that reading is accomplished by the statements:

  DO 1 I = 1, N
  1 READ (5, 100) (X(I, J), J = 1, M)
  100 FORMAT ...

This means that N separate READ statements are executed, each dealing with M observations. If, as before, the matrix being read in comprised N rows and M columns then the first READ statement would read the M observations in the first row, the second would read the M observations in the second row, and so on. Each row would have to start on a fresh card and the M observations in the row would be punched continuously on one or more cards, depending on the FORMAT and the size of M. This form of read-in is very convenient when each row of the data matrix corresponds to an individual subject of some kind, for which a set of M variables, represented by the columns, have been measured. The data may be extended by adding new individuals or new variables, or reduced by omitting individuals or variables, all without the need for repunching or complicated modifications of FORMAT.

A third and rather different method of programming matrix input is also possible, provided that the matrix to be read in occupies completely the set of locations reserved for it in a previous DIMENSION statement. Thus if this were DIMENSION X(10, 10) one could read in a 10×10 matrix by simply writing

    READ (5, 100) X

    100 FORMAT ...

This automatic method of reading in a matrix is very convenient but one must be clear that the elements are always read in *column by column* when this technique is followed.

For example, suppose one wished to read in the matrix X where

$$X = \begin{pmatrix} 3.8 & 4.2 & 2.1 \\ 2.3 & 3.2 & 8.4 \\ 7.1 & 7.2 & 6.4 \\ 1.2 & 2.1 & 1.8 \end{pmatrix}$$

The values would be read in the order 3.8, 2.3, 7.1 and 1.2 (making up the first column) 4.2, 3.2, 7.2 and 2.1 (making the second column) and so on. The data cards should therefore be punched so as to ensure that the values follow each other in this sequence. Note also, that the method works only if the matrix corresponds in size to that given in the DIMENSION statement. Attempting to read in a matrix of a different size would lead to errors.

More sophisticated methods of matrix storage are also possible, enabling certain types of matrix to be stored more economically. These require special input techniques that are indicated on pp. 451ff. Compressed storage is an important feature of the subroutines available in the IBM Scientific Subroutines Package (p. 472).

Output: Corresponding to the READ statement involving two implied DO-loops one can provide an output statement:

    WRITE (6, 100) ((X(I, J), J = 1, M), I = 1, N)

    100 FORMAT ...

which would cause the elements of the stored matrix to be printed out row by row. This, in fact, is the only way that the printer can work and a matrix can never be printed out one column after another in the way that one might write it out by hand. The above WRITE statement therefore works perfectly, provided that the number of elements in a row can be accommodated in a single line of printed output containing, say, 120 characters. Very often, however, one wishes to print out much larger matrices and for these some care is needed. The most straightforward way is to control the printing by a normal DO-loop for the rows incorporating an implied DO-loop for the columns. For example the segment

```
    DO 10 I = 1, 6
 10 WRITE (6, 100) (X(I, J) J = 1, 20)
100 FORMAT (/(10F10.4))
```

would have the following effect: The outer DO-loop would begin with I set at 1 and the implied DO-loop would immediately begin to print the first row, starting $X(1, 1)$ $X(1, 2)$ $X(1, 3)$ as far as $X(1, 10)$. This, as the FORMAT statement indicates, would be the end of the first *printed line*, and would represent the first row, first 10 columns of the stored matrix. The remaining 10 columns would then be printed in the next printed line, the format repeating as indicated by the inner parentheses. When $X(1, 20)$ had been printed the implied DO-loop would be satisfied and control would pass to the outer DO-loop with I = 2. That is, the second row of the matrix would now be printed, beginning $X(2, 1)$, $X(2, 2)$ and so on. This would, however, be separated from the previous line by a space because of the slash format. The result would be an output list of the following form:

```
 2.8  2.1  1.3  3.2  4.8  7.2  1.9  2.8  3.6  1.4
 1.3  2.4  4.1  1.2  3.7  6.7  2.8  9.3  4.1  2.7

14.8 12.7 14.3  1.2  6.1  9.2  2.7  7.3  4.8  2.8
 1.9 10.6 11.2  2.3  3.2  4.8  7.2  2.8  1.6  2.6
```

If one wished to make the numbering of rows and columns more explicit, the values assumed by I and J could be printed out at the appropriate places to give results like the following example:

```
    DO 1 I = 1, 10
  1 WRITE (6, 100) I, (J, X(I, J), J = 1, 10)
100 FORMAT (/4H ROW, I4/(5(I10, F6.1)))
```

would print out a 10 × 10 matrix in the form:

ROW 1
  1 2.8   2 4.7   3 3.9   4 4.1   5 6.2
  6 7.3   7 6.2   8 1.8   9 5.9 10 3.1

ROW 2
  1 3.5   2 7.2   3 2.1   4 4.8   5 6.2
  6 7.1   7 2.9   8 8.3   9 3.7 10 5.8

When the matrix to be printed out is a symmetric one, the preceding method is somewhat uneconomical, since most of the values are printed twice. The following technique avoids this waste by printing out a half-matrix consisting of the main diagonal and the upper triangle. This can be done by a statement of the form:

DO 2 I = 1, M
2 WRITE (6, 100) I, (J, X(I, J), J = I, M)
100 FORMAT (/4H ROW, I4/(5(I 10, F6.3)))

which leads to output of the following form (for, say a 5 × 5 matrix):

ROW 1
  1 7.200   2 8.301   3 2.105   4 3.921   5 4.830

ROW 2
  2 6.275   3 6.847   4 5.921   5 4.383

ROW 3
  3 2.921   4 7.289   5 8.311

ROW 4
  4 5.920   5 4.841

ROW 5
  5 6.212

With ingenuity alternative forms of output can be devised which may be more decorative than the ones indicated above but are no more informative or economical. One must beware of writing formats that work when a row of the matrix can be accommodated in a single line of printed output but break down when larger matrices are to be printed out.

Elementary operations on a single matrix: Once a matrix is in storage, many operations can be carried out on its component parts—rows, columns and so forth. A few of these occur regularly in a variety of programs and are outlined below for reference.

(a) *Addition of the elements of each row to give a vector of row sums.*

    DIMENSION X(50, 50), RSUM (50)
    DO 1 I = I, M
    RSUM (I) = 0.0
    DO 1 J = 1, N
  1 RSUM (I) = RSUM (I) + X(I,J)

It is assumed that there are M rows and N columns, the numerical values of these variables having previously been defined. The two statements after the DIMENSION statement set the relevant location of RSUM at zero; the next two lines accumulate the row sum by adding the elements in successive columns. The reader in doubt about the sequence of events should follow through each step of the two nested DO-loops using a simple set of numerical data.

(b) *Addition of the elements of each column to give a vector of column sums.*

    DIMENSION X(50, 50), CSUM (50)
    DO 1 J = 1, N
    CSUM (J) = 0.0
    DO 1 I = 1, M
  1 CSUM (J) = CSUM (J) + X(I, J)

(c) *Addition of all elements to give a grand total.*

    DIMENSION X(50, 50)
    GRAND = 0.0
    DO 1 I = 1, M
    DO 1 J = 1, N
  1 GRAND = GRAND + X(I, J)

It is not difficult to see how (a), (b) and (c) could be combined into a single program segment or subroutine which might be very useful in an analysis of variance program.

(d) *Interchange of two rows, say the K-th and the L-th.*

```
DIMENSION X(50, 50)
DO 1 J = 1, N
TEMP = X (K, J)
X (K, J) = X (L, J)
1 X (L, J) = TEMP
```

Note that K and L must be defined (i.e. given numerical values) earlier in the program. The interchange of the two rows occurs element by element, using a single 'holding' location TEMP to store one of the elements temporarily. This prevents it from being destroyed when the other element is read into its original location. Interchange of two columns may be accomplished by a very similar program.

(e) *Transfer of the main diagonal elements of a matrix to a vector.*

The vector chosen—we shall call it VEC—must be dimensioned with the number of elements in the main diagonal of the matrix.

```
DIMENSION X (50, 50), VEC (50)
DO 1 I = 1, M
1 VEC (I) = X (I, I)
```

This operation is, of course, only possible if the matrix is square.

(*f*) *Addition of the main diagonal elements, i.e. computation of the trace of a matrix.*

```
DIMENSION X (50, 50)
TR = 0.0
DO 1 I = I, M
1 TR = TR + X (I, I)
```

(g) *Evaluation of the continued product of the diagonal elements.*

The continued product of a set of numbers is denoted by the Greek capital letter Π, with the subscript and superscript figures indicating the first and last terms, e.g.

$$\prod_{i=1}^{n} a_i = a_1 \times a_2 \times a_3 \times \ldots a_{n-1} \times a_n$$

```
DIMENSION X (50, 50)
PI = 1.0
DO 1 I = 1, M
1 PI = PI * X (I, I)
```

Addition and subtraction of matrices. Two matrices can be added to give a third matrix provided that they contain the same numbers of rows and of columns, i.e. that they are conformable to addition. The operation is represented in matrix notation as

$$A + B = C$$

or, in more detailed form:

$$\begin{pmatrix} a_{11} & a_{12} \\ a_{21} & a_{22} \end{pmatrix} + \begin{pmatrix} b_{11} & b_{12} \\ b_{21} & b_{22} \end{pmatrix} = \begin{pmatrix} a_{11} + b_{11} & a_{12} + b_{12} \\ a_{21} + b_{21} & a_{22} + b_{22} \end{pmatrix} = C$$

That is, the sum **C** of the two matrices is found by adding their corresponding elements. Subtraction is carried out in a comparable way, e.g.

$$\begin{pmatrix} 3 & 1 \\ 2 & 5 \end{pmatrix} - \begin{pmatrix} 1 & 1 \\ 3 & 1 \end{pmatrix} = \begin{pmatrix} 2 & 0 \\ -1 & 4 \end{pmatrix}$$

In Fortran addition is achieved by:

```
DIMENSION A (50, 50), B (50, 50), C (50, 50)
DO 1 I = 1, M
DO 1 J = 1, N
1 C (I, J) = A (I, J) + B (I, J)
```

and subtraction in an exactly comparable manner.

Matrix multiplication. A matrix may be multiplied by a scalar quantity (i.e. a single number such as 3 or 5.27) or by a vector or by another matrix. In the latter two cases, however, the meaning of the term 'multiplication' is not at first obvious and the matrices or vectors being multiplied must be conformable for multiplication in a way that will emerge shortly.

First, however, consider the multiplication of a matrix by a scalar:

$$x\mathbf{A} = \mathbf{B}.$$

This consists of multiplying each element of **A** by x to give the corresponding element of **B**. For example:

$$3 \begin{pmatrix} 2 & 1 \\ 4 & 8 \end{pmatrix} = \begin{pmatrix} 6 & 3 \\ 12 & 24 \end{pmatrix}$$

In Fortran:

```
DIMENSION A (50, 50), B (50, 50)
DO 1 I = 1, M
DO 1 J = 1, N
1 B (I, J) = A (I, J) * X
```

If there is any strong motive for conserving storage space and if A is not required for further computation, then the product matrix can be stored in the same locations as contained A, either under the same variable name or through an EQUIVALENCE declaration, e.g.

```
    DIMENSION A (50, 50), B (50, 50)
    EQUIVALENCE (A (1, 1), B (1, 1))
    DO 1 I = 1, M
    DO 1 J = 1, N
  1 B (I, J) = X * A (I, J)
```

The multiplication of two matrices represents a special set of operations indicated in the following example for two 2×2 matrices:

$$\begin{pmatrix} a & b \\ c & d \end{pmatrix} \times \begin{pmatrix} e & f \\ g & h \end{pmatrix} = \begin{pmatrix} ae + bg & af + bh \\ ce + dg & cf + dh \end{pmatrix}.$$

Or, in numerical terms:

$$\begin{pmatrix} 2 & 1 \\ 3 & 4 \end{pmatrix} \times \begin{pmatrix} 0 & 1 \\ -2 & 3 \end{pmatrix} = \begin{pmatrix} -2 & 5 \\ -8 & 15 \end{pmatrix}.$$

In words, each element in the first *row* of the first matrix is multiplied by each element in the first *column* of the second matrix and the products added. This gives the element in the first row and the first column of the product matrix. The operation is then repeated for the first row of the first matrix and the *second* column of the second matrix, and so on.

Two consequences follow from this definition of multiplication in matrix algebra:

(a) Two matrices can be multiplied only if the number of columns in the first matrix is equal to the number of rows in the second, e.g.

$$\begin{pmatrix} a & b & c \\ d & e & f \end{pmatrix} \times \begin{pmatrix} x & y \\ z & m \\ n & p \end{pmatrix} = \begin{pmatrix} ax + bz + cn & ay + bm + cp \\ dx + ez + fn & dy + em + fp \end{pmatrix}.$$

Such matrices are said to be conformable to multiplication and the rule may be expressed alternatively by saying that an $m \times n$ matrix can only be multiplied by an $n \times p$ matrix and that the product matrix is of order $m \times p$. The reader should verify this using simple numerical examples.

(b) The product of two matrices generally depends on the order in which they are multiplied. Thus, if **A** and **B** are two square matrices of order m, then in general **BA** ≠ **AB**, as can be seen from the following numerical example:

$$\begin{pmatrix} 2 & 3 \\ 1 & 2 \end{pmatrix} \times \begin{pmatrix} 1 & 4 \\ 1 & 2 \end{pmatrix} = \begin{pmatrix} 5 & 14 \\ 3 & 8 \end{pmatrix}$$

but

$$\begin{pmatrix} 1 & 4 \\ 1 & 2 \end{pmatrix} \times \begin{pmatrix} 2 & 3 \\ 1 & 2 \end{pmatrix} = \begin{pmatrix} 6 & 11 \\ 4 & 7 \end{pmatrix}.$$

This peculiarity is expressed by saying that matrix multiplication is a non-commutative operation. Because of this one must distinguish between **AB** and **BA**. In the matrix equation

$$\mathbf{C} = \mathbf{AB}$$

one says that **B** is premultiplied by **A** to give **C**, whereas in the equation

$$\mathbf{D} = \mathbf{BA}$$

it is **A** which is premultiplied by **B** to give **D**.

A Fortran subroutine for matrix multiplication is given below. Note that it contains no input–output procedures, which must therefore be provided in the calling program.

```
      SUBROUTINE MATPLY (A, B, C, M, N, L)
C
C.........................................................
C
C     THIS SUBROUTINE EXECUTES A MATRIX MULTIPLICATION. PARAMETERS ARE AS
C     FOLLOWS −
C        A = FIRST INPUT MATRIX
C        B = SECOND INPUT MATRIX
C        C = OUTPUT (PRODUCT) MATRIX
C        M = NUMBER OF ROWS IN A AND C
C        L = NUMBER OF COLUMNS IN A AND ROWS IN B
C        N = NUMBER OF COLUMNS IN B AND C
C
C.........................................................
C
      DIMENSION A(50, 50), B(50, 50), C(50, 50)
      DO 1 I = 1,M
      DO 1 J = 1,N
      C(I,J) = 0.0
      DO 1 K = 1,L
    1 C(I,J) = C(I,J) + (I,K) * B(K,J)
      RETURN
      END
```

See also p. 444 for a comment on the efficiency of this method of programming a matrix multiplication.

Although vectors have been described above as special kinds of matrices the subroutine MATPLY cannot be used to multiply vectors. The reason is simply that in Fortran, vectors are normally written as singly subscripted arrays and matrices as doubly subscripted ones. Some qualifications to this statement will be discussed in Chapter 15; meanwhile we may summarise multiplication of matrices, vectors and scalars in the following table.

Premultiplier	Postmultiplier	Product
Matrix ($m \times l$)	Matrix ($l \times n$)	Matrix ($m \times n$)
Scalar	Matrix ($l \times n$)	Matrix ($l \times n$)
Scalar	Row or column vector (m)	Row or column vector (m)
Row vector (l)	Matrix ($l \times n$)	Row vector (n)
Row vector (l)	Column vector (l)	Scalar
Column vector (l)	Row vector (n)	Matrix ($l \times n$)
Matrix ($m \times l$)	Column vector (l)	Column vector (m)

The nature and dimensions of the entities being multiplied and the product obtained should be clear from this table and the reader in doubt on any point can easily verify the results with simple numerical examples. The Fortran operations involved in three of the above are given below in the form of program-segments:

(a) *Row vector VEC2 postmultiplied by matrix A to give row vector VEC1.*

    DIMENSION A (50,50), VEC1 (50), VEC2 (50)
    DO 1 I = 1,N
    VEC1 (I) = 0.0
    DO 1 J = 1,L
  1 VEC1(I) = VEC1(I) + VEC2(J) * A(J,I)

where L is the number of elements in VEC2, and N the number in VEC1; A is of order L by N.

(b) *Row vector VEC1 multiplied by column vector VEC2 to give a scalar S.*

    DIMENSION VEC1 (50), VEC2 (50)
    S = 0.0
    DO 1 I = 1, N
  1 S = S + VEC1(I) * VEC2(I)

(c) *Column vector VEC1 multiplied by row vector VEC2 to give the matrix A.*

    DIMENSION A (50,50), VEC1 (50), VEC2 (50)
    DO 1 I = 1,L
    DO 1 J = 1,N
    A(I,J) = 0.0
  1 A(I,J) = A(I,J) + VEC1(I) * VEC2(J)

where L is the number of elements in VEC1 and N the number in VEC2.

Transpose of a matrix. The transpose of a matrix, denoted by $\mathbf{A}'$ or $\mathbf{A}^T$ or $\tilde{\mathbf{A}}$, is another matrix in which the rows of the first become the columns of the second and *vice versa*. Thus the transpose of

$$\mathbf{A} = \begin{pmatrix} a_{11} & a_{12} & a_{13} \\ a_{21} & a_{22} & a_{23} \end{pmatrix}$$

would be

$$\mathbf{A}' = \begin{pmatrix} a_{11} & a_{21} \\ a_{12} & a_{22} \\ a_{13} & a_{23} \end{pmatrix}.$$

This operation is easily achieved in Fortran:

    DIMENSION A(50,50), AT(50,50)
    DO 1 I = 1,N
    DO 1 J = 1,M
  1 AT(I,J) = A(J,I)

where M is the number of rows in A and columns in AT and N is the number of columns in A and rows in AT.

Successive multiplication. When several matrices or vectors are multiplied successively, the separate multiplications are carried out in order with due regard for pre- and post-multiplication. Multiplication is associative, so that

$$(\mathbf{AB})\mathbf{C} = \mathbf{A}(\mathbf{BC}).$$

It does not matter whether one begins with the first or last matrix in the sequence, so long as one works from there to the other end of the sequence. Thus to evaluate **ABCD** one may post-multiply **A** by **B**, then postmultiply the product (**AB**) by **C**, then the resulting product (**ABC**) by **D**. Alternatively one could premultiply **D** by **C**, then premultiply the product (**CD**) by **B** and then the resulting product (**BCD**) by **A**. It must, of course, be established first that the matrices are conformable for multiplication. The two methods can be set out in stages as follows:

First method	Second method
Compute **X** = (**AB**)	Compute **X** = (**CD**)
Compute **Y** = (**XC**)	Compute **Y** = (**BX**)
Compute **Z** = (**YD**)	Compute **Z** = (**AY**)

A more useful example for statistical work is provided by the so-called quadratic form, a scalar defined as follows:

$$x = \mathbf{v}'\mathbf{A}\mathbf{v}$$

where the row vector **v**' is postmultiplied by the matrix **A** to give another row vector that is itself postmultiplied by the column vector **v** to give the scalar x. The necessary Fortran program-segment therefore consists of an amalgamation of (a) and (b) above:

```
DIMENSION A(50,50), V(50), WV(50)
DO 1 I = 1,L
WV(I) = 0.0
DO 1 J = 1, L
1 WV(I) = WV(I) + V(J) * A(J,I)
X = 0.0
DO 2 I = 1,L
2 X = X + WV(I) * V(I)
```

where WV is a 'working vector' intended to hold the product resulting from the first multiplication and then itself multiplied by V to give the scalar X.

In trying to summarise the elementary features of matrix algebra and its implementation in Fortran, the above account has necessarily become somewhat formal and separated from the biological data. It should, however, only be recalled how often these data are arranged in tabular form as matrices and vectors to realise how valuable the techniques are in manipulating the data for statistical and other numerical purposes. Examples will be found in almost every program in this book and much of the success of a programmer depends on his ability to formulate the numerical operations he uses in terms of matrix algebra. Indeed, the subject is so important in multivariate statistical analysis that it will be necessary to discuss further aspects in the following chapter. First, however, it will be convenient to provide a Fortran subroutine that constructs a dispersion or correlation matrix from relatively large bodies of data. Matrices of this kind represent a first step in the computations for a principal component analysis (p. 291) or in multiple discriminant analysis (p. 300) or in the rather more familiar process of multiple regression analysis (p. 274).

Program 23: Dispersion and correlation matrices

This program operates on a primary data matrix composed of N rows and M columns. Each row represents a set of measurements made on each of M variables and the matrix which will be formed expresses the various possible associations between these variables. It will therefore be a real symmetric matrix of order M. To understand this, consider the primary data matrix.

$$\begin{pmatrix} a_{11} & a_{12} & a_{13} & \cdots & a_{1m} \\ a_{21} & a_{22} & a_{23} & \cdots & a_{2m} \\ a_{31} & a_{32} & a_{33} & \cdots & a_{3m} \\ \vdots & \vdots & \vdots & & \vdots \\ a_{n1} & a_{n2} & a_{n3} & \cdots & a_{nm} \end{pmatrix}$$

and suppose that we wish to compute from this the matrix **R** of correlations between the m variables. To compute r_{12}, the correlation coefficient between the first and second variables, we would use the n observations in the first and the second columns. Similarly r_{13}, expressing the correlation between the first and third variables would use the observations in the first and third columns, and so on. The matrix **R** would therefore have the form:

$$\begin{pmatrix} r_{11} & r_{22} & r_{13} & \cdots & r_{1m} \\ r_{21} & r_{22} & r_{23} & \cdots & r_{2m} \\ r_{31} & r_{32} & r_{33} & \cdots & r_{3m} \\ \vdots & \vdots & \vdots & & \vdots \\ r_{m1} & r_{m2} & r_{m3} & \cdots & r_{mm} \end{pmatrix}$$

The coefficients in the main diagonal $(r_{11}, r_{22}, r_{33}, \ldots r_{mm})$ must all be 1.0 since they represent the correlation between a variable and itself. The matrix **R** is symmetric because r_{ij}, the correlation between the i-th and j-th variable, must obviously be the same as r_{ji}, the correlation between the j-th and the i-th.

In fact the program computes up to four matrices in succession, all of them stored one after the other in the array R. These matrices are: (a) that in which the elements are the uncorrected sums of squares or cross-products of the primary data, the main diagonal elements being sums of squares and the off-diagonal elements the sums of cross-products; (b) the corrected or deviation sums of squares and cross products; (c) the dispersion matrix, in which the main diagonal elements represent the variances of the first, second, third, etc., variables while the off-diagonal elements are covariances (e.g. $r_{13} = r_{31}$ = covariance of first and third variables). (d) the correlation matrix. The first matrix, referred to in (a) above, is formed as the result of an initial accumulation of sums, the other three are each formed from the previous matrix on principles similar to those outlined for Program 2. The user selects which of these four matrices he requires.

In order to economise in storage space—always at a premium in large multivariate analyses—the primary data matrix is never accommodated in core storage as a whole. Instead, each row will be read in separately as a linear array X comprising only M locations and the uncorrected sums of squares or cross-products will be accumulated row by row. The same M locations are therefore used over and over again (N times, in fact) and there is no limit on the number of rows that can be read in succession. A very large primary data matrix—larger than the computer's core-storage capacity—can therefore be dealt with using only M locations. The only serious storage limitation is that imposed by the M × M matrix R which is formed by the subroutine.

A further gain in efficiency is made possible by the fact that the matrices are all symmetric. It is therefore enough to compute the main diagonal and upper off-diagonal elements. Sequences like

DO 4 I = 1, M
DO 4 J = I, M
4 ......

ensure that only the half-matrices are computed and printed out. At the end of the subroutine, however, the other off-diagonal elements are inserted by the simple assignment procedure:

DO 13 I = 1,M
DO 13 J = I,M
13 R(J,I) = R(I,J)

so that when matrix R is returned to the calling program it contains both sets of off-diagonal elements (as it must if it is to be operated on by subsequent program-segments). When a large symmetric matrix is computed in this way the number of computational steps is almost halved: $m(m + 1)/2$ instead of m^2 elements are computed.

The subroutine, then, begins with the name SUBROUTINE CORMAT and a DIMENSION statement in respect of:

R   the variable name under which some or all of the four successive matrices will appear.

X   the successive row-vectors that make up the primary data matrix.

SX   the vector of sums of each of the M columns of the primary data matrix.

SDEV   the corresponding vector of standard deviations.

The first data card to be read gives the values of M and N for the primary data matrix and of INDEX. The latter is set at 1, 2 or 3 by the user, according to whether he requires the deviation squares and cross-products matrix, the dispersion matrix or the correlation matrix. The next segment initialises at zero the arrays SX and R, which will be used to accumulate sums, as well as NC which will be used to count the number of subsequent rows of data read.

After initialisation the first row of the data matrix is read by an implied DO-loop under the FORMAT specification 10F7.2. The number of cards on which this first row of data is punched will, of course, depend on the size of M. The DO-loops ending on 3 then begin to accumulate the sums for each variable in SX and the sums of squares and cross-products in R. Once the outer of these two nested loops is satisfied the counter NC is increased by 1 and compared with N. So long as further rows of the primary data matrix remain to be read, the sequence beginning again with statement 2 will continue. When this basic process of accumulation is complete (i.e. all rows of the data matrix have been read and processed) the array R consists of the uncorrected sums of squares and cross-products. That is, it now represents the first of the four successive matrices referred to above.

Array R is now converted into the corrected or deviation squares and cross-products matrix by the two nested DO-loops ending on the essential arithmetic statement

$$4 R(I,J) = R(I,J) - SX(I)*SX(J)/AN$$

where AN is the real equivalent of N. It will be recognised that these loops are the Fortran equivalent of the expression

$$s = \Sigma x^2 - (\Sigma x)^2/n$$

normally used to compute corrected sums of squares and cross-products but now applied to all elements of the matrix. If INDEX were set originally at 1 the deviation squares and cross products half-matrix will be printed out, the lower off-diagonal elements stored and the full matrix returned to the calling program.

If INDEX is 2 or 3, however, control passes to statement 5 and the dispersion matrix is calculated, by dividing each element of R by (AN − 1.0). The main diagonal elements of R are now the variances of the first, second,

Flowchart for Program 23

Dispersion and correlation matrices

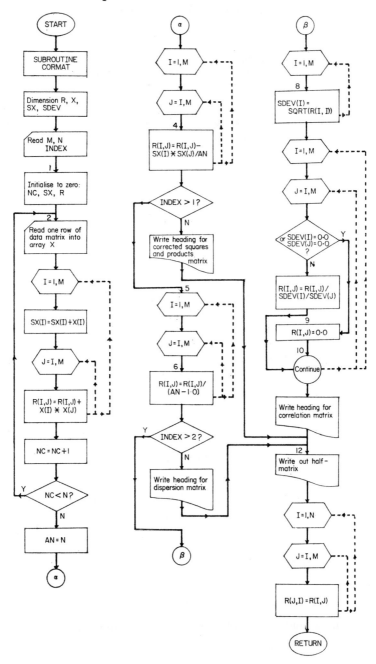

third, ... variables and the off-diagonal elements are covariances. If INDEX = 2 the dispersion half-matrix is printed out and the complete matrix returned in R. If, however, INDEX = 3 then the dispersion half-matrix is instead converted into a correlation half-matrix by dividing each element by the product of the two relevant standard deviations in the usual way. The correlation half-matrix is then printed out and the full matrix returned in R. It must, however, be recognised that occasional atypical data could result in zero values for a standard deviation. Division by zero is not allowed in any mathematical process and some computers cease execution when instructed to perform it. To guard against this the DO-loops ending on 10 contain a test for zero standard deviations; if these occur the relevant element of R is set to zero and execution proceeds normally.

The reader who has followed the logical structure of this program will see that it is based essentially on the same operations as occurred in Program 2 except that each correlation coefficient or its preceding covariance or sum of squares is computed as an element in the matrix R. The method of reading in the data matrix and operating on it row by row is, of course, a deliberate attempt to enable the program to be used with large data matrices. An alternative kind of program in which the whole data matrix is stored in core or on tape is also possible and an adaptation of SUBROUTINE CORMAT to allow taped data to be used is discussed in Program 30.

It sometimes happens that there is a need for the dispersion matrix to be computed from logarithmically transformed data. This or any other transformation can easily be accomplished by a DO-loop such as the following, which would be inserted into CORMAT immediately after statement 101:

DO 25 I = 1,M
25 X(I) = ALOG10 (X(I))

Program 23

Dispersion and correlation matrices. Data are read and computation carried out by SUBROUTINE CORMAT, for which a short calling program must be provided by the user. An example of output from SUBROUTINE CORMAT is given under Program 28 (p. 274).

```
C
      SUBROUTINE CORMAT
C
C.................................................................
C
C  THIS SUBROUTINE COMPUTES AND PRINTS A DEVIATION SUMS OF SQUARES AND
C  CROSS-PRODUCTS MATRIX OR A DISPERSION (VARIANCE-COVARIANCE) MATRIX
C  OR A CORRELATION MATRIX. PARAMETERS ARE AS FOLLOWS -
C      M    = NUMBER OF COLUMNS IN PRIMARY DATA MATRIX
C      N    = NUMBER OF ROWS IN PRIMARY DATA MATRIX
```

Program 23 (*continued*)

```
C       INDEX = 1 FOR DEVIATION SUMS OF SQUARES AND CROSS-PRODUCTS MATRIX
C               2 FOR DISPERSION MATRIX
C               3 FOR CORRELATION MATRIX
C       X     = VECTOR (LENGTH M) REPRESENTING ONE ROW OF DATA MATRIX
C       R     = OUTPUT MATRIX
C
C.................................................................
C
        DIMENSION R(50,50), X(50), SX(50), SDEV(50)
C
        READ (5,100) M, N, INDEX
    100 FORMAT (3I5)
C
C   INITIALISE FOR SUBSEQUENT ACCUMULATION OF SUMS.
        NC = 0
        DO 1 I = 1,M
        SX(I) = 0.0
        DO 1 J = 1,M
      1 R(I,J) = 0.0
C
C   READ PRIMARY DATA ONE ROW AT A TIME.
      2 READ (5,101) (X(I), I = 1,M)
    101 FORMAT (10F7.0)
C
C   COMPUTE SQUARES AND CROSS-PRODUCTS HALF-MATRIX.
        DO 3 I = 1,M
        SX(I) = SX(I) + X(I)
        DO 3 J = I,M
      3 R(I,J) = R(I,J) + X(I) * X(J)
        NC = NC + 1
        IF (NC .LT. N) GO TO 2
C
C   COMPUTE AND PRINT DEVIATION SQUARES AND CROSS-PRODUCTS HALF-MATRIX
        AN = N
        DO 4 I = 1,M
        DO 4 J = I,M
      4 R(I,J) = R(I,J) - SX(I) * SX(J)/AN
        IF(INDEX .GT. 1) GO TO 5
        WRITE (6,102)
    102 FORMAT (1H1, 23X, 42HDEVIATION SQUARES AND PRODUCTS HALF-MATRIX)
        GO TO 11
C
C   COMPUTE AND PRINT DISPERSION HALF-MATRIX.
      5 DO 6 I = 1,M
        DO 6 J = I,M
      6 R(I,J) = R(I,J)/(AN - 1.0)
        IF (INDEX.GT. 2) GO TO 7
        WRITE (6,105)
    105 FORMAT (1H1, 35X, 22HDISPERSION HALF-MATRIX)
        GO TO 11
C
C   COMPUTE AND PRINT CORRELATION HALF-MATRIX.
      7 DO 8 I = 1,M
      8 SDEV(I) = SQRT(R(I,I))
        DO 10 I = 1,M
        DO 10 J = I,M
        IF (SDEV(I) .EQ. 0.0 .OR. SDEV(J) .EQ. 0.0) GO TO 9
        R(I,J) = R(I,J) / SDEV(I) / SDEV(J)
        GO TO 10
      9 R(I,J) = 0.0
     10 CONTINUE
```

Program 23 (*continued*)

```
      WRITE (6,108)
  108 FORMAT (1H1, 35X, 23HCORRELATION HALF–MATRIX)
   11 DO 12 I = 1,M
      WRITE (6,109) I
  109 FORMAT (/1X, 3HROW, I3)
   12 WRITE (6,110) (J, R(I,J), J = I,M)
  110 FORMAT (5(2X, I3, 1PE15.7))
      DO 13 I = 1,M
      DO 13 J = I,M
   13 R(J,I) = R(I,J)
      RETURN
      END
```

Chapter 9

Further Matrix Methods

The matrix methods outlined in the previous chapter are relatively simple to follow and require little mathematical thought. There are, however, several rather more advanced ideas which are likely to be encountered by the biologist who is interested in multivariate analysis as well as in other connections. These concern three main topics: (a) the evaluation of the determinant of a large matrix; (b) the inversion of a matrix; and (c) the computation of the latent roots and vectors of a matrix. These subjects will be dealt with from an elementary standpoint and with special reference to their use in the multivariate techniques described in Chapter 10.

Determinants

The *determinant* of a matrix is a scalar quantity (i.e. a single number such as 5.38 or -2.1 or 0.0) formed by summing certain products of the elements of the matrix according to definite rules. For example, the determinant of the 2×2 matrix

$$\begin{pmatrix} a & b \\ c & d \end{pmatrix}$$

s equal to $ad - bc$. Or, to use a numerical illustration, the determinant of

$$\begin{pmatrix} 2 & 5 \\ 3 & 1 \end{pmatrix}$$

is $2 - 15 = -13$. A determinant can be evaluated for square matrices only and has no meaning for rectangular matrices. The determinant of a matrix **A** is written as det **A** or as |**A**| or by writing out the whole matrix enclosed between two vertical lines:

$$|\mathbf{A}| = \begin{vmatrix} 7 & 5 \\ 1 & 3 \end{vmatrix} = -13.$$

The determinant of a 3×3 matrix requires a little more thought. It can be evaluated in terms of the second-order determinants which it contains. For example

$$|\mathbf{X}| = \begin{vmatrix} a & b & c \\ d & e & f \\ g & h & i \end{vmatrix}$$

$$= a \begin{vmatrix} e & f \\ h & i \end{vmatrix} - b \begin{vmatrix} d & f \\ g & i \end{vmatrix} + c \begin{vmatrix} d & e \\ g & h \end{vmatrix}$$

$$= a(ei - fh) - b(di - fg) + c(dh - eg). \qquad (1)$$

The second-order determinants in the last expression are known as the *minors* of a, b and c. They are obtained as those elements of **X** which are left after crossing out in turn the row and column containing each of the elements in any one row or column of the original matrix. In the above case we chose to consider the first row, comprising a, b and c. Taking first a and crossing out the row and column containing it we are left with the minor

$$\begin{vmatrix} e & f \\ h & i \end{vmatrix}.$$

Similarly, taking b and crossing out the first row and second column we are left with the minor

$$\begin{vmatrix} d & f \\ g & i \end{vmatrix}$$

and similarly for c with its minor

$$\begin{vmatrix} d & e \\ g & h \end{vmatrix}$$

The alternating positive and negative signs in expression (1) are obtained by a rule which says that the minor of any element x_{ij} is multiplied by $(-1)^{i+j}$. Thus, in **X** element a is x_{11} and so its minor is multiplied by $(-1)^2 = 1$, while the minor of $b(x_{12})$ is multiplied by $(-1)^3 = -1$. The minors with the appropriate signs attached are the *cofactors* of the determinant.

To exemplify this procedure we may evaluate the determinant of the 3×3 matrix

$$\begin{pmatrix} 2 & 1 & 3 \\ 4 & 1 & 2 \\ 2 & 3 & 1 \end{pmatrix}.$$

This is given by

$$2\begin{vmatrix} 1 & 2 \\ 3 & 1 \end{vmatrix} - 1\begin{vmatrix} 4 & 2 \\ 2 & 1 \end{vmatrix} + 3\begin{vmatrix} 4 & 1 \\ 2 & 3 \end{vmatrix} = 2(1-6) - 1(4-4) + 3(12-2)$$

$$= -10 - 0 + 30$$

$$= 20$$

If we apply the principle just given for a 3×3 matrix to the case of a second-order determinant we shall see that it works perfectly well. Thus in evaluating

$$\begin{vmatrix} a & b \\ c & d \end{vmatrix}$$

the cofactor of a is d and the cofactor of b is $-c$ so that the determinant is $ad - bc$ as originally stated.

This method of expansion by minors is, in fact, a perfectly valid procedure for computing a determinant of any order. Unfortunately, as we saw for the 3×3 matrix, each step involves evaluating a set of the next lowest order determinants so that the whole procedure becomes impossibly tedious for a large matrix. Even a large electronic digital computer would be quite unable to evaluate high-order determinants in this way. It is therefore necessary to introduce an alternative method of computation if we are to produce a practical program, but before doing so it is worth indicating a few of the ways in which determinants enter into biological statistics. The significance tests used in conjunction with the method of principal components (p. 291) require the determinant of a variance–covariance matrix or of a correlation matrix. If a multiple discriminant analysis is to be carried out (p. 300) one should first ascertain whether the within-group dispersion matrices are equal and to perform such a test the determinants of the matrices are needed. Again, in certain circumstances a matrix may have a zero determinant; such matrices, as we shall soon see, cannot be inverted and may therefore be unsuitable for certain purposes. Even when a matrix has a non-zero determinant, this may be very small and such a condition may indicate that the matrix possesses undesirable properties which affect the accuracy of quantities computed from it (see, for example, p. 463). Thus although determinants may have no direct biological significance in themselves, and may therefore appear as unnecessarily abstract mathematical concepts, they are involved in many important operations of matrix algebra and a program to evaluate a determinant will be developed below.

Program 24: Determinant of a matrix

The first principle which we use to evaluate the determinant is that if a matrix is in triangular form (i.e. has one set of off-diagonal elements all equal to zero) the determinant is the continued product of the main diagonal elements. For example,

$$\begin{vmatrix} 3 & 9 & 8 \\ 0 & 5 & 2 \\ 0 & 0 & 1 \end{vmatrix} = 3 \times 5 \times 1 = 15.$$

The second principle is that any matrix can be reduced to triangular form by certain 'elementary row and column operations' which do not affect the value of the determinant. In particular, we can change a row (or column) by adding to it any multiple of another row (or column). Thus, consider the matrix **A**, for which we have already found $|\mathbf{A}| = 20$ through expansion by minors (p. 241).

$$\mathbf{A} = \begin{pmatrix} 2 & 1 & 3 \\ 4 & 1 & 2 \\ 2 & 3 & 1 \end{pmatrix}$$

Our object is to replace the elements below the main diagonal by zeros. We first change the second row by subtracting from it twice the first row:

$$\begin{matrix} 2 & 1 & 3 \\ 0 & -1 & -4 \\ 2 & 3 & 1 \end{matrix}$$

We then change the third row by subtracting the first row from it:

$$\begin{matrix} 2 & 1 & 3 \\ 0 & -1 & -4 \\ 0 & 2 & -2 \end{matrix}$$

We finally change the third row again by adding to it twice the second row:

$$\begin{matrix} 2 & 1 & 3 \\ 0 & -1 & -4 \\ 0 & 0 & -10 \end{matrix}$$

The matrix is now in triangular form so that det $\mathbf{A} = 2 \times -1 \times -10 = 20$ as we had previously shown.

Put more generally, the method involves the following steps:

(a) Multiply each element of the first row by a_{21}/a_{11} and subtract from the corresponding element of the second row, thus replacing a_{21} by a zero.

(b) Repeat using a_{31}/a_{11}, thus replacing a_{31} by zero, and so on until the first column of the matrix consists entirely of zeros except for a_{11}—the *pivot*—which retains its original value.

(c) Now consider the smaller matrix formed by excluding the first row and column of $\mathbf{A}$. The new pivot is the element originally denoted as a_{22}. Using this pivot repeat steps (a) and (b) so as to put zeros in all positions of the second column of $\mathbf{A}$ below the main diagonal.

(d) Continue, using a_{33}, a_{44} ... as pivots until all elements of the lower triangle are zeros.

(e) Form the determinant as the continued product of the main diagonal elements.

These operations can very easily be programmed, but certain precautions are necessary. Since many divisions are carried out by each pivot it is very desirable that these should not be small numbers, division by which would cause serious loss of accuracy (p. 462). This danger can be overcome by preliminary row and column exchanges which ensure that the absolutely largest numbers in the matrix being reduced are always selected as pivots. Each time a row or column is interchanged with another row or column the sign of the determinant changes though its absolute value remains the same. We can allow for this either by recording the number of row and column exchanges or by altering the sign of the elements in the row which will contain the new pivot. Another potential cause of accuracy-loss concerns subtractions which instead of yielding a true zero value will, through round-off errors, produce very small positive or negative numbers. It can be arranged that these are set equal to zero if the difference is below some very small tolerance. If all this sounds very complicated, the reader should find it becomes clearer when he has considered the program in detail.

The program itself is written as a function subprogram, since it returns only a single value which is associated with the name of the program—FUNCTION DET (A, N) where A is the input matrix and N its order. Note that A is destroyed in the course of the computations, which means that if subsequent use is to be made of it by the calling program it must be copied before the first use of the function DET on the right-hand side of any assignment statement in the calling program. A is the only dimensioned array in the program and we begin by defining an indexing parameter NM1, which is one less than the order of matrix A. This is then used to define the upper limit of the big outer DO-loop which ends on 7 and embraces most of the instructions in the program. This loop is responsible for selecting and using the pivots and works, so to speak, along the main diagonal up to but not including the element A(N,N).

Within the main loop the first operation consists of selecting the element of largest absolute value as the pivot. On the first pass through the loop this will, of course, be the largest element in the matrix A. On each subsequent pass it will be the largest element in the reduced matrix formed by excluding the row and column in which the previous pivot occurred. The selection of a pivot-value proceeds on the usual lines for finding a maximum element (see, for example, Program 7) using IM and JM to hold its row and column subscripts. This is done by the nested DO-loops ending on 1 and if, after they are satisfied, it transpires that the original or reduced matrix contains no value greater than 0.00001, then for reasons of numerical accuracy the determinant is set at zero and control returns to the main pcogram.

Assuming, however, that a maximum element exceeding this very small value can be found, we next carry out a row and column interchange to bring this element into the main diagonal where it can function as a pivot. Statements 2 and 4 cater for the possibility that it happens already to be in the correct position, otherwise the operation of the DO-loops ending on 3 and 5 serve to interchange the elements of the rows and columns respectively. This interchange again makes use of a holding location TEMP but also incorporates a change of sign which will ensure that the determinant ultimately computed is of the correct sign.

Having selected an element as a pivot and moved it to the correct position we are now able to begin or continue reducing the matrix to triangular form. The two nested DO-loops ending on 7 help to achieve this by carrying out the multiplication and subtraction that puts a zero in position in the pivot column below the pivot itself. The inner loop computes DIFF, whose successive values make up the reduced row. It is at this point that the second accuracy consideration of p. 243 is implemented. If the absolute value of DIFF is equal to or less than 1/100,000 of the element being reduced (i.e. if five or more significant digits have been lost through the subtraction) the result is set to zero. One could, of course, vary this tolerance according to the circumstances under which the program was being used.

When the outermost DO-loop ending on 7 is satisfied the matrix is in triangular form and the determinant is easily evaluated as the continued product of the main diagonal elements. The program shows three points of special interest. The first concerns the choice of algorithm. It can be shown that to compute the determinant of an $n \times m$ matrix through expansion by minors requires between $2n!$ and $n.n!$ multiplications. Pennington (1970)

Flowchart for Program 24

Computation of a determinant.

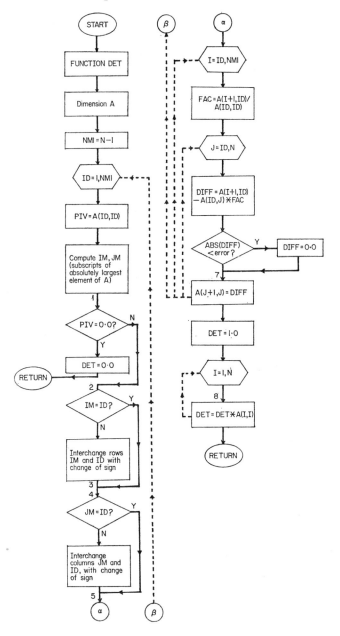

246   FURTHER MATRIX METHODS

points out that a computer with a multiplication time of 10 μsec would, using this algorithm, take about a hundred (American) billion years to evaluate a 20×20 determinant. On the other hand, the method of pivotal condensation used in Program 24 requires about $n^3/3$ multiplications, so that a 20×20 determinant can be evaluated in a fraction of a second! The second point of interest concerns the incorporation of safeguards against accuracy loss. In this program the safeguards account for about two-thirds of the Fortran instructions and emphasise the role that simple numerical analysis has to play in programming. Finally, the choice of the FUNCTION subprogram form should encourage the reader to revise the Fortran rules governing this kind of program (p. 69).

Program 24

Computation of a determinant: the listing comprises Function Sub-program DET and a short calling program which illustrates how it may be used.

```
C
C   THIS PROGRAM CONTROLS I/O AND USES THE FUNCTION SUBPROGRAM DET TO
C   COMPUTE THE DETERMINANT OF THE MATRIX X OF ORDER M.
C
      DIMENSION X(50,50)
      WRITE (6,99)
   99 FORMAT (1H1, 30X, 26HEVALUATION OF DETERMINANTS///)
    1 READ (5,100) NPROB, M
  100 FORMAT (2I5)
      IF (NPROB .EQ. 0) STOP
      DO 2 I = 1,M
    2 READ (5,101) (X(I,J), J = 1,M)
  101 FORMAT (10F7.0)
      D = DET(X,M)
      WRITE (6,102) NPROB, D
  102 FORMAT (/11X, 11HDETERMINANT, I4, 2H = , 1PE15.7)
      GO TO 1
      END


C
      FUNCTION DET(A,N)
C
C.............................................................
C
C   THIS FUNCTION SUBPROGRAM COMPUTES THE DETERMINANT (DET) OF
C   MATRIX A, OF ORDER N.  N.B. MATRIX A IS DESTROYED.
C
C.............................................................
C
      DIMENSION A(50,50)
      NM1 = N - 1
      DO 7 ID = 1,NM1
C
C   SELECT ELEMENT OF MAXIMUM ABSOLUTE VALUE AS PIVOT.
      PIV = A(ID,ID)
      IM = ID
```

Program 24 (*continued*)

```
        JM = ID
        DO 1 I = ID,N
        DO 1 J = ID,N
        IF (ABS(PIV) .GE. ABS(A(I,J))) GO TO 1
        PIV = A(I,J)
        IM = I
        JM = J
    1   CONTINUE
        IF (ABS(PIV) .GE. 0.00001) GO TO 2
        DET = 0.0
        RETURN
C
C   INTERCHANGE ROWS AND COLUMNS, BRINGING PIV TO MAIN DIAGONAL.
    2   IF (IM .EQ. ID) GO TO 4
        DO 3 I = ID,N
        TEMP = - A(IM,I)
        A(IM, I) = A(ID,I)
    3   A(ID,I) = TEMP
    4   IF (JM .EQ. ID) GO TO 6
        DO 5 I = ID,N
        TEMP = - A(I,JM)
        A(I,JM) = A(I,ID)
    5   A(I,ID) = TEMP
C
C   REDUCE MATRIX TO TRIANGULAR FORM.
    6   DO 7 I = ID,NM1
        FAC = A(I+1,ID) / A(ID,ID)
        DO 7 J = ID,N
        DIFF = A(I+1,J) - A(ID,J) * FAC
        IF (ABS(DIFF) .LE. 0.00001 * ABS(A(I+1,J))) DIFF = 0.0
    7   A(I+1,J) = DIFF
C
C   COMPUTE DETERMINANT.
        DET = 1.0
        DO 8 I = 1,N
    8   DET = DET * A(I,I)
        RETURN
        END
```

Matrix Inversion

In Chapter 8 we considered some matrix operations such as addition, subtraction and multiplication, but we said nothing of division. Just as multiplication of matrices needs to be specially defined, and does not follow obviously from the rules for scalar multiplication, so 'division' of matrices needs special attention. If we think for a moment of the ordinary scalar division of a by b we might write this as $a \div b = c$, but we could equally well express it as $1/b \times a = c$. In the latter form we have in effect replaced the process of dividing a by b by the exactly equivalent operation of multiplying a by the reciprocal of b. This is the key to the corresponding matrix operation. If we wish to 'divide' matrix **A** by matrix **B**, what we do is indicated by the matrix equation $\mathbf{B}^{-1}\mathbf{A} = \mathbf{C}$. That is, we multiply **A** by the reciprocal or *inverse* of B. This inverse, denoted by $\mathbf{D}^{-1}$ is itself a matrix and therefore, as we saw in Chapter 8, it must be conformable for multiplication by **A**. We can pursue further the parallel between the inverse matrix and the reciprocal of a scalar. If we multiply any scalar b by its reciprocal we always get the answer 1. Now the matrix equivalent of 1 is the *identity matrix* **I**, consisting of unities in the main diagonal and zeros elsewhere. It is not surprising to find, therefore, that if we multiply **B** by its inverse $\mathbf{B}^{-1}$ we obtain the identity matrix, i.e.

$$\mathbf{B} \cdot \mathbf{B}^{-1} = \mathbf{I}. \tag{2}$$

Equally, $\mathbf{B}^{-1}\mathbf{B} = \mathbf{I}$, i.e. the result of pre- or post-multiplying any matrix by its inverse is an identity matrix of the same order. We can therefore define the inverse of a matrix **B** more precisely by saying that it is another matrix, denoted by $\mathbf{B}^{-1}$, whose product with **B** is the identity matrix. It will be seen that **B** and $\mathbf{B}^{-1}$ must always be conformable for multiplication whichever is the premultiplier and that **B** and $\mathbf{B}^{-1}$ must therefore always be square and of the same order.

Perhaps a simple example, showing how an inverse matrix can be used to solve a set of simultaneous equations will help to put the subject in a more concrete form. If we had a simple equation involving only scalar quantities $bx = a$ we could solve it for x by writing $x = a/b$ i.e. by dividing a by b or, what comes to the same thing, by multiplying a by the reciprocal of b, always assuming $b \neq 0$. Now the matrix equation

$$\mathbf{Bx} = \mathbf{a}$$

is simply a short-hand form of the following set of simultaneous equations:

$$\begin{aligned} b_{11}x_1 + b_{12}x_2 + \ldots + b_{1n}x_n &= a_1 \\ b_{21}x_1 + b_{22}x_2 + \ldots + b_{2n}x_n &= a_2 \\ &\vdots \\ b_{n1}x_1 + b_{n2}x_2 + \ldots + b_{nn}x_n &= a_n. \end{aligned}$$

In other words, **x** represents the column vector $(x_1\ x_2\ x_3 \ldots x_n)$, **a** represents the column vector $(a_1\ a_2\ a_3 \ldots a_n)$ and **B** is the matrix of coefficients of **x**. A simple numerical example might be:

$$\begin{aligned} 3x_1 + 2x_2 + x_3 &= 4 \\ 4x_1 + 7x_2 - 2x_3 &= 5 \\ x_1 + 9x_2 + 4x_3 &= 6 \end{aligned}$$

where

$$\mathbf{B} = \begin{pmatrix} 3 & 2 & 1 \\ 4 & 7 & -2 \\ 1 & 9 & 4 \end{pmatrix}$$

and
$$\mathbf{a} = \begin{pmatrix} 4 \\ 5 \\ 6 \end{pmatrix}.$$

Now just as we solved $bx = a$ by writing $x = 1/b \times a$ so we can solve $\mathbf{Bx} = \mathbf{a}$ by expressing it in the form $\mathbf{x} = \mathbf{B}^{-1} \mathbf{a}$. In other words, the numerical values of $x_1\, x_2\, x_3\, \ldots\, x_n$ which satisfy the set of equations can be found by premultiplying the vector $\mathbf{a}$ by the *inverse* of the coefficient matrix $\mathbf{B}$.

The question arises, then, how can we calculate $\mathbf{B}^{-1}$ from $\mathbf{B}$—i.e. how do we invert $\mathbf{B}$? In fact a prior question should have been, how do we know that $\mathbf{B}^{-1}$ exists? A little consideration will show that an inverse matrix need not always exist for we could, say, write the following simultaneous equations:

$$2x_1 + 3x_2 + 4x_3 = 6$$
$$4x_1 + 6x_2 + 8x_3 = 12$$
$$6x_1 + 9x_2 + 12x_3 = 18$$

in which the second and third equations are simply multiples of the first and cannot be used in conjunction with it to obtain unique numerical solutions for x_1, x_2 and x_3. Clearly in this case the coefficient matrix

$$\mathbf{S} = \begin{pmatrix} 2 & 3 & 4 \\ 4 & 6 & 8 \\ 6 & 9 & 12 \end{pmatrix}$$

does not have an inverse. The position is that a matrix has an inverse if its determinant is not equal to zero. Such a matrix is said to be *non-singular*. Matrices with a zero determinant are called *singular* matrices and have no inverse. If the reader cares to evaluate the determinant of the above matrix $\mathbf{S}$ by the method of cofactors (p. 240) he will find that it is zero:

$$\mathbf{S} = 2 \begin{vmatrix} 6 & 8 \\ 9 & 12 \end{vmatrix} - 3 \begin{vmatrix} 4 & 8 \\ 6 & 12 \end{vmatrix} + 4 \begin{vmatrix} 4 & 6 \\ 6 & 9 \end{vmatrix}$$
$$= (2 \times 0) - (3 \times 0) + (4 \times 0)$$
$$= 0.$$

If we know that an inverse exists then one method of finding it is by the use of cofactors. This consists of the following steps:

(i) Starting with a matrix $\mathbf{B}$, we consider each element in turn and write down its cofactor (p. 240), so obtaining a matrix of cofactors which we shall call $\mathbf{C}$.

(ii) We then transpose **C**, giving **C′**, and

(iii) We multiply **C′** by the scalar quantity $1/|\mathbf{B}|$.

The resulting matrix is the required inverse, $\mathbf{B}^{-1}$. As a numerical example consider the matrix

$$\mathbf{B} = \begin{pmatrix} 1 & 2 & 3 \\ 4 & 5 & 6 \\ 7 & 8 & 10 \end{pmatrix}.$$

The first element 1 has as its cofactor

$$\begin{vmatrix} 5 & 6 \\ 8 & 10 \end{vmatrix} = 2.$$

The next element 2 has as a cofactor

$$-\begin{vmatrix} 4 & 6 \\ 7 & 10 \end{vmatrix} = 2$$

and so on until we have the complete matrix of cofactors:

$$\mathbf{C} = \begin{pmatrix} 2 & 2 & -3 \\ 4 & -11 & 6 \\ -3 & 6 & -3 \end{pmatrix}.$$

from which we derive the transpose:

$$\mathbf{C'} = \begin{pmatrix} 2 & 4 & -3 \\ 2 & -11 & 6 \\ -3 & 6 & -3 \end{pmatrix}$$

The determinant $|\mathbf{B}|$ will be found to be -3, whence

$$\mathbf{B}^{-1} = -\frac{1}{3}\begin{pmatrix} 2 & 4 & -3 \\ 2 & -11 & 6 \\ -3 & 6 & -3 \end{pmatrix} = \begin{pmatrix} -2/3 & -4/3 & 1 \\ -2/3 & 11/3 & -2 \\ 1 & -2 & 1 \end{pmatrix}.$$

If the reader cares to multiply this last matrix by the original matrix **B**, he will find that the product is an identity matrix, thus verifying the correctness of the calculations.

In fact the above method is not the one used when inverting a matrix by computer, since the evaluation of cofactors in a large matrix requires a very large number of multiplications. Before discussing a more suitable technique, for which a Fortran program is provided, it is worth indicating some of the uses to which matrix inversion programs are put in statistical and biometric work. As we saw above (p. 248) an inverse matrix immediately provides the solution to a set of simultaneous equations and though there are usually somewhat more efficient methods of obtaining such a solution, circumstances can occur in which the inversion procedure is desirable. Inverse matrices appear frequently when simplifying expressions in matrix algebra and may—to take only one example—find a use in evaluating the determinant of a very large matrix which has to be partitioned—i.e. split into, say, four smaller matrices. An inverse matrix also occurs in the computation of the so-called quadratic form—a scalar quantity given by the expression

$$q = \mathbf{u}' \mathbf{A}^{-1} \mathbf{u} \qquad (3)$$

where $\mathbf{u}'$ and $\mathbf{u}$ are the row and column forms of a vector and $\mathbf{A}^{-1}$ an inverse matrix. Such forms often arise in matrix algebra. In more purely statistical contexts, one of the most frequent needs for a program of matrix inversion arises in the calculation of a multiple regression equation (p. 274) in the course of which a correlation matrix has to be inverted. Some methods of multivariate analysis also require matrix inversion routines, as may be the case in multiple discriminant analysis and in the computation of factor scores in factor analysis.

Examples like the above indicate the desirability of having available a matrix inversion subroutine and we must therefore outline the principles on which an algorithm suitable for computer implementation may be constructed.

Program 25: Matrix inversion

The numerical technique on which this program is based is known as the Gauss–Jordan or exchange method, and can be programmed so as to require storage space for only the matrix to be inverted and two working vectors. Much of the computation bears a close resemblance to Program 24 and as in that program the input matrix will be destroyed. It will, however, be replaced in the same locations by the required inverse. Should the original matrix and its inverse both be needed subsequently we must copy the original in the main program before calling this subprogram.

Since the subprogram returns a matrix and not a single scalar quantity, we must plan it as a subroutine subprogram, denoted here as SUBROUTINE INVERT with the following dummy arguments which we shall transmit explicitly:

A = matrix to be inverted; on return A will be the required inverse
N = order of A.

DET = determinant of input matrix A. This is conveniently computed as a by-product of the inversion process and obviates the need to call our FUNCTION DET subprogram if, as often happens, both the inverse and the determinant are needed later.

IR = a working vector used to keep a record of the row interchanges made during part of the program.

IC = another working vector used similarly to record column interchanges.

The DIMENSION statement includes A, IR and IC and we start by initialising DET at 1.0 (since it will ultimately yield the determinant as a continued product) while NI is a set at zero. NI will be used to record the total number of row and column interchanges and thus eventually to provide the correct sign for DET (a necessary alternative to the method used in Program 24). We then begin the main loop that ends on statement 10 and includes within its range the greater part of the subroutine.

As in Program 24 this main loop is intended to operate with a series of pivots which must be identified and moved to successive main diagonal positions. The next program segment of 13 statements selects the element of maximum absolute value for use as a pivot. It is identical with the corresponding segment of Program 24 and if the potential pivot proves to be less than 0.00001 it is treated as zero and the matrix is declared singular by recording a zero determinant and returning to the main program. It this happens the array A will usually contain a partially manipulated matrix that does not correspond to $\mathbf{A}$ (nor, of course, to $\mathbf{A}^{-1}$ which does not exist under these circumstances).

Having found a suitably large element to act as the pivot we must then transfer it to the main diagonal position by row and column interchanges on the lines of Program 24. In this case, however, we must keep a record in vectors IR and IC of the rows and columns we have interchanged. The reason for this is that when, later, we carry out the inversion proper we shall find it necessary to assemble the rows and columns of the reduced matrix in an order depending on the interchanges made earlier in the interests of numerical accuracy. Each location of IR must therefore contain the number of the row in which the maximum element occurred on the corresponding pass through the outermost DO-loop. The same holds for IC in respect of columns. Thus the segment we are now considering contains (a) two logical IF statements that enable the loops ending on 3 and 5 to be bypassed if the maximum element happens already to occupy the main diagonal position; (b) the loops themselves, which bring about the row and column exchanges in the usual way; (c) a statement incrementing NI each time an exchange is made; and (d) statements 4 and 6 which record the numbers of the row and column interchanged with those originally intersecting at the main diagonal position defined by the outermost DO-loop (i.e. row ID and column ID).

Having carried out the preliminaries needed to obtain maximum accuracy we are now able to reduce the matrix. We begin with the nested DO-loops ending on 7 which, because of the logical IF they contain, operate on all elements of the matrix *except* those in the pivot row and column. The operations consists of subtracting from the element a_{ij} the quantity $(a_{ik} - a_{kj})/a_{kk}$ where a_{kk} is the pivot which has already been moved to the main diagonal and is denoted in the program as PIV (equal to the current value of A(ID,ID)). Two further operations, denoted by the DO-loops ending on 8 and 9 transform the pivot row and column by dividing each of their elements by the pivot and changing the sign of the row elements. After this we use the pivot to form a continued product (DET) with previous pivots and, as the final operation in the main DO-loop, we execute statement 10 which replaces the pivot by its reciprocal.

When this main DO-loop is satisfied we have gone much of the way towards inverting the matrix. However, we first complete our evaluation of the determinant by changing its sign if NI happens to have recorded an odd number of row or column interchanges. The IF statement incorporating a little fixed-point arithmetic does just this. Finally, to obtain the inverse we must re-arrange the rows and columns of the matrix now accommodated in array A. The rule is that if two rows were exchanged originally when bringing the pivot to the main diagonal position, we must now exchange ithe correspondingly numbered columns of A. Conversely, if two columns were originally exchanged we now exchange the correspondingly numbered rows. The original row and column interchanges were recorded in IR and IC respectively, so our last segment uses these vectors to bring about the final interchange, which must be accomplished in the reverse order to the original. Consider, therefore, the nested DO-loops ending on statement 12 which perform this interchange for the columns of the array currently stored in A. The statement $K = N - L + 1$ provides us with a subscript K which will refer to the last-recorded original row-interchange and which will work 'backwards' along the vector IR as the loop is traversed repeatedly. If the number recorded as IR(K) is the same as the location itself—i.e. IF(IR(K).EQ.K)—then there was clearly no row-interchange originally and therefore no column interchange is now needed. Control therefore passes to the 12 CONTINUE statement. Otherwise, we have an inner DO-loop ending on 11 which simply exchanges the columns corresponding to the original row-interchange recorded as IR(K). The final row-interchange using the vector IC follows a similar pattern and the array A now contains the inverse matrix we set out to obtain.

The use of pivots in this program is similar to that in Program 24, which

Flowchart for Program 25
Matrix inversion.

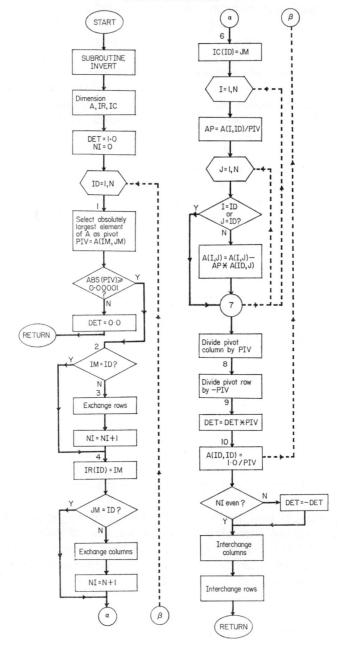

should again be followed through carefully if the reader has difficulty in understanding the present program. The rationale of this method of matrix inversion is described in text-books of numerical analysis and Orden (1960) should also be consulted. The technique requires a very much smaller number of operations than the method of cofactors described earlier, though it is not the only method suitable for computer programming. It has, however, the advantage of requiring little storage space beyond that for the original matrix. Programming it requires some facility in manipulating the elements of a matrix, many of them concerned with accuracy demands. Even so, the inversion of large matrices is always a source of accuracy-loss and there is much to be said for incorporating in the calling program a check whereby the original matrix and its inverse are multiplied and the result—which should be the identity matrix of the same order—is printed out for inspection by the user. This has not been done in the very short calling program provided.

Program 25

Matrix inversion: the inverse matrix is computed by SUBROUTINE INVERT but a short calling program is also listed to show how the subroutine may be used. Sample data and output are included.

```
C
C     THIS PROGRAM CONTROLS I/O AND CALLS SUBROUTINE INVERT.
C
      DIMENSION X(50,50), IR(50), IC(50)
   15 READ (5,100) NPROB, N
  100 FORMAT (2I5)
      IF (NPROB .EQ. 0) STOP
      WRITE (6,101) NPROB
  101 FORMAT (25H1MATRIX INVERSION PROBLEM, I4)
      DO 1 I = 1,N
    1 READ (5,102) (X(I,J), J = 1,N)
  102 FORMAT (10F7.0)
      CALL INVERT (X,N, DET, IR, IC)
      WRITE (6,103) DET
  103 FORMAT (//14H DETERMINANT = , 1PE15.7)
      IF (DET) 3,2,3
    2 WRITE (6,104)
  104 FORMAT (//39H MATRIX IS SINGULAR. NO INVERSE EXISTS)
      STOP
    3 DO 4 I = 1,N
    4 WRITE (6,105) I, (X(I,J), J = 1,N)
  105 FORMAT (// 4H ROW, I4/(1P10E12.4))
      GO TO 15
      END


C
      SUBROUTINE INVERT (A,N, DET, IR, IC)
C
```

Program 25 (*continued*)

```
C.....................................................................
C
C   THIS SUBROUTINE INVERTS A MATRIX A, OF ORDER N, USING THE
C   GAUSS-JORDAN EXCHANGE METHOD AND COMPUTES THE DETERMINANT
C   (DET). MATRIX A IS DESTROYED AND REPLACED BY ITS INVERSE.
C   IR AND IC ARE WORKING VECTORS OF LENGTH N.
C
C.....................................................................
C
      DIMENSION A(50,50), IR(50), IC(50)
C
C   INITIALISE AND BEGIN MAIN LOOP.
      DET = 1.0
      NI = 0
      DO 10 ID = 1,N
C
C   SELECT ELEMENT OF MAXIMUM ABSOLUTE VALUE AS PIVOT.
      PIV = A(ID,ID)
      IM = ID
      JM = ID
      DO 1 I = ID,N
      DO 1 J = ID,N
      IF (ABS(PIV) .GE. ABS(A(I,J))) GO TO 1
      PIV = A(I,J)
      IM = I
      JM = J
    1 CONTINUE
      IF (ABS(PIV) .GE. 0.00001) GO TO 2
      DET = 0.0
      RETURN
C
C   INTERCHANGE ROWS AND COLUMNS IF NECESSARY, BRINGING
C   PIV TO MAIN DIAGONAL. RECORD ROW AND COLUMN INTERCHANGES
C   IN IR AND IC RESPECTIVELY AND COUNT TOTAL INTERCHANGES
C   BY INCREMENTING NI.
    2 IF (IM .EQ. ID) GO TO 4
      DO 3 J = 1,N
      TEMP = A(ID,J)
      A(ID,J) = A(IM,J)
    3 A(IM,J) = TEMP
      NI = NI + 1
    4 IR(ID) = IM
      IF (JM .EQ. ID) GO TO 6
      DO 5 I = 1,N
      TEMP = A(I,ID)
      A(I,ID) = A(I,JM)
    5 A(I,JM) = TEMP
      NI = NI + 1
    6 IC(ID) = JM
C
C   REDUCE MATRIX EXCEPT FOR PIVOT ROW AND PIVOT COLUMN.
      DO 7 I = 1,N
      AP = A(I,ID) / PIV
      DO 7 J = 1,N
      IF (I .EQ. ID .OR. J .EQ. ID) GO TO 7
      A(I,J) = A(I,J) - AP * A(ID,J)
    7 CONTINUE
C
C   DIVIDE PIVOT COLUMN BY PIVOT.
      DO 8 I = 1,N
    8 A(I,ID) = A(I,ID) / PIV
```

Program 25 (*continued*)

```
C
C   DIVIDE PIVOT ROW BY MINUS PIVOT.
        DO 9 J = 1,N
      9 A(ID,J) = A(ID,J) / (-PIV)
C
C   COMPUTE DETERMINANT AND REPLACE PIVOT BY ITS RECIPROCAL.
        DET = DET * PIV
     10 A(ID,ID) = 1.0 / PIV
        IF (NI - 2*(NI/2) .GT. 0) DET = - DET
C
C   PERFORM FINAL INTERCHANGE OF COLUMNS AND ROWS.
        DO 12 L = 1,N
        K = N - L + 1
        IF (IR(K) .EQ. K) GO TO 12
        INT = IR(K)
        DO 11 I = 1,N
        TEMP = A(I,K)
        A(I,K) = A(I,INT)
     11 A(I,INT) = TEMP
     12 CONTINUE
C
        DO 14 L = 1,N
        K = N - L + 1
        IF (IC(K) .EQ. K) GO TO 14
        INT = IC(K)
        DO 13 J = 1,N
        TEMP = A(K,J)
        A(K,J) = A(INT,J)
     13 A(INT,J) = TEMP
     14 CONTINUE
        RETURN
        END
```

DATA

```
   1    4
 10.0    7.0    8.0    7.0
  7.0    5.0    6.0    5.0
  8.0    6.0   10.0    9.0
  7.0    5.0    9.0   10.0
   0
```

OUTPUT

       MATRIX INVERSION PROBLEM   1

       DETERMINANT =   1.0000063E 00

ROW 1
 2.5000E 01   -4.1000E 01   9.9999E 00   -6.0000E 00

ROW 2
-4.1000E 01   6.8000E 01   -1.7000E 01   9.9999E 00

ROW 3
 9.9999E 00   -1.7000E 01   5.0000E 00   -3.0000E 00

ROW 4
-6.0000E 00   9.9999E 00   -3.0000E 00   2.0000E 00

Latent Roots and Vectors

Many biometric and statistical applications of matrix algebra involve the concept of the latent roots and vectors of a matrix. In order to understand something of these mathematical ideas, one may begin by posing the question whether there exists, for a square matrix **A**, a scalar λ and a vector **u** such that

$$\mathbf{Au} = \lambda \mathbf{u}. \tag{4}$$

In words, is there a vector **u** such that when premultiplied by the matrix the result is equal to the product of the vector and a scalar quantity λ. If so, then the scalar λ is referred to as a latent root of the matrix **A** and the vector **u** as the corresponding latent vector. These are not the only terms used, λ sometimes being known as a characteristic root or eigenvalue of the matrix and **u** as a characteristic vector or eigenvector. The equation (1) can be rewritten as $\mathbf{Au} - \lambda \mathbf{u} = 0$ or, more usefully, in the form

$$(\mathbf{A} - \lambda \mathbf{I})\mathbf{u} = 0. \tag{5}$$

Now apart from the trivial case when $\mathbf{u} = 0$ this equation may be solved only if the determinant of the matrix $(\mathbf{A} - \lambda \mathbf{I})$ is zero, i.e. if

$$|\mathbf{A} - \lambda \mathbf{I}| = 0. \tag{6}$$

Equation (5) is known as the *characteristic equation* of the matrix and is a polynomial in λ whose degree is equal to the order of **A**.

A simple numerical example will perhaps make matters clear. Consider the 2×2 matrix

$$\mathbf{A} = \begin{pmatrix} 1 & 8 \\ 2 & 1 \end{pmatrix}.$$

If from this we subtract the matrix

$$\lambda \mathbf{I} = \begin{pmatrix} \lambda & 0 \\ 0 & \lambda \end{pmatrix}$$

we get

$$(\mathbf{A} - \lambda \mathbf{I}) = \begin{pmatrix} 1 - \lambda & 8 \\ 2 & 1 - \lambda \end{pmatrix}$$

and the determinant of the latter matrix is $(1 - \lambda)^2 - 16$. Expanding this we obtain the characteristic equation

$$\lambda^2 - 2\lambda - 15 = 0$$

which we can easily solve to give $\lambda_1 = 5$ and $\lambda_2 = -3$. These, then are the two latent roots of the original matrix **A**. Each latent root will have a latent vector associated with it. Let us call these $\mathbf{u}_1$ and $\mathbf{u}_2$, with elements $\alpha_1 \beta_1$ and $\alpha_2 \beta_2$ respectively. Considering first the latent root $\lambda_1 = 5$ we can rewrite $(\mathbf{A} - \lambda \mathbf{I})$ of equation (5) in the form:

$$\begin{pmatrix} 1 & 8 \\ 2 & 1 \end{pmatrix} - \begin{pmatrix} 5 & 0 \\ 0 & 5 \end{pmatrix} = \begin{pmatrix} -4 & 8 \\ 2 & -4 \end{pmatrix}.$$

whence (5) becomes

$$\begin{pmatrix} -4 & 8 \\ 2 & -4 \end{pmatrix} \begin{pmatrix} \alpha_1 \\ \beta_1 \end{pmatrix} = 0$$

i.e.

$$-4\alpha_1 + 8\beta_1 = 0$$
$$2\alpha_1 - 4\beta_1 = 0.$$

As in all such cases, these equations are not independent and therefore will not provide unique solutions for α_1 and β_1 but if we put $\alpha_1 = 1$ then $\beta_1 = 0.5$, so that a latent vector associated with the first latent root $\lambda_1 = 5$ is

$$\begin{pmatrix} 1 \\ 0.5 \end{pmatrix}.$$

Similarly for the second latent root $\lambda_2 = -3$ we can show that there is an associated latent vector

$$\begin{pmatrix} 1 \\ -0.5 \end{pmatrix}.$$

To sum up, therefore, the original matrix

$$\mathbf{A} = \begin{pmatrix} 1 & 8 \\ 2 & 1 \end{pmatrix}$$

has two latent roots $\lambda_1 = 5$ and $\lambda_2 = -3$, with which are respectively associated the latent vectors

$$\begin{pmatrix} 1 \\ 0.5 \end{pmatrix} \text{ and } \begin{pmatrix} 1 \\ -0.5 \end{pmatrix}.$$

It is left as a simple exercise for the reader to verify that these numerical values satisfy the equation (4) from which we started.

One point only requires further comment, namely the fact that we chose arbitrarily a value 1 as the first element of each latent vector. This, as we saw, was necessary because a latent vector of a matrix is only determined up to multiplication by an arbitrary constant. The two latent vectors we have just calculated could equally well be represented as

$$\begin{pmatrix} 2.0 \\ 1.0 \end{pmatrix} \text{ and } \begin{pmatrix} 2.0 \\ -1.0 \end{pmatrix}$$

or, for that matter, as

$$\begin{pmatrix} -3.0 \\ -1.5 \end{pmatrix} \text{ and } \begin{pmatrix} -3.0 \\ 1.5 \end{pmatrix}.$$

Any such solutions would equally well satisfy equations (4) and (5) above. In practice, therefore, it is usual to standardise the elements of a latent vector in some way. One method (used above) is to take the largest element as equal to 1.0 and scale the others accordingly, but a more usual and more useful method is to normalise the vector by arranging for the sum of the squares of its elements to be equal to 1. This procedure is further discussed below (p. 262) in connection with a FORTRAN program for computing latent roots and vectors, but it is not difficult to see that the two latent vectors just calculated could be normalised as

$$\begin{pmatrix} 2/\sqrt{5} \\ 1/\sqrt{5} \end{pmatrix} \text{ and } \begin{pmatrix} 2/\sqrt{5} \\ -1/\sqrt{5} \end{pmatrix}.$$

The preceding discussion will inevitably seem very formal to the biologist but if he has grasped the essential features of the explanation it will be possible to indicate something of the importance of latent roots and vectors in biological work. They form a major feature of various kinds of multivariate analysis, notably in principal component studies, factor analysis, multiple discriminant analysis and canonical correlation analysis, in all of which the vectors provide systems of weights associated with the original variables. Many of these multivariate techniques have assumed a special importance in recent years, being employed as methods of cluster-analysis in numerical taxonomy and in related classificatory problems such as statistical zoogeography and medical diagnosis. The existence of stable age-distributions in demographic work and in the dynamics of animal populations can also be discussed in terms of latent roots and vectors and they may arise in connection with the probability transition matrices employed in some kinds of genetical research.

Program 26: Latent roots and vectors of a real symmetric matrix

The method indicated above for obtaining the latent roots and vectors from the characteristic equation is not very suitable for computer implementation. Several other techniques are, however, available, of which the simplest is the iterative method used in this program. It is due to Hotelling (1933) and explained clearly by Lawley and Maxwell (1963). The method involves two main processes: (a) an iterative approach to the largest latent root and associated vector, followed by (b) reduction of the matrix. The sequence is then repeated to find the next largest latent root and vector, and so on. For most biological purposes it is rarely necessary to compute more than the 10 largest roots and their vectors and this is the maximum set in the present program. Some limitations of the method are discussed at the end of the program description, together with alternative techniques.

The program is written as a subroutine subprogram—SUBROUTINE LATENT—and is accompanied by a short calling program which controls input/output and is not described here. The parameters of SUBROUTINE LATENT are:

A = input matrix, which must be real symmetric and which is destroyed during computation so that it may have to be copied before LATENT is called.

N = order of A

NROOT = maximum number of latent roots and vectors required. This must not exceed 10. If, for reasons indicated below, it is not possible to extract all the latent roots required, NROOT will on return contain the number which have been extracted.

LAMBDA = vector of latent roots, extracted in descending order of size.

VEC = a matrix composed of the latent vectors arranged in not more than 10 columns with each normalised so that the sum of the squares of its elements is equal to 1.

U1, UN = working vectors which will hold successive estimates of the vectors being computed. When these agree within certain limits of error the vector has 'converged' and the program can proceed to compute the next root and vector.

The subroutine begins with a type declaration REAL LAMBDA; the latent root of a matrix is invariably denoted in mathematical works by the Greek symbol λ so that it seems worth using it as a Fortran variable and over-riding the implicit typing. Dimensions are then established for A, LAMBDA, VEC and the two working vectors U1 and UN. It is an advantage of this method that it uses relatively little storage space beyond that needed for the input matrix.

The first program segment begins the main loop which will control the computation of successive roots and vectors and which embraces all the remaining statements in the program. It also sets up the first trial vector U1. This is done by summing the elements in each column of the input matrix (using the DO-loop ending on 1), selecting the largest column sum (U1MAX) and dividing the vector of column sums by U1MAX to give the first trial vector. When this has been done we initialise the counter NCYC to zero; this will enable us to count subsequent iterative cycles and end the program if a vector has not converged after, say, 100 cycles of computation.

The long segment beginning with statement 4 and ending on 11 carries out the iteration and begins with a test of cycles completed, followed by a diagnostic warning message if convergence has not been accomplished in 100 cycles. In this case NROOT is set equal to (K − 1), thus indicating the number of roots and vectors that had already been extracted. The iteration begins by computing a new trial vector UN from the relation:

$$\mathbf{u}_n = \mathbf{A}\mathbf{u}_1. \tag{7}$$

Once again, we standardise UN in terms of its largest element UNMAX but before doing so we make an accuracy check. If UNMAX is very small (we use the arbitrary value of 0.00005), the division may cause considerable loss of significant digits and it is therefore wiser to terminate the iteration with a diagnostic message. Provided UNMAX is sufficiently large, however, we use it to standardise UN in the DO-loop ending on statement 9. We are now ready to compare UN with the previous estimate U1. This we do by the logical IF statement, arranging for the iteration to stop if all elements of UN and U1 agree to within 0.00005. This means that each element of the vector will be accurate to 4 decimal places—enough for biological applications of the technique. If agreement is not sufficiently close, however, we take UN as the new value of U1 and return to statement 4 to compute a new UN, repeating the complete cycle. Provided convergence occurs without undue accuracy loss, then, this segment will give us values for UNMAX, which is the latent root currently being extracted, and for UN, its associated vector, standardised with the largest element equal to 1.

The final program segment therefore begins by placing UNMAX in the k-th location of LAMBDA and normalising the vector UN (i.e. rescaling it so that the sum of its squared elements will be equal to 1). This operation is performed by the successive DO-loops ending on 12 and 13: effectively these square each element of UN, sum the squares, take the positive square root (DEN) and divide each element by that number. The normalised vector is then placed in the k-th column of VEC by the DO-loop ending on 14 and if all NROOT vectors have been computed control is returned to the main program. If, however, the outermost loop is still unsatisfied we reduce the matrix so that a further root and vector can be extracted. This reduction is done by the two nested DO-loops ending on 15 which compute a reduced matrix, which

Flowchart for Program 26

Latent roots and vectors of a real symmetric matrix.

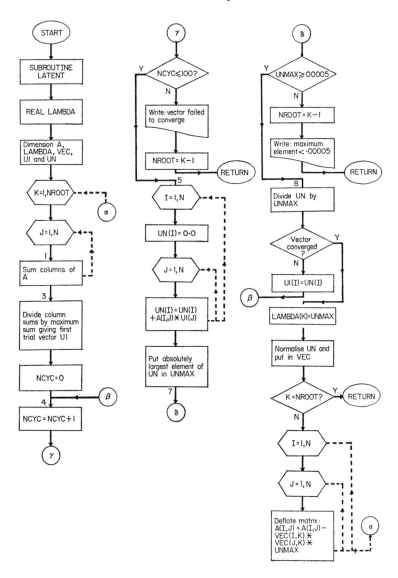

we may call A_{k+1}, from the present matrix A_k by the relation:

$$A_{k+1} = A_k - \lambda u_k u_k' \tag{8}$$

where u_k and u_k' are respectively the column and row forms of the k-th latent vector. This product of column and row vectors is itself a matrix (the so-called outer product of two vectors).

The iterative method used in this subroutine has the advantage that it requires only simple arithmetical operations which can be understood without specialised knowledge of matrix algebra. It also requires little storage space and if only a few latent roots are being extracted from a large matrix it can be quicker than some more sophisticated methods. It has the disadvantage that convergence will not occur if two latent roots are identical and may be very slow if they are close in value. For this reason it is sometimes recommended that one should raise the input matrix to the power of, say, 8 before applying the iterative procedure since the latent roots of the powered matrix would be more widely separated. However, the same result can be achieved by increasing the number of iterative cycles (by a factor of 8 in the example cited; Kendall, 1957: p. 24). The program sets a maximum of 100 which should be enough for all but the most awkward cases. If satisfactory convergence does not occur the program will not, of course, extract all the roots and vectors required, though those which have been successfully computed will be accurate and their number is returned to the main program in NROOT. When *all* the latent roots of a large matrix are required, the iterative method cannot compete in speed or accuracy with other techniques. Most of these are conveniently reviewed by White (1958) and in works by Fox (1964) and Wilkinson (1965). The Jacobi technique is widely used for real symmetric matrices in biological problems: a suitable version in Fortran II is listed by Cooley and Lohnes (1962) and a faster Fortran IV program is included in the Scientific Subroutine Package (p. 472). Both programs depend on methods discussed by Greenstadt (1960). The Jacobi method extracts all the roots and vectors together and generally requires rather more storage space than the method given here. Another widely advocated technique for real symmetric matrices is the Givens–Householder method, fully discussed and written as a complete Fortran IV program by Ortega (1967).

Program 26

Latent roots and vectors of a real symmetric matrix: SUBROUTINE LATENT with short calling program, sample data and output.

```
C
C    CALLING PROGRAM FOR SUBROUTINE LATENT.
C
     REAL LAMBDA
     DIMENSION A(50,50), LAMBDA(10), VEC(50,10), U1(50), UN(50)
```

Program 26 (*continued*)

```
   15 READ (5,99) NPROB, N, NROOT
   99 FORMAT (3I5)
      IF (NPROB .EQ. 0) STOP
      WRITE (6,100) NPROB
  100 FORMAT (44H1EIGENANALYSIS OF SYMMETRIC MATRIX PROBLEM, I4)
      DO 1 I = 1,N
    1 READ (5,101) (A(I,J), J = 1,N)
  101 FORMAT (10F7.0)
      CALL LATENT (A,N,NROOT,LAMBDA,VEC,U1,UN)
      IF (NROOT .EQ. 0) GO TO 15
      WRITE (6,102) (LAMBDA(I), I = 1,NROOT)
  102 FORMAT (//13H LATENT ROOTS//10F12.3)
      WRITE (6,103)
  103 FORMAT (//15H LATENT VECTORS/)
      DO 2 I = 1,N
    2 WRITE (6,104) (VEC(I,J), J = 1,NROOT)
  104 FORMAT (10F12.3)
      GO TO 15
      END

C
      SUBROUTINE LATENT (A, N, NROOT, LAMBDA, VEC, U1, UN)
C
C.......................................................
C
C THIS PROGRAM COMPUTES THE LATENT ROOTS AND VECTORS OF A REAL SYMMETRIC
C MATRIX BY ITERATION. FOR COMPUTATIONAL DETAILS SEE LAWLEY AND
C MAXWELL (1963) FACTOR ANALYSIS AS A STATISTICAL METHOD. PARAMETERS
C ARE AS FOLLOWS -
C    A       = INPUT MATRIX, DESTROYED IN COMPUTATION
C    N       = ORDER OF INPUT MATRIX
C    NROOT   = NUMBER OF LATENT ROOTS AND VECTORS REQUIRED (MAXIMUM 10).
C              ON RETURN NROOT CONTAINS NUMBER OF ROOTS COMPUTED.
C    LAMBDA  = VECTOR OF LATENT ROOTS (IN DESCENDING ORDER OF SIZE).
C    VEC     = MATRIX OF NORMALISED LATENT VECTORS ARRANGED COLUMNWISE
C              (MAXIMUM OF 10 COLUMNS)
C    U1, UN  = WORKING VECTORS (LENGTH N)
C
C.......................................................
C
      REAL LAMBDA
      DIMENSION A(50,50), LAMBDA(10), VEC(50,10), U1(50), UN(50)
C
C START MAIN LOOP WITH FIRST TRIAL VECTOR.
      DO 14 K = 1,NROOT
      DO 1 J = 1,N
      U1(J) = 0.0
      DO 1 I = 1,N
    1 U1(J) = U1(J) + A(I,J)
      U1MAX = ABS(U1(1))
      DO 2 I = 2,N
      IF (U1MAX .GE. ABS(U1(I))) GO TO 2
      U1MAX = ABS(U1(I))
    2 CONTINUE
      DO 3 I = 1,N
    3 U1(I) = U1(I) / U1MAX
      NCYC = 0
C
C COMPUTE LATENT ROOT AND VECTOR.
    4 NCYC = NCYC + 1
```

Program 26 (*continued*)

```
          IF (NCYC .LE. 100) GO TO 5
          WRITE (6,100) K
    100   FORMAT (//7H VECTOR, I3, 19H FAILED TO CONVERGE)
          NROOT = K – 1
          RETURN
      5   DO 6 I = 1,N
          UN(I) = 0.0
          DO 6 J = 1,N
      6   UN(I) = UN(I) + A(I,J)*U1(J)
          UNMAX = ABS(UN(1))
          DO 7 I = 2,N
          IF (UNMAX .GE. ABS(UN(I))) GO TO 7
          UNMAX = ABS(UN(I))
      7   CONTINUE
          IF (UNMAX .GE. 0.00005) GO TO 8
          WRITE (6,101) K
    101   FORMAT(//15H WORKING VECTOR,I3,39H WITH MAXIMUM ELEMENT LESS THAN
         A0.00005)
          NROOT = K – 1
          RETURN
      8   DO 9 I = 1,N
      9   UN(I) = UN(I) / UNMAX
          DO 11 I = 1,N
          IF (ABS(U1(I) – UN(I)) .LT. 0.00005) GO TO 11
          DO 10 J = 1,N
     10   U1(J) = UN(J)
          GO TO 4
     11   CONTINUE
C
C   PLACE LATENT ROOT AND NORMALISED LATENT VECTOR IN ARRAYS LAMBDA AND
C   VEC, AND REDUCE MATRIX.
          LAMBDA (K) = UNMAX
          DEN = 0.0
          DO 12 I = 1,N
     12   DEN = DEN + UN(I) * UN(I)
          DEN = SQRT(DEN)
          DO 13 I = 1,N
     13   VEC(I,K) = UN(I) / DEN
          IF (K .EQ. NROOT) GO TO 15
          DO 14 I = 1,N
          DO 14 J = 1,N
     14   A(I,J) = A(I,J) – VEC(I,K)*VEC(J,K)*UNMAX
     15   RETURN
          END
```

DATA

```
    1      5      5
  1.000   0.438  −0.137   0.205  −0.178
  0.438   1.000   0.031   0.180  −0.304
 −0.137   0.031   1.000   0.161   0.372
  0.205   0.180   0.161   1.000  −0.013
 −0.178  −0.304   0.372  −0.013   1.000
    0
```

OUTPUT
    EIGENANALYSIS OF SYMMETRIC MATRIX PROBLEM

LATENT ROOTS

      1.757   1.331   0.781   0.709   0.422

Program 26 (*continued*)

LATENT VECTORS

0.556	0.186	0.216	0.641	−0.447
0.565	0.247	0.439	−0.308	0.576
−0.270	0.662	0.320	−0.399	−0.477
0.236	0.556	−0.787	−0.014	0.123
−0.494	0.395	0.195	0.579	0.475

Program 27: Latent roots and vectors of a non-symmetric matrix

Some of the commoner biological applications of latent roots and vectors relate to symmetric matrices. There are, however, two major methods of multivariate analysis—multiple discriminant analysis and canonical correlation analysis—which require the latent roots and vectors of a non-symmetric matrix. For various reasons, this entails a more elaborate program. The problem is also complicated by the fact that whereas a real symmetric matrix has no negative or complex roots, these may occur in real non-symmetric matrices. The many techniques available for dealing with non-symmetric matrices are discussed by the authors cited above; here we shall develop only one method, an iterative procedure due originally to Hotelling. As in Program 26 this involves the deflation of the matrix after each root and vector has been calculated. In order to accomplish this deflation, however, we now need the *two* vectors associated with each latent root. One of these is the so-called right latent vector or column vector **u** which we computed in Program 26 and which is derived from the relationships expressed in equation (4) above:

$$\mathbf{Au} = \lambda \mathbf{u}.$$

The other is a left latent vector or row vector **v′** determined by the equation

$$\mathbf{v'A} = \lambda \mathbf{v'}. \qquad (9)$$

If **A** is symmetric, the right and left vectors are identical, but in a non-symmetric matrix they will differ. The program therefore involves three essential steps:

(a) Calculation of a latent root and right vector by the method used in Program 26, from the iterative sequence

$$\mathbf{u}_{i+1} = \mathbf{Au}_i.$$

(b) Calculation of the left vector by an almost exactly similar process in which, however, we use the sequence

$$\mathbf{v'}_{i+1} = \mathbf{v'}_i \mathbf{A}.$$

(c) After both vectors have converged satisfactorily we deflate the matrix by the relation

$$\mathbf{A}_{k+1} = \mathbf{A}_k - \lambda\mathbf{uv}'. \tag{10}$$

This should be compared with equation (8), to which it reduces in the symmetric case when $\mathbf{u} = \mathbf{v}$.

Program 27 therefore begins in a manner almost identical with Program 26, and continues so until we have computed the right vector. The only difference between the two programs up to this point is that in the present one the warning diagnostic messages refer specifically to the right vector which we are calculating. It is this normalised vector which will eventually be reported as *the* latent vector and which is therefore stored as before in VEC.

Starting with statement 14, however, we go on to calculate the left (row) vector, using V1, VN, V1MAX and VNMAX in place of U1, UN, U1MAX and UNMAX. The instructions needed to compute the left vector differ in three respects from those used to compute the right one: (a) in the iterative sequence whereby a new trial vector is calculated from an old one (statements 19 to 20 as compared with 5 to 6); (b) in the warning diagnostics which now refer specifically to the left vector; and (c) should this fail to converge we now set NROOT = K (instead of the (K − 1) used when computing the right vector). The reason for the latter change should be clear: if the k-th left vector fails to converge we shall be unable to deflate the matrix and to compute the next vector, but we already have the k-th right vector stored in VEC and can therefore return K latent roots and vectors to the main program.

Having obtained both right and left vectors for a given root we must normalise them (actually, renormalise the right vector) before deflating the matrix. This normalisation is accomplished by the two consecutive DO-loops ending on statement 27 and involves adjusting the vectors so that the scalar product $\mathbf{v}'\mathbf{u}$ is equal to unity. The last step is then taken by the nested loops ending on 28 and consists simply of deflating the matrix by subtracting from it the product $\lambda\mathbf{uv}'$. The latter, involving the premultiplication of a row vector by a column vector, is itself a matrix. Note also that the normal return of this subroutine is by the statement immediately after 13—there is no need to compute the left vector corresponding to the last required latent root.

Consideration of the iterative process programmed here shows immediately that only positive latent roots can be extracted by means of this subroutine; control returns to the main program if the root being extracted is or approaches zero. This is not a serious drawback when the subroutine is used as part of a multiple discriminant analysis (see Program 31) or of a canonical correlation analysis, the two main statistical uses to which it is put; the matrices involved in these techniques do not yield negative or complex roots and only the positive roots are of any importance. A rather more troublesome

Outline flowchart for Program 27

Latent roots and vectors of a real non-symmetric matrix.

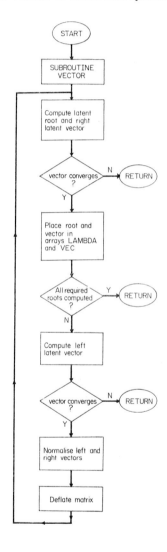

feature of the method—as also of the method used in Program 26—is that the vector fails to converge when the root being extracted is the same or only slightly larger than the following one. Fortunately such cases are uncommon in the statistical analyses just mentioned and the large number of iterations (100 is the maximum set in the program) will usually deal with cases where the roots are close. For both reasons given, however, the method may be quite unsatisfactory if applied to certain non-symmetric matrices arising in genetics or population dynamics, where zero, negative, complex or multiple roots may occur (see, for example, some of the illustrations given by Searle, 1966). It would take us beyond the scope of this book to consider methods appropriate in these cases; it is enough to note, in general, that the extraction of the latent roots and vectors of non-symmetric matrices is more troublesome than the same process for symmetric matrices.

Since this and Program 26 are so similar, there is no need to provide a detailed flowchart. The one given on p. 269 simply summarises the main steps in the procedure. It should be clear to the reader that this subroutine can also be used for computing the latent roots and vectors of a real symmetric matrix (the right and left vectors of a root being identical in that case). It takes nearly twice as long to execute as Program 26, however, and is therefore obviously not a method one would use in practice.

Program 27

Latent roots and vectors of a real non-symmetric matrix. The listing comprises only SUBROUTINE VECTOR; a short calling program almost identical with that used in Program 26 (p. 264) will suffice.

```
C
          SUBROUTINE VECTOR (A,N,NROOT,LAMBDA,VEC,U1,UN,V1,VN)
C
C.........................................................................
C
C     THIS PROGRAM COMPUTES THE LATENT ROOTS AND VECTORS OF A REAL
C     NON-SYMMETRIC MATRIX BY ITERATION, USING HOTELLING'S MATRIX
C     DEFLATION TECHNIQUE. FOR COMPUTATIONAL DETAILS SEE WHITE (1958)
C     J.SOC.INDUST.APPL.MATH. VOL.6, P.414. PARAMETERS ARE AS FOLLOWS -
C        A        = INPUT MATRIX, DESTROYED IN COMPUTATION
C        N        = ORDER OF INPUT MATRIX
C        NROOT    = NUMBER OF LATENT ROOTS AND VECTORS REQUIRED (MAXIMUM 10).
C                   ON RETURN NROOT CONTAINS NUMBER OF ROOTS COMPUTED.
C        LAMBDA   = VECTOR OF LATENT ROOTS (IN DESCENDING ORDER OF SIZE).
C        VEC      = MATRIX OF NORMALISED LATENT VECTORS ARRANGED COLUMNWISE
C                   (MAXIMUM OF 10 COLUMNS)
C        U1, UN   = WORKING VECTORS (LENGTH N)
C        V1, VN   = WORKING VECTORS (LENGTH N)
C
C.........................................................................
C
          REAL LAMBDA
          DIMENSION A(50,50), LAMBDA(50), VEC(50,10), U1(50), UN(50),
         A  V1(50), VN(50)
```

```
C
C   START MAIN LOOP WITH FIRST TRIAL VECTOR.
        DO 28 K = 1,NROOT
        DO 1 J = 1,N
        U1(J) = 0.0
        DO 1 I = 1,N
    1   U1(J) = U1(J) + A(I,J)
        U1MAX = ABS(U1(1))
        DO 2 I = 2,N
        IF (U1MAX .GE. ABS(U1(I))) GO TO 2
        U1MAX = ABS(U1(I))
    2   CONTINUE
        DO 3 I = 1,N
    3   U1(I) = U1(I) / U1MAX
        NCYC = 0
C
C   COMPUTE LATENT ROOT AND RIGHT (COLUMN) VECTOR.
    4   NCYC = NCYC + 1
        IF (NCYC .LE. 100) GO TO 5
        WRITE (6,100) K
  100   FORMAT (//13H RIGHT VECTOR, I3, 19H FAILED TO CONVERGE)
        NROOT = K − 1
        RETURN
    5   DO 6 I = 1,N
        UN(I) = 0.0
        DO 6 J = 1,N
    6   UN(I) = UN(I) + A(I,J)*U1(J)
        UNMAX = ABS(UN(1))
        DO 7 I = 2,N
        IF (UNMAX .GE. ABS(UN(I))) GO TO 7
        UNMAX = ABS(UN(I))
    7   CONTINUE
        IF (UNMAX .GE. 0.00005) GO TO 8
        WRITE (6,101) K
  101   FORMAT (//21H RIGHT WORKING VECTOR, I3, 39H WITH MAXIMUM ELEMENT
       ALESS THAN 0.00005)
        NROOT = K − 1
        RETURN
    8   DO 9 I = 1,N
    9   UN(I) = UN(I) / UNMAX
        DO 11 I = 1,N
        IF (ABS(U1(I) − UN(I)) .LT. 0.00005) GO TO 11
        DO 10 J = 1,N
   10   U1(J) = UN(J)
        GO TO 4
   11   CONTINUE
C
C   PLACE LATENT ROOT AND NORMALISED RIGHT VECTOR IN ARRAYS LAMBDA AND VEC.
        LAMBDA (K) = UNMAX
        DEN = 0.0
        DO 12 I = 1,N
   12   DEN = DEN + UN(I) * UN(I)
        DEN = SQRT(DEN)
        DO 13 I = 1,N
   13   VEC(I,K) = UN(I) / DEN
        IF (K .EQ. NROOT) RETURN
C
C   COMPUTE LEFT (ROW) VECTOR.
   14   DO 15 J = 1,N
        V1(J) = 0.0
        DO 15 I = 1,N
```

Program 27 (*continued*)

```
    15  V1(J) = V1(J) + A(I,J)
        V1MAX = ABS(V1(1))
        DO 16 I = 2,N
        IF (V1MAX .GE. ABS(V1(I))) GO TO 16
        V1MAX = ABS(V1(I))
    16  CONTINUE
        DO 17 I = 1,N
    17  V1(I) = V1(I) / V1MAX
        NCYC = 0
    18  NCYC = NCYC + 1
        IF (NCYC .LE. 100) GO TO 19
        WRITE (6,102) K
   102  FORMAT (//12H LEFT VECTOR, I3, 19H FAILED TO CONVERGE)
        NROOT = K
        RETURN
    19  DO 20 J = 1,N
        VN(J) = 0.0
        DO 20 I = 1,N
    20  VN(J) = VN(J) + V1(I) * A(I,J)
        VNMAX = ABS(VN(1))
        DO 21 I = 2,N
        IF (VNMAX .GE. ABS(VN(I))) GO TO 21
        VNMAX = ABS(VN(I))
    21  CONTINUE
        IF (VNMAX .GE. 0.00005) GO TO 22
        WRITE (6,103) K
   103  FORMAT (//20H LEFT WORKING VECTOR, I3, 39H WITH MAXIMUM ELEMENT
       A LESS THAN 0.00005)
        NROOT = K
        RETURN
    22  DO 23 I = 1,N
    23  VN(I) = VN(I) / VNMAX
        DO 25 I = 1,N
        IF (ABS(V1(I) – VN(I)) .LT. 0.00005) GO TO 25
        DO 24 J = 1,N
    24  V1(J) = VN(J)
        GO TO 18
    25  CONTINUE
C
C   NORMALISE LEFT AND RIGHT VECTORS AND DEFLATE MATRIX.
        DEN = 0.0
        DO 26 I = 1,N
    26  DEN = DEN + VN(I)*UN(I)
        IF (DEN .LT. 0.0) GO TO 29
        DEN = SQRT(DEN)
        DO 27 I = 1,N
        UN(I) = UN(I) / DEN
    27  VN(I) = VN(I) / DEN
        DO 28 I = 1,N
        DO 28 J = 1,N
    28  A(I,J) = A(I,J) – LAMBDA(K)*UN(I)*VN(J)
    29  WRITE (6,104) K
   104  FORMAT (//41H SCALAR PRODUCT OF LEFT AND RIGHT VECTORS, I3,
       A  9H NEGATIVE)
        NROOT = K
        RETURN
        END
```

Chapter 10

Multiple Regression and Multivariate Analysis

The methods discussed in this chapter all involve relatively heavy computation and for that reason were not widely applied before the advent of of electronic digital computers. They are, however, particularly relevant to the biological sciences since they concern the treatment of many variables and are therefore often appropriate when a complex of causal processes underlies the system being studied. Four programs will be described in detail, the last three being examples of multivariate analysis in the strict sense of this term.

Program 28: Multiple linear regression analysis.

Program 29: Hotelling's T^2 and discriminant function for two groups.

Program 30: Principal Component Analysis.

Program 31: Multiple Discriminant Analysis.

The first of these methods is an extension of the regression techniques studied in Chapter 7, while the second can be regarded as a generalisation to several variables of Students' t-test (program 3, p. 97). The third program represents an attempt to express concisely the structure inherent in a complex of inter-correlated variables and is one form of what is loosely known as 'factor analysis'. The last technique provides an effective method of discriminating between several groups, all characterised by measurements on the same set of variables. Programs 30 and 31 are typical of those methods of multivariate analysis used in studying the classificatory or descriptive problems of taxonomy, faunistics or biogeography, but they have also been employed in studies of growth, development and structural relationships. A survey of the biological applications of various forms of multivariate analysis is not possible here and the reader is referred to Rao (1952) and Seal (1964) for many examples. Hope (1968) provides a very useful introduction to the theory, with the absolute minimum of mathematics; Morrison (1967) has a balanced account of the whole field, while Anderson (1958) is the major theoretical work in this field of statistics.

Program 28: Multiple linear regression analysis

A not infrequent problem in biological statistics consists in attempting to relate a single dependent variable to a linear function of several independent variables. Thus the abundance of an animal species (the dependent variable) might be related to mean seasonal temperatures, percentage parasitism, laboratory estimates of fecundity at various times or places, and so on. The data comprise several sets of observations, each consisting of a measurement of the dependent variable and one for each of the independent variables. As in the simple linear regression problem considered earlier, it is assumed that the dependent variable is normally distributed with constant variance while the independent variables may have some possibly irregular

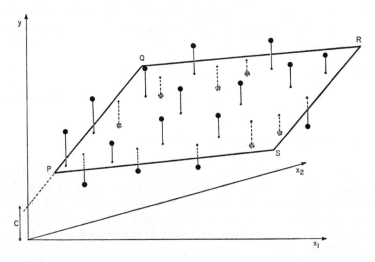

FIG. 29. Representation of a multiple regression equation involving a linear relationship between the two independent variables x_1 and x_2 and the dependent variable y. The regression equation $y = b_1x_1 + b_2x_2 + C$ is depicted by the plane PQRS. Individual observations (large circles) usually lie above or below this plane, which is located so as to minimise the sum of the squares of the distances from each point to the plane; the distances are measured parallel to the y-axis.

or arbitrary distribution. From such data one would hope to derive an equation of the form

$$y = b_1 x_1 + b_2 x_2 + b_3 x_3 + \ldots + b_m x_m + C \quad (1)$$

where y is the predicted value of the dependent variable and $b_1, b_2, b_3 \ldots b_m$ form a set of coefficients for the values of the m independent variables $x_1, x_2, x_3, \ldots x_m$. C is the 'intercept' a constant added to the other terms on the right-hand side of equation (1). If there are only two independent variables the equation represents a plane in 3-dimensional space (Fig. 29). If there are more than two independent variables the situation cannot be visualised. Given values of $x_1, x_2, x_3, \ldots x_m$, however, it is possible to predict a value of y from a set of observations $x_1, x_2, x_3 \ldots$. For this reason the x's are sometimes referred to as the predictor variables and y as the criterion variable. The equation may also be spoken of as the linear regression of the dependent variable on a set of independent variables, and the coefficients $b_1, b_2, b_3, \ldots$ form a set of partial regression coefficients.

The values of these partial regression coefficients depend on the units in which the original variables were measured. For some purposes it is therefore an advantage to standardise the measurements by expressing each one as the deviation from its mean, measured in units of 1 standard deviation:

$$z = \frac{(x - \bar{x})}{s}. \quad (2)$$

The regression equation may then be written:

$$y' = \beta_1 z_1 + \beta_2 z_2 + \beta_3 z_3 + \ldots + \beta_m z_m. \quad (3)$$

We no longer require an intercept since the regression plane denoted by (3) must pass through the origin of the axes. The β-coefficients—sometimes known technically as standard partial regression coefficients—provide a measure of the contribution of each independent variable to the predicted value of the dependent variable. They are, for this reason, sometimes referred to as β-weights.

The problem we have set ourselves, therefore, is to estimate from several sets of observations the following quantities:

(a) the set of β-weights of equation (3)
(b) the set of b-coefficients of equation (1)
(c) the intercept C of equation (1)
(d) the multiple correlation coefficient R which is a measure of the over-

all effectiveness of the multiple regression analogous to the simple product–moment correlation coefficient r. Actually we calculate R^2, which is an estimate of the proportion of total variance in the dependent variable which can be accounted for by the known variance in the independent variables.

(e) an assessment of the significance of R^2 by a variance–ratio test.

The procedure for achieving these ends is explained more fully in many elementary statistical text-books, but the rationale of the method is often obscured by a good deal of tedious, though simple, algebra. This can be greatly simplified by the use of matrix notation. We begin by computing the matrix of correlation coefficients between each pair of independent variables and between each of the latter and the dependent variable. The observations on the variables constitute the primary data matrix, each row of which comprises the observed values $x_1, x_2, x_3, \ldots$ etc. of the above discussion and ends with the observed value of y. In Fortran, each row is represented by the singly subscripted variable X(I) with I varying from 1 to M, where there are (M–1) independent variables, X(M) being the dependent variable. The successive rows of the data matrix are read one after another by the SUBROUTINE CORMAT which is called almost immediately after the main program has started. CORMAT will return the following variables:

R,   the matrix of correlation coefficients

SDEV, a vector of standard deviations, each corresponding to the variables specified in the data matrix

SX,   the corresponding vector of sums for the variables.

Note that communication between the main program and subroutine CORMAT is established through a COMMON statement. The matrix of correlation coefficients **R** is printed out, in half-matrix form. This matrix can be represented in full as follows:

$$\mathbf{R} = \left(\begin{array}{ccccc|c} r_{11} & r_{12} & r_{13} & \cdots r_{1(m-1)} & r_{1m} \\ r_{21} & r_{22} & r_{23} & \cdots r_{2(m-1)} & r_{2m} \\ \vdots & \vdots & \vdots & \vdots & \vdots \\ r_{(m-1)1} & r_{(m-1)2} & r_{(m-1)3} & \cdots r_{(m-1)(m-1)} & r_{(m-1)m} \\ \hline r_{m1} & r_{m2} & r_{m3} & \cdots r_{m(m-1)} & r_{mm} \end{array} \right)$$

We can divide the elements of this matrix into four groups ('partition the matrix') as shown by the horizontal and vertical lines drawn on the matrix. The four component parts could be given the names R_{11} R_{12} R_{21} and R_{22}, i.e.

$$R = \left(\begin{array}{c|c} R_{11} & R_{12} \\ \hline R_{21} & R_{22} \end{array} \right)$$

and have the following meaning:

R_{11} represents the matrix of correlations between the independent variables; r_{12}, for example, is the correlation coefficient between the observed values of x_1 and those of x_2.

R_{12} is a column vector—i.e. an $(m-1) \times 1$ matrix containing the correlation coefficients between the observed values of y and those of each of the $(m-1)$ independent variables.

R_{21} is a row vector containing exactly the same elements as those in R_{12}. This must be so since R is symmetric and therefore $r_{ij} = r_{ji}$.

R_{22} is the scalar 1.0, being the correlation of $X(M)$—i.e. the Fortran variable under which y appears—with itself.

The relation between β (the vector of β-weights we require) and the above component matrices is a simple one:

$$R_{11}\beta = R_{12}. \tag{4}$$

This equation can be solved by premultiplying throughout by the inverse of R_{11}:

$$R_{11}^{-1} R_{11} \beta = R_{11}^{-1} R_{12}$$

i.e.
$$\beta = R_{11}^{-1} R_{12} \tag{5}$$

since the product of R_{11} and its inverse is the identity matrix I which leaves β unchanged when used as multiplier.

Our Fortran program therefore proceeds by obtaining the inverse of the matrix R_{11} by calling subroutine INVERT. We specify R_{11} simply by using the first $(M-1)$ rows and columns of the full correlation matrix R. Moreover, since we shall not require R_{11} again we shall not be disturbed by the fact that INVERT will destroy it in order to obtain the inverse. We need, however, to guard against the possibility that our data are unsuitable, in the sense that R_{11} is a singular matrix (i.e. one which has no inverse). This we do by branching on a zero determinant (DET, returned by INVERT) to a warning message followed by the termination of execution for that problem. Even if the determinant is non-zero, we also arrange for it to be printed out since it provides the user with an indication of the 'condition' of the matrix and hence the reliability of the analysis (see p.463 for a discussion of accuracy problems in relation to ill-conditioned matrices.)

Having computed the inverse it is a simple matter to obtain the vector of β-weights, which we have denoted in Fortran as BETA, by applying a form of equation (5) above. This is done by the nested DO-loops ending on statement 3. Note that in Fortran we can refer to both the matrix and the vector involved simply as elements of R by selecting the appropriate subscripts. The reader who has worked out multiple regression equations by hand will realise that the processes described in this and the preceding paragraph are merely the equivalent in matrix algebra and Fortran array manipulation of solving a set of simultaneous equations for $\beta_1, \beta_2, \beta_3, \ldots \beta_m$. The so-called Doolittle method described in many text-books of statistics as a technique for calculating a multiple regression equation is a systematic, tabular way of achieving the same results.

After computing BETA we are able to calculate the vector of partial regression coefficients b_1, b_2, b_3, etc. from the relationship

$$b_i = \frac{\beta_i s_m}{s_i} \tag{6}$$

where s_m and s_i are respectively the standard deviations of the dependent variable and the i-th independent variable. These, of course, have already been returned in the vector SDEV by subroutine CORMAT. The DO-loop ending on 4 accomplishes the calculation of b-coefficients in the vector B and the following four statements complete the computation of the regression equation by computing the intercept from the relation:

$$C = \bar{x}_m - (\mathbf{b}'\mathbf{x}) \tag{7}$$

where $\bar{x}_m$ is the mean for the dependent variable (i.e. SX(M)/AN in Fortran) and $\mathbf{x}$ is the vector of the $(m-1)$ independent variable means. In the program the vector product is first computed as CEPT and then subtracted from SX(M)/AN; both operations make use of the vector SX which had previously been returned by CORMAT.

It now remains to compute R^2 (i.e. the squared multiple correlation coefficient), which we do from the relation:

$$R^2 = \boldsymbol{\beta}'\mathbf{r} \tag{8}$$

where $\mathbf{r}$ is the vector of correlation coefficients between the dependent variable and each independent variable. These are, of course, given as $\mathbf{R}_{12}$ or $\mathbf{R}_{21}$ of the partitioned $\mathbf{R}$-matrix from which we started and the appropriate multiplication is performed by the DO-loop ending on statement 6. The significance of R^2 is tested by a variance-ratio test in which we compare the

Outline flowchart for Program 28

Multiple linear regression analysis.

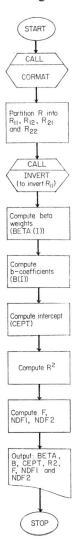

variance accounted for by the multiple regression with the residual variance. The quantities needed are already present and the computation is a simple division

$$F = \frac{R^2(N-m)}{(1-R^2)(m-1)} \qquad (9)$$

where N is the total number of sets of observations (i.e. the number of rows in the primary data matrix). The appropriate number of degrees of freedom for F are:

$$n_1 = (m-1), \qquad n_2 = (N-m) \qquad (10)$$

and the significance of R^2 is then assessed by consulting a table of variance-ratios using the computed values of F and its degrees of freedom.

Program 28

Multiple linear regression analysis. The main program calls SUBROUTINE CORMAT (see p. 236) and SUBROUTINE INVERT (see p. 255). The listing gives only the first few lines of these subroutines to show how their parameters are transmitted by a COMMON statement (p. 72). Sample data and output are included.

```
C
C......................................................................
C
C   THIS PROGRAM EXECUTES A MULTIPLE REGRESSION ANALYSIS.
C   INPUT PARAMETERS ARE AS FOLLOWS -
C       NPROB = PROBLEM REFERENCE NUMBER (ZERO ON LAST DATA CARD)
C       M     = NUMBER OF VARIABLES (1 DEPENDENT + (M-1) INDEPENDENT)
C       N     = NUMBER OF OBSERVATIONS ON EACH VARIABLE
C       INDEX = 3 (FOR CORRELATION MATRIX)
C       X     = PRIMARY DATA MATRIX
C   N.B. THIS PROGRAM CALLS SUBROUTINES CORMAT AND INVERT.
C       M, N, INDEX AND X ARE READ BY SUBROUTINE CORMAT
C
C......................................................................
C
      DIMENSION R(50,50), IR(50), IC(50), BETA(50), SDEV(50), SX(50),
     A X(50), A(50), B(50)
      COMMON R, M, N, MM, IR, IC, SDEV, SX, AN, DET, X
C
    1 READ (5,100) NPROB
  100 FORMAT (I5)
      IF (NPROB .EQ. 0) STOP
      CALL CORMAT
C
C   PARTITION CORRELATION MATRIX AND INVERT R11.
      MM = M - 1
      CALL INVERT
      WRITE (6,101) DET
  101 FORMAT (///14H DETERMINANT = , 1PE15.7)
      IF (DET .NE. 0.0) GO TO 2
      WRITE (6,102)
```

Program 28 (*continued*)

```
  102 FORMAT (///39H MATRIX SINGULAR. EXECUTION TERMINATED.)
      GO TO 1
C
C  COMPUTE BETA WEIGHTS.
    2 DO 3 I = 1,MM
      BETA(I) = 0.0
      DO 3 J = 1,MM
    3 BETA(I) = BETA(I) + R(M,J) * R(J,I)
C
C  COMPUTE B COEFFICIENTS.
      DO 4 I = 1,MM
      A(I) = SDEV(M) / SDEV(I)
    4 B(I) = BETA(I) * A(I)
C
C  COMPUTE INTERCEPT.
      CEPT = 0.0
      DO 5 I = 1,MM
    5 CEPT = CEPT + B(I) * SX(I) / AN
      CEPT = SX(M) / AN - CEPT
C
C  COMPUTE R-SQUARED.
      RSQ = 0.0
      DO 6 I = 1,MM
    6 RSQ = RSQ + BETA(I) * R(M,I)
C
C  COMPUTE VARIANCE RATIO FOR R-SQUARED.
      AM = M
      F = RSQ * (AN - AM) / ((1.0 - RSQ) * (AM - 1.0))
      NDF1 = AM - 1.0
      NDF2 = AN - AM
C
C  OUTPUT RESULTS.
      WRITE (6,103)
  103 FORMAT (1H1, 30X, 28HMULTIPLE REGRESSION ANALYSIS///)
      WRITE (6,104) (I, BETA(I), I = 1,MM)
  104 FORMAT (1X, 12HBETA WEIGHTS//(5(I5, 1PE15.7)))
      WRITE (6,105) (I, B(I), I = 1,MM)
  105 FORMAT (//1X, 14HB COEFFICIENTS//(5(I5, 1PE15.7)))
      WRITE (6,106) CEPT, RSQ, F, NDF1, NDF2
  106 FORMAT (//1X, 11HINTERCEPT = , 1PE15.7//1X, 11HR-SQUARED = , 1PE15.7
     A//1X, 16HVARIANCE RATIO = , 1PE15.7//1X, 6HNDF1 = , I3//
     B1X, 6HNDF2 = , I3)
      GO TO 1
      END

      SUBROUTINE CORMAT
C
C.......................................................................
C
C  THIS SUBROUTINE COMPUTES AND PRINTS A DEVIATION SUMS OF SQUARES AND
C  CROSS-PRODUCTS MATRIX OR A DISPERSION (VARIANCE-COVARIANCE) MATRIX
C  OR A CORRELATION MATRIX. PARAMETERS ARE AS FOLLOWS -
C     M     = NUMBER OF COLUMNS IN PRIMARY DATA MATRIX
C     N     = NUMBER OF ROWS IN PRIMARY DATA MATRIX
C     INDEX = 1 FOR DEVIATION SUMS OF SQUARES AND CROSS-PRODUCTS MATRIX
C             2 FOR DISPERSION MATRIX
C             3 FOR CORRELATION MATRIX
C     X     = VECTOR (LENGTH M) REPRESENTING ONE ROW OF DATA MATRIX
C     R     = OUTPUT MATRIX
C
C.......................................................................
```

Program 28 (*continued*)

```
C
      DIMENSION R(50,50), SDEV(50), SX(50), X(50), IR(50), IC(50)
      COMMON R, M, N, MM, IR, IC, SDEV, SX, AN, DET, X
      READ (5,100) M, N, INDEX
  100 FORMAT (3I5)
C
C     INITIALISE FOR SUBSEQUENT ACCUMULATION OF SUMS.

      SUBROUTINE INVERT
C
C.............................................................
C
C     THIS SUBROUTINE INVERTS A MATRIX A, OF ORDER N, USING THE
C     GAUSS-JORDAN EXCHANGE METHOD AND COMPUTES THE DETERMINANT
C     (DET). MATRIX A IS DESTROYED AND REPLACED BY ITS INVERSE.
C     IR AND IC ARE WORKING VECTORS OF LENGTH N.
C
C.............................................................
C
      DIMENSION A(50,50), IR(50), IC(50), SX(50), X(50), SDEV(50)
      COMMON A, M, NSUB, N, IR, IC, SDEV, SX, AN, DET, X
C
C     INITIALISE AND BEGIN MAIN LOOP.
      DET = 1.0
```

DATA

```
    1
    4    30    3
  3.05   1.45   5.67   0.34
  4.22   1.35   4.86   0.11
  3.34   0.26   4.19   0.38
  3.77   0.23   4.42   0.68
  3.52   1.10   3.17   0.18
  3.54   0.76   2.76   0.00
  3.74   1.59   3.81   0.08
  3.78   0.39   3.23   0.11
  2.92   0.39   5.44   1.53
  3.10   0.64   6.16   0.77
  2.86   0.82   5.48   1.17
  2.78   0.64   4.62   1.01
  2.22   0.85   4.49   0.89
  2.67   0.90   5.59   1.40
  3.12   0.92   5.86   1.05
  3.03   0.97   6.60   1.15
  2.45   0.18   4.51   1.49
  4.12   0.62   5.31   0.51
  4.61   0.51   5.16   0.18
  3.94   0.45   4.45   0.34
  4.12   1.79   6.17   0.36
  2.93   0.25   3.38   0.89
  2.66   0.31   3.51   0.91
  3.17   0.20   3.08   0.92
  2.79   0.24   3.98   1.35
  2.61   0.20   3.64   1.33
  3.74   2.27   6.50   0.23
  3.13   1.48   4.28   0.26
  3.49   0.25   4.71   0.73
  2.94   2.22   4.58   0.23
    0
```

Program 28 (*continued*)

```
OUTPUT
    CORRELATION HALF-MATRIX

ROW  1
  1   9.9999996E-01    2  2.0940064E-01    3  9.2564464E-02    4  -7.1772755E-01

ROW  2
  2   1.0000000E 00    3  4.0739921E-01    4  -4.9963826E-01

ROW  3
  3   1.0000000E 00    4  1.7936699E-01

ROW  4
  4   1.0000000E 00

DETERMINANT =  7.9740232E-01
    MULTIPLE REGRESSION ANALYSIS

BETA WEIGHTS
    1  -6.4474572E-01    2  -5.5395864E-01    3  4.6472986E-01

B COEFFICIENTS
    1  -5.3145380E-01    2  -4.3963615E-01    3  2.0897581E-01

INTERCEPT =  1.8110354E 00

R-SQUARED =  8.2288789E-01

VARIANCE RATIO =  4.0266557E 01

NDF1 =  3

NDF2 = 26
```

From a programming standpoint, the main interest of this program is that it represents a relatively complicated statistical procedure which we can easily carry out using two of our standard matrix algebra subroutines (CORMAT and INVERT). The ease with which multiple regressions can now be calculated by digital computers should not mislead the biologist into thinking that they solve all problems in which several variables are present. The method used to compute the regression coefficients tends to capitalise on the errors of observation involved. Predictions made with their aid, using a new set of observations on the independent variables may therefore be misleading. The problem of assessing the relative importance of the different independent variables is also a difficult one and it is sometimes advocated that a stepwise procedure be adopted, adding or removing variables in turn and noting their contributions to the prediction of the dependent variable (Efroymson, 1960).

The flowchart for this program indicates only the major steps, details of which are readily available from the program listing.

Program 29: Hotelling's T^2 and discriminant function for two groups.

The simple t-test described in connection with Program 3 enables one to decide whether two samples come from the same population or from populations with different means. It related to observations on a single normally distributed variable and raises the question of whether an analogous test can be made between two samples, each individual in which has been measured in respect of *several* variables. Hotelling's T^2 provides this multivariate generalisation of Student's t-test (Hotelling, 1931). The problem is closely connected with the question of how best to discriminate between the two samples, a solution to which is given by Fisher's discriminant function. This provides a linear function of the measurements on each variable such that an individual can be assigned to one or other of the two groups with the least chance of being misclassified. The discriminant function can be written as

$$z = a_1 x_1 + a_2 x_2 + \ldots + a_m x_m \tag{11}$$

where **a** is the vector of discriminant coefficients and **x** the vector of observations made on an individual which is to be assigned to one or other of the two groups under study. Both T^2 and the linear discriminant function are clearly of great interest in taxonomy, where the numerical data typically consist of measurements made on several characters and the problem is often either to decide to which of two groups an individual belongs or to decide whether one group (considered in terms of the totality of its measured attributes) is significantly different from another.

The method used to answer these questions is mathematically very similar to the multiple regression analysis considered in Program 28 but we shall deal with it here as a problem in its own right and consider that we have n_1 individuals in the first group and n_2 in the second group. Both groups will have been measured in respect of the same m variables so that the primary data may be regarded as arranged in two matrices, the first of order $n_1 \times m$ and the second of order $n_2 \times m$. In fact we shall read in all the data as though they constituted a single matrix of order $(n_1 + n_2) \times m$. The Fortran input variables are therefore the doubly subscripted X (denoting the data matrix) and N1, N2 and M, denoting n_1, n_2 and m of the above discussion. The initialisation procedure carried out in the first program segment is intended to provide the real and integer numbers needed for subsequent manipulation. It is followed by reading of the primary data matrix.

The next segment begins computation by evaluating the vector of variable means for each group. Note that this is done by first summing the columns of X over the rows 1 to N1 and then by repeating the procedure over rows (N1 + 1) to (N1 + N2), the latter quantities having been calculated in the initialisation segment. Having obtained the means we can easily calculate the pooled covariance matrix. The DO-loops ending on statement 9 first calculate the pooled sum of products of deviations from the respective group means, again making use of N1P (equal to N1 + 1) and N (equal to N1 + N2) for delimiting the appropriate rows of the matrix X. From the pooled sum of products matrix we obtain the covariance matrix by dividing each element by the appropriate number of degrees of freedom (equal to N1 + N2 − 2). The covariance matrix is stored in S and since it is symmetric all the above computations have been carried out on its upper triangle only. The nested DO-loops ending on 10 then copy the upper triangular elements into the appropriate locations of the lower triangle.

We are now in a position to compute the vector of discriminant coefficients **a**. This can be done by solving for **a** the set of simultaneous equations that are compactly expressed in matrix notation as

$$\mathbf{Sa} = \mathbf{d} \tag{12}$$

where **S** is the pooled covariance matrix and **d** the difference between the mean vectors of the two groups. To solve this equation we premultiply both sides by the inverse of the covariance matrix, obtaining

$$\mathbf{a} = \mathbf{S}^{-1}\mathbf{d}. \tag{13}$$

These calculations are performed by the next program segment which begins by calling subroutine INVERT to obtain the inverse of the covariance matrix. If the pooled covariance matrix is based on too small a number of individuals, or if for other reasons it is singular (we choose a value of det **S** less than 0.0001) the 'inverse' returned will be meaningless or likely to produce inaccurate solutions for **a**. A diagnostic message is then produced and execution is terminated. If, as one hopes, however, the data are such that an accurate inverse of the covariance matrix is available, we compute the vector of mean differences (XDIFF)—corresponding to **d** above—and permultiply it by the inverse to obtain the vector of discriminant coefficients A. This last operation is carried out by the DO-loops ending on statement 13.

Having obtained **a**, the vector of discriminant coefficients, we compute T^2 from the relation:

$$T^2 = \frac{n_1 n_2}{(n_1 + n_2)} \mathbf{d}'\mathbf{a}. \tag{14}$$

The vector product **d'a** is, of course, the so-called inner product, a scalar resulting from the multiplication of a row vector by a column vector. The

Outline flowchart for Program 29
Hotelling's T^2 and discriminant function for two groups.

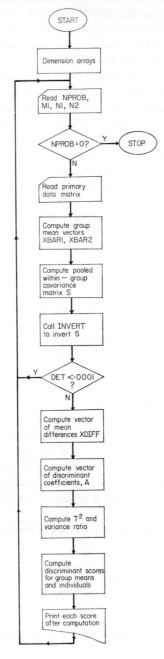

significance of T^2 can be assessed by a variance-ratio test, and the program goes on to compute F from the expression

$$F = \frac{n_1 + n_2 - m - 1}{(n_1 + n_2 - 2)m} \cdot T^2 \qquad (15)$$

with degrees of freedom m and $(n_1 + n_2 - m - 1)$. This test (which the user must complete using tabulated values of F in the usual way) allows one to decide whether the groups differ significantly from each other.

At this point the program prints out T^2, F, its degrees of freedom and the vector of discriminant coefficients. It is also of interest, however, to compute the so-called discriminant scores (i.e. the value of z in equation (11)) for each individual in the two groups and also for the mean values ('centroids') for each group. This is done by the last program segment, after which control returns to statement 1 for another problem to be analysed.

Program 29

Hotelling's T^2 and discriminant function for two groups; with sample data and output. The main program calls SUBROUTINE INVERT (see p. 255).

```
C
C........................................................................
C
C   THIS PROGRAM COMPUTES HOTELLING'S T-SQUARED AND THE LINEAR
C   DISCRIMINANT FUNCTION FOR TWO GROUPS. FOR THEORY SEE
C   MORRISON, D.F.(1967) MULTIVARIATE STATISTICAL METHODS, CHAP.4.
C   INPUT PARAMETERS ARE AS FOLLOWS -
C       NPROB = PROBLEM REFERENCE NUMBER (ZERO ON LAST DATA CARD)
C       M     = NUMBER OF VARIABLES
C       N1    = NUMBER OF INDIVIDUALS IN FIRST GROUP
C       N2    = NUMBER OF INDIVIDUALS IN SECOND GROUP
C       X     = PRIMARY DATA MATRIX
C
C........................................................................
C
      DIMENSION X(100,25), S(25,25), XBAR1(25), XBAR2(25), XDIFF(25),
     A A(25), IR(25), IC(25)
C
C   READ PROBLEM SPECIFICATION, INITIALISE AND READ DATA.
    1 READ (5,100) NPROB, M, N1, N2
  100 FORMAT (4I5)
      IF (NPROB .EQ. 0) STOP
      N = N1 + N2
      AN = N
      AM = M
      AN1 = N1
      AN2 = N2
      NM2 = N - 2
      ANM2 = NM2
      N1P = N1 + 1
      DO 2 I = 1,N
```

Program 29 (*continued*)

```
    2 READ (5,101) (X(I,J), J = 1,M)
  101 FORMAT (10F7.0)
C
C   COMPUTE GROUP MEAN VECTORS.
      DO 3 J = 1,M
      XBAR1(J) = 0.0
      DO 3 I = 1,N1
    3 XBAR1(J) = XBAR1(J) + X(I,J)
      DO 4 I = 1,M
    4 XBAR1(I) = XBAR1(I) / AN1
C
      DO 5 J = 1,M
      XBAR2(J) = 0.0
      DO 5 I = N1P, N
    5 XBAR2(J) = XBAR2(J) + X(I,J)
      DO 6 I = 1,M
    6 XBAR2(I) = XBAR2(I) / AN2
C
C   COMPUTE POOLED WITHIN–GROUP COVARIANCE MATRIX.
      DO 9 I = 1,M
      DO 9 J = I,M
      SUM = 0.0
      DO 7 K = 1,N1
    7 SUM = SUM + (X(K,I)–XBAR1(I)) * (X(K,J)–XBAR1(J))
      DO 8 K = N1P, N
    8 SUM = SUM + (X(K,I)–XBAR2(I)) * (X(K,J)–XBAR2(J))
    9 S(I,J) = SUM
      DO 10 I = 1,M
      DO 10 J = I,M
      S(I,J) = S(I,J) / ANM2
   10 S(J,I) = S(I,J)
C
C   INVERT COVARIANCE MATRIX AND COMPUTE DISCRIMINANT COEFFICIENTS.
      CALL INVERT (S,M,DET,IR,IC)
      WRITE (6,102) NPROB, DET
  102 FORMAT (60H1HOTELLING'S T–SQUARED AND DISCRIMINANT FUNCTION.
     APROBLEM, I4///14H DETERMINANT = , F12.5)
      IF(DET .GE. 0.0001) GO TO 11
      WRITE (6,103)
  103 FORMAT (//21H EXECUTION TERMINATED)
      GO TO 1
   11 DO 12 I = 1,M
   12 XDIFF(I) = XBAR1(I) – XBAR2(I)
      DO 13 I = 1,M
      A(I) = 0.0
      DO 13 J = 1,M
   13 A(I) = A(I) + S(I,J)*XDIFF(J)
C
C   COMPUTE T–SQUARED AND TEST SIGNIFICANCE.
      T2 = 0.0
      DO 14 I = 1,M
   14 T2 = T2 + XDIFF(I) * A(I)
      T2 = T2 * AN1*AN2/AN
      NDF1 = M
      NDF2 = N – M – 1
      DF2 = NDF2
      F = T2 * DF2 / ANM2 / AM
      IF (F .GE. 1.0) GO TO 15
      F = 1.0 / F
      NDF1 = NDF2
      NDF2 = M
```

Program 29 (continued)

```
   15 WRITE (6,104) T2, F, NDF1, NDF2, (A(I), I = 1,M)
  104 FORMAT (//12H T-SQUARED = , F12.3//4H F = , F12.3//
     A     21H DEGREES OF FREEDOM = , I4, 4H AND, I4///
     B     26H DISCRIMINANT COEFFICIENTS//(10F12.3))
C
C  COMPUTE AND PRINT DISCRIMINANT SCORES.
      D1 = 0.0
      D2 = 0.0
      DO 16 I = 1,M
      D1 = D1 + XBAR1(I)*A(I)
   16 D2 = D2 + XBAR2(I) * A(I)
      WRITE (6,105) D1
  105 FORMAT (//28H DISCRIMINANT SCORES GROUP 1//11H CENTROID = ,
     A F12.3)
      DO 18 I = 1,N1
      D1 = 0.0
      DO 17 J = 1,M
   17 D1 = D1 + X(I,J) * A(J)
   18 WRITE (6,106) I, D1
  106 FORMAT (I9, 2H = , F12.3)
      WRITE (6,107) D2
  107 FORMAT (//28H DISCRIMINANT SCORES GROUP 2//11H CENTROID =
     A F12.3)
      IX = 0
      DO 20 I = N1P,N
      IX = IX + 1
      D2 = 0.0
      DO 19 J = 1,M
   19 D2 = D2 + X(I,J) * A(J)
   20 WRITE (6,108) IX, D2
  108 FORMAT (I9, 2H = , F12.3)
      GO TO 1
      END
```

DATA

```
   1    3    6    4
  7.0   9.0   8.0
  9.0   8.0   7.0
  4.0  10.0   5.0
  7.0   7.0   7.0
  8.0   5.0   5.0
  7.0   9.0   4.0
  5.0   6.0   6.0
  6.0   8.0   3.0
  4.0   3.0   6.0
  3.0   7.0   5.0
   0
```

OUTPUT

     HOTELLING'S T-SQUARED AND DISCRIMINANT FUNCTION.   PROBLEM   1

DETERMINANT =   17.06055

T-SQUARED =   14.386

F =   3.596

DEGREES OF FREEDOM =   3 AND   6

Program 29 (continued)

DISCRIMINANT COEFFICIENTS

    1.320   0.972   0.749

DISCRIMINANT SCORES GROUP 1

CENTROID =	21.514
1 =	23.984
2 =	24.903
3 =	18.748
4 =	21.291
5 =	19.168
6 =	20.988

DISCRIMINANT SCORES GROUP 2

CENTROID =	15.520
1 =	16.929
2 =	17.946
3 =	12.692
4 =	14.512

The program itself has no features of very great technical interest, though the method of computing the pooled covariance matrix is worth noting. The direct use of deviations from the group means promotes accuracy; the method of handling the two submatrices of which X is composed makes it very compact; and the fact that only the upper triangular elements of the covariance matrix are computed is a further gain in efficiency.

A few further comments may be made on the statistical nature and uses of this program. The computation of the discriminant scores allows one to view the individuals originally measured as two clusters of points, each centred on its group centroid scores. These points are, of course, dispersed along a line. The main value of the linear discriminant function for taxonomic work, however, lies in the application of the discriminant coefficient in computing a discriminant score for any new individual which has yet to be assigned to one of the two groups (or even set aside as not, or only doubtfully, belonging to either). If the two groups from which the discriminant coefficients were calculated are equally represented in the population from which our samples are drawn, then the mid-point between the group centroid scores may be used as a 'cutting point' to assign a previously unclassified individual to its group. Thus, if the discriminant scores for the group centroids were 10.5 and 15.5, then the mid-point would be 13.0 and previously unclassified individuals with a score of <13.0 could be assigned to the first group while those with a score of >13.0 could be placed in the second one. If, however, the two groups are unequally represented in the population then the cutting point must be shifted from the midpoint towards the less numerous group by an amount equal to $(\log_e R)/D$ where R is the ratio of the more numerous group to the less numerous one and D is a quantity referred to in the next paragraph. Program 29 could easily be extended to assign unclassified individuals to their groups on the basis of the previously computed discriminant function. A good example of this use of a linear discriminant function in systematic entomology is reported by Barlow, Graham and Adisoemarto (1969). The method deserves wider use by taxonomists who, in the past, seem to have been deterred by the amount of desk computation needed.

The difference between the group centroids, as calculated in Program 29, is the so-called 'generalised distance' between the two groups, often referred to as the Mahalanobis D^2-statistic, after the Indian statistician responsible for developing it. This distance is itself of value in expressing the mutual relationships between several groups in taxonomic work, though we cannot pursue this aspect of the subject further here. We may, however, note that D^2 is also obtainable directly from T^2 by the relationship

$$D^2 = \frac{n_1 + n_2}{n_1 n_2} \cdot T^2 \tag{16}$$

or from the more general relationship

$$D^2 = \mathbf{d'S^{-1}d}. \tag{17}$$

We may also use D^2 to obtain a measure of the effectiveness with which our calculated discriminant function can be used to classify further individuals. To do this we evaluate $D/2$, which is a standardised normal deviate, and consult tabulated values of the cumulative normal distribution function, e.g. a value of $D/2$ equal to 1.5 would mean that about 93% of the individuals would be correctly assigned to their group on the basis of the discriminant function.

One final statistical point is worth making: the methods used in this program assume that the covariance matrices of the two groups are equal within the limits of sampling error, so that one is justified in using the pooled matrix. This assumption should, strictly speaking, be examined before T^2 and the discriminant function are calculated. Suitable tests are discussed in works on multivariate analysis. If the matrices are not homogeneous it may be possible to achieve this by transforming the original data (e.g. by taking logarithms).

Program 30: Principal component analysis

This method of multivariate analysis has its origins in work by Karl Pearson in 1901, though it was Hotelling (1933) who first established its use for analysing the structure of a set of interrelated variables in terms of a simpler set of orthogonal, i.e. uncorrelated, components. Suppose, for example, that our data comprise measurements of m variables on each of a set of individuals. The measurements made on an individual can be represented as $x_1, x_2, \ldots x_m$ and we shall seek a set of up to m principal components of the form

$$c = a_1x_1 + a_2x_2 + a_3x_3 + \ldots + a_mx_m. \tag{18}$$

In other words, the principal components are variously weighted sums of the original variates and our main task is to determine the numerical values of the coefficients $a_1, a_2, a_3, \ldots a_m$ for each principal component. It can be shown that the values we require are the elements of successive latent vectors of the covariance matrix S derived from the set of multivariate observations on all individuals in the sample. The principal component (or component score as it is sometimes called) is then computed by multiplying each of the original measurements by the corresponding element of the appropriate latent vector and summing the products. Since the covariance matrix will have $k(\leqslant m)$ non-zero latent roots, there will be k orthogonal vectors and k different principal components. The sum of the latent roots is equal to the variance of the whole system of principal components and the proportion of that variance which is associated with the j-th component can be calculated as the fraction which the j-th latent root forms of the trace of S. Further, the size of the elements of a given latent vector indicates the importance of the corresponding variable in determining the principal component concerned.

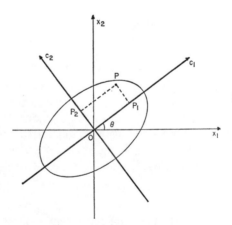

FIG. 30. Diagrammatic representation of the method of principal components for two variables. For explanation see text. θ represents the angle through which the original axes x_1 and x_2 have been rotated to give the component axes c_1 and c_2.

Geometrically, the ideas underlying principal component analysis can easily be visualised for the case of two variables (Fig. 30). Each individual described by measurements on these two variables (say x_1 the height and x_2 the length of leg) can be denoted by a point P in the system of two perpendicular axes along which x_1 and x_2 are measured. If a large number of individuals are so measured, the resulting points will take on the appearance of an elliptical swarm. The first principal component of P is then given by the projection OP_1 of the point P on to the major axis of the ellipse. The second principal component is given by the projection OP_2 on to the minor axis of the ellipse. If the individuals were characterised by measurements on three variables, the ellipse of Fig. 30 would be replaced by an ellipsoid and a third principal component would be represented by a projection on to the third principal axis of the ellipsoid. Systems involving 4, 5, 6, ... m variables require additional dimensions and so cannot be visualised, but they are all amenable to the same mathematical treatment. In calculating the principal components we are, in effect, using a method which determines the axes of the ellipse of Fig. 30 by minimising the squares of the distances corresponding to PP_1 and PP_2 for each point that has been observed. It is also clear from the figure that we have in effect replaced the original axes x_1, x_2 by the new ones c_1 and c_2. The matrix of latent vectors that arises in the course of the computations is therefore a mathematical operator that transforms the coordinates of any point P into the new set of coordinates in which the principal components are expressed.

As an example of the way in which the method of principal components is used in biometrics, we may consider its application to the analysis of the dimensions of an animal or plant. A sample of several individuals of a given species is described by measurements of m more or less highly correlated dimensions. The $m \times m$ covariance matrix may have up to m non-zero latent roots and associated vectors, each of which may be regarded as expressing an independent component or aspect of the form of the organism. The relative importance of each component may be assessed by the size of the corresponding latent root and one may expect to describe all the major features in terms of an appreciably smaller number of components than the original m variables, thus bringing about a simplification or, as it is often put, a more parsimonious description of the form. The importance of any character in determining the size of a component may be estimated from the size of the relevant vector element and in favourable cases this information may help in providing a biological interpretation of the components. Principal component analysis may also provide a method of 'cluster analysis' since the points representing individuals may fall into two or more easily discernible groups when plotted in a co-ordinate system involving any two or three of the principal component axes.

We may note finally that principal component analysis does not assume any particular distributions for the variables (though significance tests may require them to have been sampled from a multivariate normal population). If the components are to have a physical meaning, however, it is necessary that the dimensions of all m variables should be the same and expressed in the same units. When they are not the same (one being a weight, for example, another an area and a third the molar concentration of a chemical substance), one can nevertheless standardise the original data, expressing them all as deviations from the means of the variables in units of 1 standard deviation. The covariance matrix S is then replaced by a correlation matrix.

The above very brief introduction to principal component analysis should be enough to make intelligible the Fortran program given below. This will employ subroutines CORMAT and LATENT and function subprogram DET so that we begin the main program by a type-declaration for LAMBDA (which will accommodate the latent roots returned by LATENT) and by suitable DIMENSION and EQUIVALENCE statements. Multivariate analysis usually makes considerable demands on storage space and we should always make the larger matrices equivalent if that is at all possible. The first program segment then causes a small group of parameters to be read: NPROB, as usual, is a problem reference number; NROW and NCOL define the number of rows (individuals) and columns (variables) in the primary data matrix; INDEX provides the option of calling CORMAT so as to provide a covariance matrix (INDEX = 2) or a correlation matrix (INDEX = 3); NTAPE is the number of a binary device (tape or disk) which will be used to store the data and which must be numbered in conformity with procedure at the computer installation where the job is to be run. NTAPE must not be 5 or 6 since these are numbers used in all our programs to denote the formatted input and output tapes. We may note in passing that we use tape or disk storage for the data because it will be required twice—once to compute the covariance matrix and a second time to compute the principal component scores. In many ways it would be simpler to have stored the data in central memory but we already require a good deal of central memory space for the covariance matrix, latent vectors and so on and we are also envisaging that the program may be used with large bodies of data (up to perhaps 50 variables and several hundred individuals). We therefore have no alternative but to arrange for this to be held in an auxiliary store. For further details of the Fortran statements governing tape or disk manipulation see p. 449. The first program segment then continues by reading a variable input format, after which we are ready to read the main bulk of the data. Since this is to be stored on tape or disk we rewind to the starting point ('load point') and then by a DO-loop ending on statement 1 we read the cards, on each of which a row of the primary data matrix (DATA) has been punched. After each row has been read we also write it on tape, the latter process being carried out in binary and therefore needing no format. When the DO-loop is satisfied we close the file (END FILE NTAPE) and rewind the tape for subsequent use.

The next program-segment computes and prints the covariance or correlation matrix by calling subroutine CORMAT. In the present program CORMAT is slightly different from the version given as Program 23 on p. 236 since the subroutine now entails reading successive values of DATA from the tape or disk on which they have been stored. The differences are as follows:

(a) The first statement is
SUBROUTINE CORMAT (M, N, INDEX, R, NTAPE),
the parameters being transmitted explicitly.

(b) The statement
READ (5,100) M, N, INDEX
and its associated format declaration are now omitted.

(c) Statement 2 of Program 23 is now replaced by
2 READ (NTAPE) (X(I), I = 1, M)
and the format declaration numbered 101 is omitted.

When R, the required covariance or correlation matrix, has been returned, we compute its trace (TR) by summing the diagonal elements and its determinant by using FUNCTION DET. We need to bear in mind, however, that DET causes destruction of the input matrix so that before calling it we copy R into another array RR and use the latter at this point. The determinant and trace of the matrix are printed out and thus made available for significance tests described in Lawley and Maxwell (1963, pp. 51–54). The particular tests used will vary according to the user's requirements and it is better for him to select them rather than provide instructions in the program.

The extraction of latent roots and vectors of the matrix R is carried out in the next program-segment by calling subroutine LATENT. This returns the roots (arranged in descending order of size) in LAMBDA and the corresponding vectors, arranged columnwise, in VEC. It is first necessary to define NROOT, the number of non-zero roots and associated vectors to be extracted. This follows the principles indicated above, with an arbitrary maximum of 10 which is sufficient for practically all biological work and enables one to achieve a compact output. On return, NROOT will contain the number of latent roots which have actually been extracted (thus allowing for failure to converge during iteration). The returned value of NROOT is needed to control the output of latent roots and vectors. This is

Outline flowchart for Program 30
Principal component analysis.

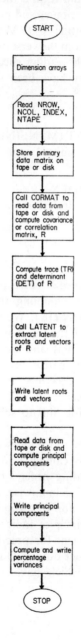

arranged so that under each latent root is a column consisting of the elements of the corresponding latent vector. These elements are the weights $a_1, a_2, a_3, \ldots a_m$ of equation (18) above. Each vector is normalised so that the sum of its squared elements is unity. In some versions of principal component analysis it is usual to scale the vectors so that this sum is equal to the corresponding latent root, and an addition of this sort could easily be made to the program if desired.

Having computed and printed the latent vectors it is possible to obtain the principal components for each individual. This requires the primary data matrix to be read again from tape, one row at a time. Each row-vector of observations is then multiplied by each latent vector to give the corresponding principal component. The DO-loops ending on 6 achieve this and write out the principal components as a matrix in which each row corresponds to one of the individuals in the primary data and successive columns correspond to the first, second, third, etc., principal component.

Finally the last short program segment allows one to calculate, from the latent roots and the trace, what proportion of the total variance is associated with each component.

Program 30

Principal component analysis. The listing gives the main program which calls subroutines CORMAT and LATENT (p. 265) and Function Sub-program DET (p. 246). Subroutine CORMAT is amended to read data from tape and the first few lines are given here. For the remainder of CORMAT see p. 236. Sample data and output are also listed.

```
C
C.................................................
C
C  THIS PROGRAM CARRIES OUT A PRINCIPAL COMPONENT ANALYSIS OF A
C  COVARIANCE OR CORRELATION MATRIX. THE LATTER IS DERIVED FROM
C  A PRIMARY DATA MATRIX OF UP TO 50 COLUMNS AND ANY NUMBER OF
C  ROWS, CORRELATIONS BEING COMPUTED BETWEEN COLUMNS. INPUT PARAMETERS
C  ARE AS FOLLOWS -
C     NROW = NUMBER OF ROWS IN DATA MATRIX
C     NCOL = NUMBER OF COLUMNS IN DATA MATRIX
C     DATA = VECTOR OF PRIMARY OBSERVATIONS (LENGTH NCOL)
C     INDEX = 2 FOR COVARIANCE MATRIX, 3 FOR CORRELATION MATRIX
C     NTAPE = NUMBER OF SYMBOLIC BINARY DEVICE (TAPE OR DISK)
C     INFMT = VECTOR HOLDING VARIABLE INPUT FORMAT
C  FUNCTION SUBPROGRAM DET AND SUBROUTINES CORMAT AND LATENT ARE REQUIRED
C  REFERENCE - MORRISON (1967) MULTIVARIATE STATISTICAL METHODS, CHAP.7.
C
C.................................................
C
      REAL LAMBDA
      DIMENSION R(50,50), RR(50,50), VEC(50,10), LAMBDA(10),
```

Program 30 (continued)

```
      A   U1(50), UN(50), DATA(50), PRIN(10), INFMT(20)
          EQUIVALENCE (RR(1,1), VEC(1,1))
C
C     READ PROBLEM SPECIFICATION AND STORE PRIMARY DATA ON BINARY DEVICE.
          READ (5,100) NROW, NCOL, INDEX, NTAPE
  100     FORMAT (4I5)
          READ (5,101) (INFMT(I), I = 1,20)
  101     FORMAT (20A4)
          REWIND NTAPE
          DO 1 I = 1,NROW
          READ (5,INFMT) (DATA(J), J = 1,NCOL)
    1     WRITE (NTAPE) (DATA(J), J = 1,NCOL)
          END FILE NTAPE
          REWIND NTAPE
C
C     COMPUTE COVARIANCE OR CORRELATION MATRIX, TRACE AND DETERMINANT.
          CALL CORMAT (NCOL,NROW,INDEX,R,NTAPE)
          TR = 0.0
          DO 2 I = 1 NCOL
    2     TR = TR + R(I,I)
          DO 3 I = 1,NCOL
          DO 3 J = 1,NCOL
    3     RR(I,J) = R(I,J)
          D = DET(RR,NCOL)
          WRITE (6,102) TR, D
  102     FORMAT (//8H TRACE = , 1PE15.7, 16H DETERMINANT = , 1PE15.7)
C
C     COMPUTE LATENT ROOTS AND VECTORS.
          NROOT = 10
          MIN = MIN0(NCOL,NROW)
          IF (MIN .LT. 10) NROOT = MIN
          CALL LATENT (R,NCOL,NROOT,LAMBDA,VEC,U1,UN)
          IF (NROOT .EQ. 0) GO TO 8
          WRITE (6,103) (LAMBDA(I), I = 1,NROOT)
  103     FORMAT (//13H LATENT ROOTS//10F12.3)
          WRITE (6,104)
  104     FORMAT (//15H LATENT VECTORS/)
          DO 4 I = 1,NCOL
    4     WRITE (6,105) (VEC(I,J), J = 1,NROOT)
  105     FORMAT (10F12.3)
C
C     COMPUTE PRINCIPAL COMPONENTS.
          REWIND NTAPE
          WRITE (6,106)
  106     FORMAT (//21H PRINCIPAL COMPONENTS/)
          DO 6 I = 1,NROW
          READ (NTAPE) (DATA(J), J = 1,NCOL)
          DO 5 J = 1,NROOT
          PRIN(J) = 0.0
          DO 5 K = 1,NCOL
    5     PRIN(J) = PRIN(J) + DATA(K)*VEC(K,J)
    6     WRITE (6,107) (PRIN(J), J = 1,NROOT)
  107     FORMAT (10F12.3)
C
C     COMPUTE PERCENTAGE VARIANCES.
          DO 7 I = 1,NROOT
    7     LAMBDA(I) = LAMBDA(I) * 100.0 / TR
          WRITE (6,108) (LAMBDA(I), I = 1,NROOT)
  108     FORMAT (//21H PERCENTAGE VARIANCES//10F12.3)
    8     STOP
          END
```

Program 30 (*continued*)

```
      SUBROUTINE CORMAT (M,N,INDEX,R,NTAPE)
C
C.........................................................................
C
C     THIS SUBROUTINE COMPUTES AND PRINTS A DEVIATION SUMS OF SQUARES AND
C     CROSS-PRODUCTS MATRIX OR A DISPERSION (VARIANCE-COVARIANCE) MATRIX
C     OR A CORRELATION MATRIX. PARAMETERS ARE AS FOLLOWS -
C        M     = NUMBER OF COLUMNS IN PRIMARY DATA MATRIX
C        N     = NUMBER OF ROWS IN PRIMARY DATA MATRIX
C        INDEX = 1 FOR DEVIATION SUMS OF SQUARES AND CROSS-PRODUCTS MATRIX
C                2 FOR DISPERSION MATRIX
C                3 FOR CORRELATION MATRIX
C        X     = VECTOR (LENGTH M) REPRESENTING ONE ROW OF DATA MATRIX
C        R     = OUTPUT MATRIX
C
C.........................................................................
C
      DIMENSION R(50,50), X(50), SX(50), SDEV(50)
C
C     INITIALISE FOR SUBSEQUENT ACCUMULATION OF SUMS.
      NC = 0
      DO 1 I = 1,M
      SX(I) = 0.0
      DO 1 J = 1,M
    1 R(I,J) = 0.0
C
C     READ PRIMARY DATA ONE ROW AT A TIME.
    2 READ (NTAPE) (X(I), I = 1,M)
C
C     COMPUTE SQUARES AND CROSS-PRODUCTS HALF-MATRIX
      DO 3 I = 1,M
      SX(I) = SX(I) + X(I)
      DO 3 J = I,M
```

DATA

```
    5    7    3    4
(10F7.2)
    3.1   14.2   7.3   20.1   2.2   13.1   3.8
   16.0   21.2   5.8   30.0   4.0   21.1   7.8
   20.8   30.5   7.6   40.1   3.9   25.2   8.8
    5.2   17.3   8.1   18.8   3.7   11.1   6.2
    5.0   15.1   1.9   28.2   1.7    7.2   4.0
```

OUTPUT

CORRELATION HALF-MATRIX

ROW 1
  1 1.0000000E 00   2 9.4758192E-01   3 2.2051479E-01   4 8.8689353E-01   5 7.2267816E-01
  6 9.3193036E-01   7 9.2852542E-01

ROW 2
  2 1.0000000E 00   3 3.5400160E-01   4 8.6823834E-01   5 7.0221788E-01   6 8.9087273E-01
  7 9.0894378E-01

ROW 3
  3 9.9999998E-01   4 -1.2751488E-01   5 6.4236388E-01   6 4.8263591E-01   7 4.3574628E-01

ROW 4
  4 9.9999996E-01   5 3.4740972E-01   6 7.3168578E-01   7 6.9027752E-01

Program 12 (*continued*)

```
ROW 5
    5  1.0000000E 00   6  7.5512982E-01   7  9.1785858E-01

ROW 6
    6  1.0000000E 00   7  8.8120506E-01

ROW 7
    7  9.9999998E-01


TRACE = 6.9999999E 00   DETERMINANT = 0.


LATENT ROOTS
       5.231        1.354       0.302       0.112

LATENT VECTORS

       0.424       -0.199      -0.057      -0.241
       0.421       -0.124       0.234       0.558
       0.194        0.727       0.524       0.185
       0.341       -0.523       0.235       0.201
       0.366        0.372      -0.604       0.002
       0.417        0.025       0.312      -0.729
       0.425        0.063      -0.387       0.155


PRINCIPAL COMPONENTS

      23.450       -6.196      12.976       3.610
      40.646      -14.773      15.269       0.914
      52.488      -20.726      21.436       4.477
      26.094       -5.078      11.236       6.555
      23.789      -15.172      10.534       8.619


PERCENTAGE VARIANCES

      74.730       19.341       4.319       1.604
```

Program 31: Multiple discriminant analysis.

Program 29 enables one to discriminate between two groups; the present program extends the analysis to a situation in which one may wish to discriminate between several groups. It is therefore of considerable interest to those wishing to classify specimens, on each of which a number of measurements have been made and which are to be allocated to previously defined groups. The primary data will consist of k groups containing n_1, n_2, n_3, ... n_k individuals respectively. Each of the individuals will have been defined in terms of p variables. These need not have been measured in the same units—one may be a length, another an area, a third a weight, and so on. Our object is to derive a set of discriminant functions of the form

$$d = a_1 x_1 + a_2 x_2 + a_3 x_3 + \ldots + a_p x_p \tag{19}$$

where $a_1, a_2, a_3, \ldots$ are discriminant coefficients computed so as to minimise the 'overlap' between one group and another. The problem can be considered in terms of the between-group variation as compared with the within-group variation. If we denote the between-group sums of squares and products matrix as **B** and the pooled within-group sums of squares and products matrix as **W**, then we shall require the latent roots and vectors of the product matrix $\mathbf{W}^{-1}\mathbf{B}$. The elements of the normalised latent vectors will provide the weights $a_1, a_2, a_3, \ldots$ of (19) while the latent roots will provide a measure of the discriminatory power associated with each of the 'canonical variates' or discriminant axes, in terms of which the data are now represented. Both **B** and $\mathbf{W}^{-1}$ are symmetric matrices, but their product $\mathbf{W}^{-1}\mathbf{B}$ is in general non-symmetric. We shall therefore require a program in which the subroutine VECTOR can be called to extract the latent roots and vectors. The maximum number of non-zero latent roots which can be extracted will be whichever is the smaller of p and $(k-1)$. We shall, however, impose an arbitrary maximum of 10, which is more than enough for all biological applications of the technique.

While the extraction of the latent roots and vectors of $\mathbf{W}^{-1}\mathbf{B}$ is the essential procedure in a multiple discriminant analysis, the program is somewhat more elaborate than this since it computes certain other results, i.e.

(a) the means and standard deviations of each variable for each group.

(b) the means and standard deviations of each variable for all groups combined.

(c) a chi-squared test of the equality of group means.

(d) chi-squared tests of the significance of each latent root extracted.

(e) the discriminant scores (i.e. the value of d in equation (19) for each individual in every group).

(f) the discriminant scores for each group centroid (i.e. the mean of the discriminant scores of all the individuals in a given group).

In general the program follows the computational techniques set out in Hope (1968; Chapter 7).

The reader should by now require a less full commentary on the Fortran program and the following explanation will concentrate only on the important or especially interesting features. After suitable DIMENSION and EQUIVALENCE declarations the program reads a specification card defining the number of the problem (NPROB), number of groups (NK), number of variables (NP), number of the symbolic binary unit used as a backing store (NTAPE) and the number of individuals in each of the NK groups (N). In the subsequent initialisation, note particularly the two matrices

T and W which will be used eventually to accumulate the total sums of squares and products matrix and the within-groups sums of squares and products matrix.

The main loop ending on 7 carries out a number of functions:

(a) It reads the data one row at a time, writing each row on tape and accumulating the sums of each variable both for a given group (SX) and for all groups together (SXT).

(b) It also accumulates the uncorrected sums of squares and cross-products, both for a given group (in array B, used simply as a working matrix) and for all groups together (T).

(c) It then obtains the corrected sums of squares and cross-products for a given group (in B) and pools these corrected sums in W.

(d) Finally it uses the information in SX and B to produce means and standard deviations for each group and writes the group means on tape. The same array SX is used successively to hold the sums and the means derived from them.

The next program segment computes similar statistics for the entire sample (i.e., all groups together) and prints the total sample mean and standard deviations. At this point in the program, therefore, (statement 106) we have in central memory the total sums of cross-products matrix (T) and the pooled within-groups sums of cross-products matrix (W) while on the tape or disk backing store we have the data for each individual in a group followed by the means for that group.

We now deviate from the straightforward multiple discriminant analysis to carry out some significance tests and accuracy checks. We first require the determinant of the pooled within-groups matrix W. We obtain this as DW, using FUNCTION DET and remembering to apply the function to a *copy* of W. This copy is made in array B, which is still fulfilling its role as a working matrix. If DW proves to be very small (we use the arbitrary criteria of being less than 0.0001 in absolute value) it suggests that W is singular or ill-conditioned, so that its inverse either does not exist or is likely to yield inaccurate results. Under these circumstances it would be wiser to abandon the analysis and the program therefore returns control to statement 1 after printing a suitable diagnostic message. If, however, the determinant of W is sufficiently large, we next compute the determinant of T (DT), again using B as a working matrix for copying purposes. DT will form the divisor in a test of the equality of group means and the test will obviously fail if DT is zero. The program allows for this by printing a diagnostic message but execution is continued nevertheless. Assuming that the two determinants DT and DW are available in suitable form, however, we can use them in a test of the null hypothesis that the group means are equal. This involves the computation of a chi-squared approximation:

$$\chi^2 = -\left[\sum_{i=1}^{k} n_i - 1 - \tfrac{1}{2}(k + p)\right] \ln L \qquad (20)$$

where
$$L = \frac{|\mathbf{W}|}{|\mathbf{T}|}$$

and where the number of degrees of freedom is given by $p(k - 1)$. The test cannot be made if $L \leq 0$ and this possibility is allowed for in the program.

Whether or not the test is performed the next segment, starting with statement 18, continues the essential procedure of a multiple discriminant analysis. We first compute the between groups matrix B from the simple relationship $B = T - W$, then we invert W by calling subroutine INVERT and multiply to obtain the product matrix $\mathbf{W}^{-1}\mathbf{B}$ which we store in T as that array is no longer needed for its original purpose. These operations are completed by statement 21. The trace of $\mathbf{W}^{-1}\mathbf{B}$ is then computed and subroutine VECTOR is called to provide the latent roots and vectors which are printed out in the next segment, together with the percentage discriminatory power associated with each root.

A test of the significance of the dispersion of group means along each axis is now carried out, computing a chi-squared for each root by the formula

$$\chi^2 = \left[\sum_{i=1}^{k} n_i - 1 - \tfrac{1}{2}(k + p)\right] \ln(1 + \lambda_i) \qquad (21)$$

where λ_i is the i-th latent root. The number of degrees of freedom associated with the χ^2 for λ_i is given by $(p + k - 2i)$. It may be noted that these χ^2 values sum to the χ^2 calculated earlier from equation (20) and that the degrees of freedom are also additive.

The last program segment computes the discriminant scores ('canonical variates') for each individual and for the centroid of every group. For this to be done it is only necessary to read the individual observations and group means from tape and post-multiply by the latent vectors in turn. The working vector WORK is used to accumulate the discriminant scores and X is used as a working vector to hold each set of observations as it is read from tape.

Outline flowchart for Program 31

Multiple discriminant analysis.

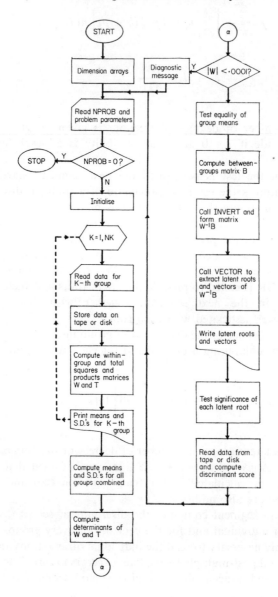

Program 31

Multiple discriminant analysis: The listing gives the main program which calls subroutines INVERT (p. 255) and VECTOR (p. 270) and Function Sub-program DET (p. 246). Sample data and output are also listed.

```
C
C........................................................................
C
C   THIS PROGRAM CARRIES OUT A MULTIPLE DISCRIMINANT ANALYSIS
C   (CANONICAL ANALYSIS). INPUT PARAMETERS ARE AS FOLLOWS –
C       NPROB = PROBLEM REFERENCE NUMBER (ZERO ON SENTINEL CARD, TO BE
C               FOLLOWED BY ONE BLANK CARD BEFORE END–OF–FILE CARD)
C       NK    = NUMBER OF GROUPS
C       NP    = NUMBER OF VARIABLES
C       NTAPE = NUMBER OF SYMBOLIC BINARY UNIT (TAPE OR DISK)
C       N     = VECTOR (LENGTH NK) GIVING NUMBERS OF INDIVIDUALS
C               IN EACH GROUP
C       X     = VECTOR (LENGTH NP) CORRESPONDING TO ONE ROW OF
C               PRIMARY DATA MATRIX
C   FOR THEORY AND DETAILS OF COMPUTATION, SEE HOPE, K., (1968) METHODS OF
C   MULTIVARIATE ANALYSIS. CHAP.7. THIS PROGRAM CALLS SUBROUTINES
C   INVERT AND VECTOR AND FUNCTION SUBPROGRAM DET.
C
C........................................................................
C
        REAL LAMBDA
        DIMENSION N(25), SXT(25), W(25,25), B(25,25), T(25,25), SX(25),
       A X(25), SDEV(25), IR(25), IC(25), U1(25), UN(25), V1(25),
       B VN(25), WORK(10), LAMBDA(10), VEC(25,10), NDF(10)
        EQUIVALENCE (B(1,1),VEC(1,1)), (B(1,11),U1(1)), (B(1,12),UN(1)),
       A (B(1,13),V1(1)), (B(1,14),VN(1)), (SXT(1),IR(1)),(SDEV(1),IC(1))
C
C   READ PROBLEM SPECIFICATION AND INITIALISE.
      1 READ (5,100) NPROB, NK, NP, NTAPE, (N(I), I = 1,NK)
    100 FORMAT (4I5/(24I3))
        IF (NPROB .EQ. 0) STOP
        WRITE (6,101) NPROB
    101 FORMAT (39H1MULTIPLE DISCRIMINANT ANALYSIS PROBLEM, I4)
        DO 2 I = 1,NP
        SXT(I) = 0.0
        DO 2 J = 1,NP
        W(I,J) = 0.0
      2 T(I,J) = 0.0
        P = NP
        ANK = NK
        REWIND NTAPE
C
C   BEGIN MAIN LOOP FOR K–TH GROUP. FROM HERE TO STATEMENT 14 B IS
C   USED AS A WORKING MATRIX ONLY.
        NT = 0.0
        DO 7 K = 1,NK
        NG = N(K)
        G = NG
        NT = NT + N(K)
        DO 3 I = 1,NP
        SX(I) = 0.0
        DO 3 J = 1,NP
      3 B(I,J) = 0.0
C
```

Program 31 (*continued*)

```
C   READ DATA ONE ROW AT A TIME AND FORM TOTAL AND POOLED WITHIN-GROUP
C   DEVIATION SQUARES AND CROSS-PRODUCTS MATRICES.
        DO 4 I = 1,NG
        READ (5,102) (X(J), J = 1,NP)
  102   FORMAT (10F7.0)
        WRITE (NTAPE) (X(J), J = 1,NP)
        DO 4 J = 1,NP
        SXT(J) = SXT(J) + X(J)
        SX(J) = SX(J) + X(J)
        DO 4 L = J,NP
        B(J,L) = B(J,L) + X(J) * X(L)
    4   T(J,L) = T(J,L) + X(J) * X(L)
        DO 5 I = 1,NP
        DO 5 J = I,NP
        B(I,J) = B(I,J) - SX(I)*SX(J)/G
    5   W(I,J) = W(I,J) + B(I,J)
        DO 6 I = 1,NP
        SX(I) = SX(I) / G
    6   SDEV(I) = SQRT(B(I,I)/(G-1.0))
        WRITE (NTAPE) (SX(I), I = 1,NP)
        WRITE (6,103) K, (SX(I), I = 1,NP)
  103   FORMAT (//6H GROUP, I3//8H    MEANS//(10F12.3))
    7   WRITE (6,104) (SDEV(I), I = 1,NP)
  104   FORMAT (/22H   STANDARD DEVIATIONS//(10F12.3))
        END FILE NTAPE
C
C   COMPUTE STATISTICS FOR TOTAL SAMPLE.
        ANT = NT
        DO 8 I = 1,NP
        DO 8 J = I,NP
    8   T(I,J) = T(I,J) - SXT(I)*SXT(J)/ANT
        DO 9 I = 1,NP
        SXT(I) = SXT(I) / ANT
    9   SDEV(I) = SQRT(T(I,I)/(ANT-1.0))
        WRITE (6,105) (SXT(I), I = 1,NP)
  105   FORMAT (//13H TOTAL SAMPLE//8H    MEANS//(10F12.3))
        WRITE (6,106) (SDEV(I), I = 1,NP)
  106   FORMAT (/22H   STANDARD DEVIATIONS//(10F12.3))
C
C   COMPUTE DETERMINANTS NEEDED FOR SIGNIFICANCE TESTS.
        DO 10 I = 1,NP
        DO 10 J = I,NP
   10   W(J,I) = W(I,J)
        DO 11 I = 1,NP
        DO 11 J = 1,NP
   11   B(I,J) = W(I,J)
        DW = DET(B,NP)
        IF (ABS(DW) .GE. 0.0001) GO TO 12
        WRITE (6,107)
  107   FORMAT (//100H DETERMINANT OF WITHIN-GROUPS MATRIX LESS THAN 0.000
       A1.   EXECUTION TERMINATED. PASS TO NEXT PROBLEM.)
        GO TO 1
C
   12   DO 13 I = 1,NP
        DO 13 J = I,NP
   13   T(J,I) = T(I,J)
        DO 14 I = 1,NP
        DO 14 J = 1,NP
   14   B(I,J) = T(I,J)
        DT = DET(B,NP)
```

Program 31 (*continued*)

```
         IF (DT .NE. 0.0) GO TO 15
         WRITE (6,108)
     108 FORMAT (//85H TOTAL SUM OF PRODUCTS MATRIX SINGULAR. NO TEST OF
        A EQUALITY OF GROUP MEANS POSSIBLE.)
         GO TO 18
      15 EL = DW / DT
         IF (EL) 17,17,16
      16 FAC = ANT - 1.0 - 0.5*(P+ANK)
         CHI2 = - FAC * ALOG(EL)
         NF = NP * (NK-1)
         WRITE (6,109) CHI2, NF
     109 FORMAT (//48H TEST OF EQUALITY OF GROUP MEANS. CHI-SQUARED = ,
        A F10.3, 5H WITH, I4, 19H DEGREES OF FREEDOM)
         GO TO 18
      17 WRITE (6,110)
     110 FORMAT (//86H RATIO OF DETERMINANTS ZERO OR NEGATIVE. NO TEST OF
        A EQUALITY OF GROUP MEANS POSSIBLE.)
C
C   FORM PRODUCT MATRIX W-1B, STORE IN T AND EXTRACT LATENT ROOTS
C   AND VECTORS.
      18 DO 19 I = 1,NP
         DO 19 J = 1,NP
      19 B(I,J) = T(I,J) - W(I,J)
         CALL INVERT (W,NP,DW,IR,IC)
         DO 21 I = 1,NP
         DO 21 J = 1,NP
         S = 0.0
         DO 20 K = 1,NP
      20 S = S + W(I,K) * B(K,J)
      21 T(I,J) = S
         TR = 0.0
         DO 22 I = 1,NP
      22 TR = TR + T(I,I)
         NROOT = 10
         MIN = MIN0(NP,(NK-1))
         IF (MIN .LT. 10) NROOT = MIN
         CALL VECTOR (T,NP,NROOT,LAMBDA,VEC,U1,UN,V1,VN)
         IF (NROOT .GE. 1) GO TO 23
         WRITE (6,111)
     111 FORMAT (//45H EXECUTION TERMINATED. PASS TO NEXT PROBLEM.)
         GO TO 1
C
C   PRINT LATENT ROOTS AND VECTORS.
      23 WRITE (6,112) (LAMBDA(I), I = 1,NROOT)
     112 FORMAT (//13H LATENT ROOTS//10F12.3)
         DO 24 I = 1,NROOT
      24 WORK(I) = LAMBDA(I) / TR * 100.0
         WRITE (6,113) (WORK(I), I = 1,NROOT)
     113 FORMAT (//26H PERCENTAGE DISCRIMINATION//10F12.3)
         WRITE (6,114)
     114 FORMAT (//15H LATENT VECTORS/)
         DO 25 I = 1,NP
      25 WRITE (6,115) (VEC(I,J), J = 1,NROOT)
     115 FORMAT (10F12.3)
C
C   TEST SIGNIFICANCE OF EACH LATENT ROOT.
         DO 26 I = 1,NROOT
      26 WORK(I) = FAC * ALOG(1.0+LAMBDA(I))
         WRITE (6,116) (WORK(I), I = 1,NROOT)
     116 FORMAT (//12H CHI-SQUARED//10F12.3)
```

Program 31 *(continued)*

```
      DO 27 I = 1,NROOT
   27 NDF(I) = NP + NK - 2*I
      WRITE (6,117) (NDF(I), I = 1,NROOT)
  117 FORMAT (//19H DEGREES OF FREEDOM//10I2)
C
C  COMPUTE DISCRIMINANT SCORES, USING X AS WORKING VECTOR.
      WRITE (6,118)
  118 FORMAT (//20H DISCRIMINANT SCORES)
      REWIND NTAPE
      DO 31 K = 1,NK
      WRITE (6,119) K
  119 FORMAT (//8H GROUP, I3/)
      NG = N(K)
      DO 29 I = 1,NG
      READ (NTAPE) (X(J), J = 1,NP)
      DO 28 J = 1,NROOT
      WORK(J) = 0.0
      DO 28 L = 1,NP
   28 WORK(J) = WORK(J) + X(L)*VEC(L,J)
   29 WRITE (6,115) (WORK(J),J = 1,NROOT)
      READ (NTAPE) (X(J), J = 1,NP)
      DO 30 J = 1,NROOT
      WORK(J) = 0.0
      DO 30 L = 1,NP
   30 WORK(J) = WORK(J) + X(L)*VEC(L,J)
   31 WRITE (6,120) K, (WORK(J), J = 1,NROOT)
  120 FORMAT (/8H    GROUP, I3, 9H CENTROID//(10F12.3))
      GO TO 1
      END
```

DATA

```
    1    3    3    4
 3  2    4
    2.0  3.0  1.0
    1.0  2.0  3.0
    2.0  2.0  1.0
    3.0  1.0  1.0
    2.0  2.0  3.0
    2.0  1.0  3.0
    1.0  2.0  3.0
    3.0  2.0  4.0
    2.0  1.0  2.0
 0
```

OUTPUT

MULTIPLE DISCRIMINANT ANALYSIS PROBLEM   1

GROUP   1

MEANS

    1.667   2.333   1.667

STANDARD DEVIATIONS

    0.577   0.577   1.155

Program 31 (*continued*)

GROUP 2

  MEANS

    2.500   1.500   2.000

  STANDARD DEVIATIONS

    0.707   0.707   1.414

GROUP 3

  MEANS

    2.000   1.500   3.000

  STANDARD DEVIATIONS

    0.816   0.577   0.816

TOTAL SAMPLE

  MEANS
    2.000   1.778   2.333

  STANDARD DEVIATIONS

    0.707   0.667   1.118

TEST OF EQUALITY OF GROUP MEANS. CHI–SQUARED =   6.233 WITH   6 DEGREES OF FREEDOM

LATENT ROOTS

    1.873   0.211

PERCENTAGE DISCRIMINATION

    89.880   10.119

LATENT VECTORS

     0.347   0.887
    −0.809   −0.152
     0.475   −0.436

CHI–SQUARED

    5.277   0.957

DEGREES OF FREEDOM

        4   2

Program 31 (*continued*)

DISCRIMINANT SCORES

GROUP 1

−1.258	0.883
0.154	−0.725
−0.450	1.035

GROUP 1 CENTROID

−0.518	0.398

GROUP 2

0.706	2.073
0.501	0.162

GROUP 2 CENTROID

0.603	1.118

GROUP 3

1.310	0.314
0.154	−0.725
1.322	0.613
0.834	0.750

GROUP 3 CENTROID

0.905	0.238

This program is the most elaborate in the present book, partly because a multiple discriminant analysis is a relatively complicated statistical undertaking and partly because we have included tests of significance and measures of accuracy and of the condition of the matrices being handled. Alternative methods of programming an analysis of this kind are available. For example it is possible to obtain the latent roots and vectors of $W^{-1}B$ directly from W and B. This obviates the need for inversion and deals only with symmetric matrices that are amenable to efficient extraction procedures such as the Jacobi and Givens–Householder techniques (p. 264). A more radical alternative has been found by Gower (1966), based on a $k \times k$ symmetric matrix formed in a way he discusses from the mean vectors of k groups and the pooled within-group dispersion matrix. When, as is often the case, $k < p$ this method has the advantage of needing the roots and vectors of a smaller matrix; it is also closely related to a method of computing the generalised distances (Mahalanobis' D^2) between any pair of groups.

In displaying the results of a multiple discriminant analysis it is usual to plot the individual and/or the group discriminant scores on two axes at a time, as illustrated, for example, by Ashton *et al.* (1965) in their interesting application of the method to the functional anatomy of the primate pectoral girdle. Plots of this kind can readily be made by computer with the aid of suitable programs or off-line plotting facilities. It is also possible to construct three-dimensional graphs by hand or using more elaborate plotting programs. As with the similar graphical representation of principal component analyses, however, one must beware of the misleading effects of perspective when studying such drawings. A more satisfactory method of display is the construction of 3-dimensional models in which clusters of points are represented by differently coloured small spheres mounted on rigid vertical rods.

Chapter 11

Nonparametric Statistics

The statistical methods discussed so far are all parametric, in the sense that they assume the data to belong to certain definite distributions characterised by certain parameters. This is so, for example, in the t-tests of Program 3, where the data are assumed to be normally distributed, while in the analysis of variance programs of Chapter 6 it is assumed that the residuals are normally distributed with constant variance. If the data do not conform, at least approximately, to the required distribution they cannot validly be analysed by methods based on it. In practice we may be faced with data which we know or suspect to be non-normal or otherwise unusual and one common way of dealing with this situation is to transform the data to normality or uniform variances. This possibility is discussed briefly on p. 346; in the present chapter we consider an alternative, namely the use of nonparametric or distribution-free statistical tests, in which fewer and less restrictive assumptions are made about the data. The best known of these tests are the chi-squared tests for association or goodness of fit, but many others are available and a few will be programmed here.

In general, nonparametric tests do not require the data to conform to any particular distribution, and they may be used with large or small samples. They are often applicable not only to data measured on an ordinary numerical scale (e.g. lengths, weights, times, etc.) but also to data set out as ranks on an *ordinal* scale or to those where qualitatively defined groups (e.g. those animals with certain colours or those which react in several quite different ways) are assigned numerical codes (i.e. are recorded on a *nominal* scale). Of course, if a parametric test is appropriate one would not deliberately apply nonparametric procedures, since the latter utilize only part of the available information. Nonparametric methods may, however, be compared for 'power-efficiency' with parametric techniques and are commonly found to be 90–95% efficient, so that a quite modest increase in sample size enables one to use them to compare populations just as precisely as by parametric methods. A comprehensive account of nonparametric procedures, requiring very little mathematical knowledge, is given by Siegel (1956), who also provides a good selection of the tables needed for nonparametric significance tests. Campbell (1967) gives a useful introduction to some of the better known tests.

The following nonparametric methods are discussed here:

Program 32: Chi-squared test of association in two-way contingency tables.

Program 33: Chi-squared test of goodness of fit.

Program 34: Spearman's Rank Correlation coefficient.

Program 35: Wilcoxon's signed ranks test for two related samples.

Program 36: Mann–Whitney Test for two independent samples.

Program 37: Kruskal–Wallis Test for several independent samples.

Program 32: Chi-squared test of association in two-way contingency tables

The construction of two-way contingency tables is described in connection with Program 7 (p. 121). If the two variables concerned show some form of association, then some cells of the table will contain higher frequencies than might otherwise be expected, while other cells will contain correspondingly lower frequencies. The chi-squared test enables one to assess the probability of the observed distribution of frequencies in relation to that expected. If in a table with r rows and c columns the observed frequency for the cell in the i-th row and j-th column is denoted as O_{ij} while the expected frequency for the same cell is E_{ij}, then the quantity

$$\sum_{i=1}^{r} \sum_{j=1}^{c} \frac{(O_{ij} - E_{ij})^2}{E_{ij}} \qquad (1)$$

is distributed as a chi-squared variable with $(r-1)(c-1)$ degrees of freedom. The corresponding probability can then be obtained from tabulated values of the chi-squared distribution. The expected frequency E_{ij} is obtained as follows: sum the frequencies for each of the rows and each of the columns in the table, denoting the sums for the i-th row and the j-th column as S_i and S_j respectively, and the grand total number of individuals as N. Then

$$E_{ij} = \frac{S_i \cdot S_j}{N}. \qquad (2)$$

We can now apply these principles in constructing a suitable program, bearing in mind a number of other considerations discussed by Cochran (1954), which will affect certain parts of the program and are mentioned as they arise. We shall in fact calculate the χ^2 value for the table; check for the occurrence of expected cell frequencies of 5 or less, identifying the cells concerned; tabulate the contribution made by each cell towards the total χ^2 for the whole table; and compute a so-called 'coefficient of contingency'.

We first dimension an array X of observed frequencies, a corresponding array EXP which will accommodate the expected frequencies we shall calculate, and arrays SUMC and SUMR for the row and column sums. Although the observed frequencies are normally recorded in a contingency table as integer values we require them to be read in eventually under an F-format since they will be used immediately in floating-point arithmetic. The problem specification card includes values for NPROB (problem reference number) and the number of rows and columns (NR and NC) in the contingency table. After the usual arrangements for stopping execution when all problems have been dealt with, we read in the array of observed frequencies (X) using a double implied DO-loop. Remember that this means the data must be punched continuously beginning with the first row and ending with the last row, each card having 10 numbers (except the last card of a problem, which of course bears a variable number of observations depending on the total number of cells).

The next segment calculates row and column sums for the array X and these are used in the following segment to compute expected frequencies (EXP). At this point we must bear in mind that the χ^2 test is unreliable if the expected frequencies of many of the cells are low. Opinions differ as to how low they may be before the test becomes unreliable so we shall simply record the number of cells with expected frequencies <5 and arrange for these cells and their frequencies to be identified in the output. It is then left to the user to decide whether he will combine certain rows or columns in order to increase the expected frequencies in the pooled cells. Guidance on this point is given in the paper by Cochran (1954). The method used for locating the cells with low expected frequencies is simply to scan EXP by the nested DO-loops ending on 7, incorporated in which is an arithmetic IF statement that causes the row and column numbers of cells with frequencies of 5 or less to be printed out with the frequency.

Having checked for low expected frequencies the program considers the special case of 2×2 contingency tables. If such a table is being analysed the cell contributions to χ^2 are calculated not from equation (1) but by a somewhat different method embodying Yates' correction for continuity. This consists of reducing the absolute value of each difference by 0.5 before squaring it. Statement 9 accomplishes this, whereas if the contingency table is larger than 2×2 the cell-contributions are computed from (1) by statement

Flowchart for Program 32

Chi-squared test of association in two-way contingency tables.

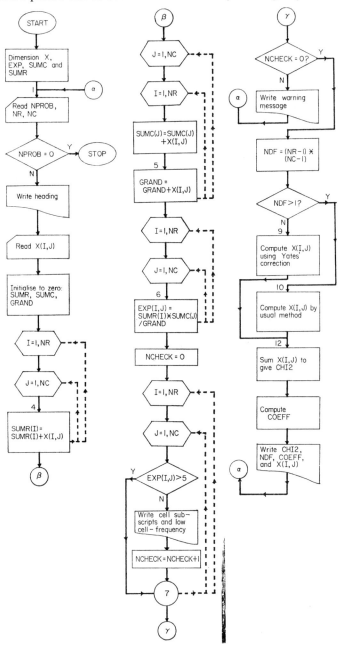

11. In either case these contributions are summed by the DO-loops ending on 13 to give χ^2, the individual contributions being retained in the array X. Finally the program prints a 'coefficient of contingency'—a simple function of χ^2 which indicates the extent of the association. There are several such contingency coefficients available; the one used here takes the form

$$C = \frac{\chi^2}{N(t - 1.0)} \qquad (3)$$

where N is the total number of observations and t is the number of rows or columns, whichever is the smaller. The program prints out the value of χ^2, its degrees of freedom, the coefficient of contingency and the contributions to χ^2 made by each cell of the table. Examination of these enables the user to judge what form the association takes.

Program 32

Chi-squared test of association in two-way contingency tables; with sample data and output.

```
C
C.............................................................
C
C   THIS PROGRAM COMPUTES CHI-SQUARED VALUES FOR ANY NUMBER OF TWO-WAY
C   CONTINGENCY TABLES. OUTPUT ALSO INCLUDES COEFFICIENT OF CONTINGENCY
C   AND CHI-SQUARED CONTRIBUTION FOR EACH CELL OF CONTINGENCY TABLE.
C   INPUT PARAMETERS ARE AS FOLLOWS -
C       NPROB = PROBLEM REFERENCE NUMBER (ZERO ON LAST DATA CARD)
C       NR    = NUMBER OF ROWS IN TABLE
C       NC    = NUMBER OF COLUMNS IN TABLE
C       X     = CELL FREQUENCY MATRIX (PUNCHED AS REAL NUMBERS)
C
C.............................................................
C
        DIMENSION X(40,40), EXP(40,40), SUMC(40), SUMR(40)
C
      1 READ (5,100) NPROB, NR, NC
    100 FORMAT (3I5)
        IF (NPROB .EQ. 0) STOP
        WRITE (6,101) NPROB
    101 FORMAT (20H1CHI-SQUARED PROBLEM, I4)
        READ (5,102) ((X(I,J), J = 1,NC), I = 1,NR)
    102 FORMAT (10F7.0)
C
C   SUMS ROWS AND COLUMNS AND FIND GRAND TOTAL
        DO 2 I = 1,NR
      2 SUMR(I) = 0.0
        DO 3 J = 1,NC
      3 SUMC(J) = 0.0
        GRAND = 0.0
        DO 4 I = 1,NR
        DO 4 J = 1,NC
      4 SUMR(I) = SUMR(I) + X(I,J)
        DO 5 J = 1,NC
        DO 5 I = 1,NR
```

Program 32 (continued)

```
      SUMC(J) = SUMC(J) + X(I,J)
    5 GRAND = GRAND + X(I,J)
C
C  COMPUTE EXPECTED FREQUENCIES
      DO 6 I = 1,NR
      DO 6 J = 1,NC
    6 EXP(I,J) = SUMR(I) * SUMC(J) / GRAND
C
C  CHECK FOR LOW EXPECTED CELL-FREQUENCIES
      WRITE (6,103)
  103 FORMAT(/10X,50HCHECKED FOR EXPECTED CELL-FREQUENCIES OF 5 OR LESS)
      NCHECK = 0
      DO 7 I = 1,NR
      DO 7 J = 1,NC
      IF (EXP(I,J) .GT. 5.0) GO TO 7
      WRITE (6,104) I, J, EXP(I,J)
  104 FORMAT (/10X, 4HCELL, 2I3, 5X, 11HFREQUENCY = , F6.2)
      NCHECK = NCHECK + 1
    7 CONTINUE
      IF (NCHECK .EQ. 0) GO TO 8
      WRITE (6,105)
  105 FORMAT (/10X,102HWARNING. CHI-SQUARED MAY BE INACCURATE. POOLING
     A OF CELLS WITH LOW EXPECTED FREQUENCIES MAY BE NEEDED)
C
C  APPLY YATES CORRECTION FOR 2 x 2 TABLES.
    8 NDF = (NR - 1) * (NC - 1)
      IF (NDF .GT. 1) GO TO 10
      DO 9 I = 1,NR
      DO 9 J = 1,NC
    9 X(I,J) = (ABS(X(I,J) - EXP(I,J)) - 0.5) ** 2 / EXP(I,J)
      GO TO 12
   10 DO 11 I = 1,NR
      DO 11 J = 1,NC
   11 X(I,J) = (X(I,J) - EXP(I,J)) ** 2 / EXP(I,J)
C
C  COMPUTE CHI-SQUARED.
   12 CHI2 = 0.0
      DO 13 I = 1,NR
      DO 13 J = 1,NC
   13 CHI2 = CHI2 + X(I,J)
C
C  COMPUTE COEFFICIENT OF CONTINGENCY.
      T = AMIN0(NR,NC)
      COEFF = CHI2 / (GRAND * (T - 1.0))
C
C  OUTPUT RESULTS.
      WRITE (6,106) CHI2, NDF, COEFF
  106 FORMAT (//10X, 13HCHI-SQUARED = , 1PE15.7//
     A10X, 20HDEGREES OF FREEDOM = , I5//
     B10X, 28HCOEFFICIENT OF CONTINGENCY = , 0PF7.4)
      WRITE (6,107)
  107 FORMAT (//10X, 33HCELL CONTRIBUTIONS TO CHI-SQUARED)
      DO 14 I = 1,NR
      WRITE (6,108) I
  108 FORMAT (/ 1X, 3HROW, I3)
   14 WRITE (6,109) (J, X(I,J), J = 1,NC)
  109 FORMAT (5(3X, I3, 1PE15.7))
C
C  READ NEXT SAMPLE
      GO TO 1
      END
```

Program 32 (continued)

DATA

```
    1   4   3
  335.   60.   18.   300.   57.   19.   294.   51.   16.   290.
   42.   12.
    0
```

OUTPUT

```
CHI-SQUARED PROBLEM    1
       CHECKED FOR EXPECTED CELL-FREQUENCIES OF 5 OR LESS

       CHI-SQUARED =  2.7195009E 00

       DEGREES OF FREEDOM =    6

       COEFFICIENT OF CONTINGENCY = 0.0009

       CELL CONTRIBUTIONS TO CHI-SQUARED

ROW  1
     1   1.1625116E-02    2   6.5353103E-02    3   5.5079015E-05

ROW  2
     1   1.5027137E-01    2   3.2564574E-01    3   4.2644441E-01

ROW  3
     1   1.0302351E-03    2   1.3019279E-03    3   5.4974084E-03

ROW  4
     1   3.0946791E-01    2   8.3480897E-01    3   5.8799966E-01
```

Program 33: Chi-squared test for goodness of fit

Suppose that we have a set of observed frequencies $O_1, O_2, O_3, \ldots O_m$. We may use a chi-squared test to decide whether these frequencies depart significantly from those expected on some hypothesis. The latter may postulate that all the frequencies are equal, or it may suppose that they follow some arbitrary or statistical distribution which the experimenter has in mind. This time we compute

$$\chi^2 = \sum_{i=1}^{m} \frac{(O_i - E_i)^2}{E_i}. \qquad (4)$$

The number of degrees of freedom associated with this χ^2 will normally be $(m - 1)$. If, however, our expected frequencies have been calculated from a relation involving parameters that must first be estimated from the data, then for each such parameter estimated we subtract one further degree of freedom. For example, suppose we wished to compare a set of 10 observed frequencies with those expected from a normal distribution. To calculate the expected frequencies we need to estimate two parameters—the mean and standard

Flowchart for Program 33

Chi-squared test of goodness of fit.

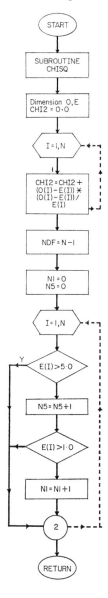

deviation—from the data (cf. p. 344). The degrees of freedom for χ^2 in such a test would therefore be $10 - 1 - 2 = 7$.

The program is written in the form of a short calling program that controls input/output while the calculation of chi-squared itself is carried out by the subroutine CHISQ (which we shall use later, in Chapter 12, when testing the goodness of fit of some theoretical distributions). The calling program reads the observed frequencies into the array DATA and then provides two options, depending on the value of INDEX. If INDEX has been set at 1 the program will read in a set of expected frequencies punched by the user as the array EXP. If INDEX has been set as 2 the calling program will itself calculate a set of expected frequencies (stored in EXP) such that each cell has the same frequency and the total of expected frequencies is the same as the total of observed frequencies. This second option therefore tests the null hypothesis that the frequencies are the same in all cells of the array.

The subroutine CHISQ is very simple: given the vectors of observed and expected frequencies χ^2 is computed by the DO-loop ending on 1. The second loop—ending on 2—scans the expected frequency array and records the number of cells with expected frequencies of less than 5 and less than 1 in N5 and N1 respectively. Note that the number of degrees of freedom calculated and printed out does not allow for the estimation of parameters from the data: this must be done by the user, depending on the expected frequency distribution that he employed.

Program 33

Chi-squared test of goodness of fit, including SUBROUTINE CHISQ; with sample data and output.

```
C
C.............................................................
C
C   THIS PROGRAM CONTROLS I/O AND CALLS SUBROUTINE CHISQ TO
C   EXECUTE TESTS OF GOODNESS OF FIT. INPUT PARAMETERS ARE AS FOLLOWS -
C       NPROB = PROBLEM REFERENCE NUMBER (ZERO ON LAST DATA CARD)
C       N     = NUMBER OF CLASSES
C       INDEX = 1 IF EXPECTED FREQUENCIES ARE TO BE READ IN
C               2 IF EXPECTED FREQUENCIES ARE TO BE COMPUTED AS ALL
C                 EQUAL WITH SAME SUM AS OBSERVED FREQUENCIES
C       DATA  = OBSERVED FREQUENCIES (PUNCHED AS REAL NUMBERS)
C       EXP   = EXPECTED FREQUENCIES (IF INDEX = 1)
C
C.............................................................
C
      DIMENSION DATA(1000), EXP(1000)
C
    1 READ (5,100) NPROB, N, INDEX
  100 FORMAT (3I5)
      IF (NPROB .EQ. 0) STOP
C
C   READ OBSERVED FREQUENCIES AND SELECT FOR EXPECTED FREQUENCIES.
```

Program 33 (continued)

```
C   READ OBSERVED FREQUENCIES (DATA) AND SELECT FOR EXPECTED ONES.
    READ (5,101) (DATA(I), I = 1,N)
101 FORMAT (10F7.0)
    GO TO (2,3), INDEX
  2 READ (5,101) (EXP(I), I = 1,N)
    GO TO 6
  3 FREQ = 0.0
    AN = N
    DO 4 I = 1,N
  4 FREQ = FREQ + DATA(I)
    FREQ = FREQ / AN
    DO 5 I = 1,N
  5 EXP(I) = FREQ
C
  6 CALL CHISQ (DATA, EXP, N, CHI2, NDF, N1, N5)
C
C   OUTPUT RESULTS.
    WRITE (6,102) NPROB, CHI2, NDF, N1, N5
102 FORMAT (/// 8H PROBLEM, I5//11X, 13HCHI-SQUARED = , F10.3,
   A   6H (WITH, I5, 6H D.F.)// 11X, 51HNO OF CELLS WITH EXPECTED FREQU
    BENCIES LESS THAN 1 = , I5/11X, 51HNO OF CELLS WITH EXPECTED FREQUEN
    CCIES LESS THAN 5 = , I5)
    GO TO 1
    END


    SUBROUTINE CHISQ (O, E, N, CHI2, NDF, N1, N5)
C
C.......................................................
C
C   THIS SUBROUTINE COMPUTES CHI-SQUARED (CHI2) FOR VECTORS OF OBSERVED
C   AND EXPECTED FREQUENCIES, O AND E, BOTH OF LENGTH N. ALSO RETURNED
C   ARE DEGREES OF FREEDOM (NDF) AND NUMBER OF CELLS WITH EXPECTED
C   FREQUENCIES OF LESS THAN 1 (N1) AND LESS THAN 5 (N5)
C
C.......................................................
C
    DIMENSION O(N), E(N)
    CHI2 = 0.0
    DO 1 I = 1,N
  1 CHI2 = CHI2 + (O(I) - E(I)) * (O(I) - E(I)) / E(I)
    NDF = N - 1
    N1 = 0
    N5 = 0
    DO 2 I = 1,N
    IF (E(I) .GE. 5.0) GO TO 2
    N5 = N5 + 1
    IF (E(I) .GE. 1.0) GO TO 2
    N1 = N1 + 1
  2 CONTINUE
    RETURN
    END

DATA

   1     8    1
  10    12   14   10    2    3    9   12
  12    12    4   12   12    4   12    0
   2    14    2
  10     9    8    7    8    9   10   10   12   13
  10    11   12   13
   0
```

Program 33 (*continued*)

OUTPUT

PROBLEM 1

    CHI–SQUARED =   35.000 (WITH   7 D.F.)

    NO OF CELLS WITH EXPECTED FREQUENCIES LESS THAN 1 =   1
    NO OF CELLS WITH EXPECTED FREQUENCIES LESS THAN 5 =   3

PROBLEM 2

    CHI–SQUARED =   4.507 (WITH   13 D.F.)

    NO OF CELLS WITH EXPECTED FREQUENCIES LESS THAN 1 =   0
    NO OF CELLS WITH EXPECTED FREQUENCIES LESS THAN 5 =   0

Program 34: Spearman's rank correlation coefficient

The well-known product-moment correlation coefficient r, discussed in Program 2, should be employed only when the observations used in computing it come from a bivariate normal distribution. If this assumption is not fulfilled it is preferable to compute a non-parametric measure of association, of which Spearman's rank correlation coefficient is probably the best known. As its name implies this coefficient requires only that we know the order in which the members of the two sets of observations can be ranked. It is not necessary that the observations have been recorded on an ordinary numerical scale, though such observations can always be used provided that they are first ranked. Ranking a set of observations is, in fact, a common procedure in many nonparametric tests and will appear again in subsequent programs in this chapter. It is a troublesome operation to perform by hand when the observations are at all numerous and it is here carried out by two short subroutines RANK and TIE.

    SUBROUTINE RANK accomplishes essentially the same thing as Program 5 (p. 111) but does it in a rather simpler way. The original unranked observations are stored in the vector X of dimension N and our object is to place the ranks in a second vector RX so that for any value X(I), its rank is given by RX(I). The subroutine comprises two nested DO-loops ending on statement 2. Every time the value X(J), which is being ranked, proves to be greater than the other elements of X, a counter NX (initially set at 1) is incremented by 1. When the inner DO-loop is satisfied NX gives the rank of X(J) and this value is accordingly placed in RX(J) by statement 2. By the time the outer loop is satisfied all the elements of X will have been ranked in this way, i.e. in ascending order, the smallest number being ranked 1. If the observations 2.0, 4.0, 4.0, 6.3 and 7.5 were being ranked by this subroutine the successive values of RX would be 1, 2, 2, 4 and 5. Though a perfectly valid method of ranking, this is not the form required by nonparametric tests. These demand that when ties occur, each tied observation is given the average of the ranks they would have occupied had no ties occurred. Thus in the above example the ranks should be adjusted to 1, 2.5, 2.5, 4, 5. This adjustment is made by the second subroutine TIE.

Though TIE is a short program the logic involved is rather complicated and the reader may find it easier to follow in terms of the above numerical example where the ranks 1, 2, 2, 4, 5 are to be adjusted to 1, 2.5, 2.5, 4, 5. The vector RX—typed explicitly as an integer variable—contains the unadjusted ranks and X will be used to accommodate the adjusted ones. The situation at the start of the program will therefore be

RX	1	2	2	4	5

X					

We are not concerned at first with the contents of X but begin by setting up the intermediate variable NP1 = N + 1 (=6 in this case). This is intended as a 'marker'; its only property is that it should be larger than any value in RX and the simplest way to ensure this is to define it as N + 1. We now begin the outer DO-loop and compare RX(I), i.e. 1, with NP1 (=6). The two are not equal so we set NX = 1 and initialise NTIE at −1. We then start the first inner DO-loop, comparing each of the values RX(I) with 1, i.e. we are comparing RX(1) with each of the elements RX(1) to RX(5) in turn. In all but one of these comparisons we pass direct to 2 CONTINUE but in one comparison (where RX(1) is compared with itself) we execute all the statements within the ranks of the loop, incrementing NTIE so that it becomes zero, writing RX(1) into X(1) and setting RX(1) at zero. The position is therefore now

RX	0	2	2	4	5

X	1.0				

The subroutine now executes the second inner DO-loop, which scans the elements of RX, ignoring RX(2) to RX(5) because they are all greater than zero. For RX(1), however, the program can initiate an adjustment of X(1) based on the value of NTIE. Since this is in fact zero at present no change of X(1) is needed. We have therefore completed our examination of the first rank in RX and to 'mark' it we replace the zero in RX(1) by a 6 (the value of

NP1). The situation is therefore now as follows at the end of the first pass through the outermost DO-loop:

RX	6	2	2	4	5

X	1.0				

Execution now proceeds by passing through the outermost DO-loop a second time, during which the vectors RX and X are successively modified as follows:

(i) at start of second passage through outermost DO-loop.

RX	6	2	2	4	5

X	1.0				

(ii) After first inner DO-loop.

RX	6	0	0	4	5

X	1.0	2.0	2.0		

(iii) At end of second passage through outer DO-loop.

RX	6	6	6	4	5

X	1.0	2.5	2.5		

What has happened here is that on passage through the first inner DO-loop a tie is recognised through NTIE being incremented twice to equal $+1$ and the two tied observations are transferred to X. Subsequently, in the passage through the second inner DO-loop, each of these is increased by 1.0×0.5 to give the appropriate adjusted values.

During the third pass through the outer DO-loop control passes directly to statement 4 since RX(3) has already been 'marked' by being set equal to 6. On the fourth and fifth passes RX(4) and RX(5) are treated as was RX(1)—i.e. the ranks are transferred to X but are not further adjusted since in both cases NTIE is zero when the second inner DO-loop is executed.

In general, should there be k tied values for any particular rank, the value of NTIE will rise to $(k-1)$ and each tied rank will be adjusted by the addition of $(k-1)/2$. Obviously RANK and TIE could be combined into a single subroutine but we have not done this since the present form allows one

PROGRAM 34

to carry our nonparametric tests on observations consisting only of the ranks from Program 5 or from other data already ranked in that simple way. The flow-charts for RANK and TIE are given below.

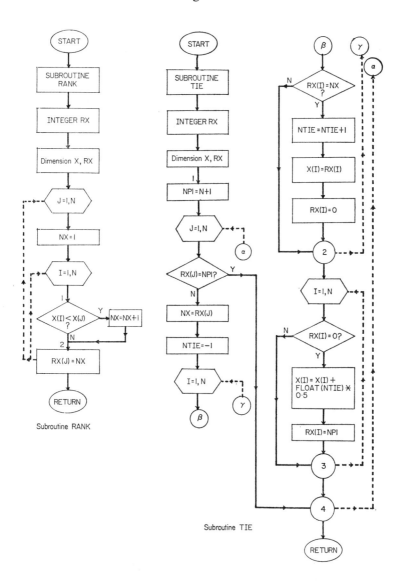

Subroutine RANK

Subroutine TIE

From a programming point of view these two subroutines are the most interesting feature of the program. With their aid the rank correlation coefficient can be computed in a straightforward fashion. We can, however, make the program somewhat more general by arranging for it to cope with data which may or may not have been ranked already and also for data which may be punched either as paired observations or as the two separate arrays to be correlated. The input variables, therefore, are

NPROB   the problem number

N   the number of pairs of observations

INDEX   an option number set at 1 if the data are not already ranked and at 2 if they are ranked by the method of Program 5 (i.e. as they would be by SUBROUTINE RANK)

MODE   a second option number set at 1 if the data are punched in pairs and at 2 if one complete array (A or RA) is read in before the other (B or RB).

Computed GO TO statements enable these options to be implemented and whichever have been selected the program ultimately executes calls to SUBROUTINE TIE (statement 9 and the following one). From the adjusted ranks returned by TIE we can compute the rank correlation coefficient r_s from the formula:

$$r_s = 1 - \frac{6 \sum_{i=1}^{N} d_i^2}{(N-1)N(N+1)} \tag{5}$$

where d_i (represented in Fortran by DIFF) is the difference between the adjusted ranks A(I) and B(I). The DO-loop ending on 10 carries out the summation and the following arithmetic statement completes the calculation of the required coefficient RS. It is necessary to emphasise that formula (5) assumes that ties are not very numerous. If many ties occur an alternative method is required (see Siegel, 1956).

After computing the rank correlation coefficient it is desirable to test its significance. Critical values of r_s for $P = 0.05$ and $P = 0.01$ are given by Siegel (1956) for N between 4 and 30. If N is greater than 10, however, one may use a test criterion that is distributed as Student's t with $(N-2)$ degrees of freedom. This criterion is therefore computed by a simple arithmetic statement and is printed out, together with the correlation coefficient and degrees of freedom.

Outline flowchart for Program 34

Spearman's rank correlation coefficient.

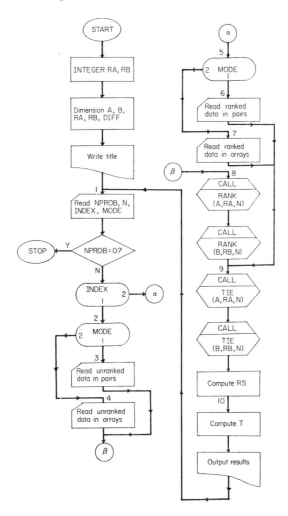

Program 34

Spearman's rank correlation coefficient, including Subroutines RANK and TIE; with sample data and output.

```
C
C.............................................................
C
C   THIS PROGRAM COMPUTES SPEARMAN'S RANK CORRELATION COEFFICIENT.
C   INPUT PARAMETERS ARE AS FOLLOWS -
C      NPROB = PROBLEM REFERENCE NUMBER (ZERO ON LAST DATA CARD)
C      N     = NUMBER OF PAIRED OBSERVATIONS
C      INDEX = 1 IF DATA ARE NOT RANKED
C              2 IF DATA ARE ALREADY RANKED
C      MODE  = 1 IF DATA ARE TO BE READ IN PAIRS
C              2 IF DATA ARE TO BE READ AS SEPARATE ARRAYS
C      A     = FIRST ARRAY OF OBSERVATIONS
C      B     = SECOND ARRAY OF OBSERVATIONS
C
C.............................................................
C
      INTEGER RA, RB
      DIMENSION A(1000), B(1000), RA(1000), RB(1000), DIFF(1000)
C
C   READ PROBLEM SPECIFICATION.
    1 READ (5,100) NPROB, N, INDEX, MODE
  100 FORMAT (4I5)
      IF (NPROB .EQ. 0) STOP
      GO TO (2,5), INDEX
    2 GO TO (3,4), MODE
C
C   READ IN DATA.
    3 READ (5,101) (A(I), B(I), I = 1,N)
  101 FORMAT (10F7.0)
      GO TO 8
    4 READ (5,101) (A(I), I = 1,N)
      READ (5,101) (B(I), I = 1,N)
      GO TO 8
    5 GO TO (6,7), MODE
    6 READ (5,102) (RA(I), RB(I), I = 1,N)
  102 FORMAT (10I7)
      GO TO 9
    7 READ (5,102) (RA(I), I = 1,N)
      READ (5,102) (RB(I), I = 1,N)
      GO TO 9
C
C   RANK ARRAYS A AND B.
    8 CALL RANK (A, RA, N)
      CALL RANK (B, RB, N)
C
C   ADJUST RANKING FOR TIES IN A AND B.
    9 CALL TIE (A,RA, N)
      CALL TIE (B, RB, N)
C
C   COMPUTE SPEARMAN'S RANK CORRELATION COEFFICIENT.
      AN = N
      RS = 0.0
      DO 10 I = 1,N
      DIFF (I) = A(I) − B(I)
   10 RS = RS + DIFF(I) * DIFF(I)
      RS = 1.0 − 6.0 * RS / ((AN − 1.0) * AN * (AN + 1.0))
C
```

Program 34 (*continued*)

```
C   COMPUTE SIGNIFICANCE CRITERION.
      T = RS * SQRT((AN - 2.0) / (1.0 - RS*RS))
      NDF = N - 2
C
C   OUTPUT RESULTS.
      WRITE (6,103) NPROB, N, RS, T, NDF
  103 FORMAT (///8H PROBLEM, I4, 13H (CONTAINING, I5,
     A  23H PAIRS OF OBSERVATIONS)// 11X,
     B  30HRANK CORRELATION COEFFICIENT = , F8.4//11X, 3HT = ,
     C  F7.3, 5H WITH, I5, 19H DEGREES OF FREEDOM)
      GO TO 1
      END


      SUBROUTINE RANK (X, RX, N)
C
C.............................................................
C
C   THIS SUBROUTINE RANKS THE N ELEMENTS OF THE VECTOR X IN
C   ASCENDING ORDER (SMALLEST = 1) AND PLACES THE RANKS IN THE
C   VECTOR RX.
C
C.............................................................
C
      INTEGER RX
      DIMENSION X(N), RX(N)
      DO 2 J = 1,N
      NX = 1
      DO 1 I = 1,N
    1 IF (X(I) .LT. X(J)) NX = NX + 1
    2 RX(J) = NX
      RETURN
      END


      SUBROUTINE TIE (X, RX, N)
C
C.............................................................
C
C   THIS PROGRAM ADJUSTS THE RANKINGS IN VECTOR RX FOR TIES AND PLACES
C   THE ADJUSTED RANKS IN VECTOR X.
C
C.............................................................
C
      INTEGER RX
      DIMENSION X(N), RX(N)
    1 NP1 = N + 1
      DO 4 J = 1,N
      IF (RX(J) .EQ. NP1) GO TO 4
      NX = RX(J)
      NTIE = -1
      DO 2 I = 1,N
      IF (RX(I) .NE. NX) GO TO 2
      NTIE = NTIE + 1
      X(I) = RX(I)
      RX(I) = 0
    2 CONTINUE
      DO 3 I = 1,N
      IF (RX(I) .NE. 0) GO TO 3
      X(I) = X(I) + FLOAT (NTIE) * 0.5
      RX(I) = NP1
```

Program 34 (continued)

```
   3 CONTINUE
   4 CONTINUE
     RETURN
     END
```

DATA

```
 1  18   1    2
49.0  44.0  32.0  42.0  32.0  53.0  36.0  39.0  37.0  45.0
41.0  48.0  45.0  39.0  40.0  34.0  37.0  35.0
27.0  24.0  12.0  22.0  13.0  29.0  14.0  20.0  16.0  21.0
22.0  25.0  23.0  18.0  20.0  15.0  20.0  13.0
 0
```

OUTPUT

PROBLEM   1 (CONTAINING 18 PAIRS OF OBSERVATIONS)

RANK CORRELATION COEFFICIENT = 0.9613

T = 13.957 WITH 16 DEGREES OF FREEDOM

Program 35: Wilcoxon's signed ranks test

This is one of a number of nonparametric tests that can be used to examine a difference between two samples and is therefore analogous to the parametric t-test of Program 3. Essentially, Wilcoxon's signed ranks test is a test of significance of the difference of the medians of two populations with identical distributions. The precise form of the distribution is not important. The test is therefore usually applicable to two *related* samples such as a biologist might expect to find when treatments can be assessed by measurements on individuals 'before and after', or by comparisons between sets of identical twins (one of each pair of twins acting as a control for the other) or by comparisons between corresponding members of otherwise matched samples. Unlike some of the other nonparametric tests that may be used in these situations, however, Wilcoxon's test is applied when the measurements are made on an ordinary numerical scale (ranks alone are not enough).

As in Program 34, we type the intermediate variable RDIFF explicitly as an integer and arrange for the two arrays A and B, each of size N, to be read in separately. It is these arrays which constitute the matched samples we are studying. The test criterion is then computed in a series of steps:

(a) The algebraic difference between each pair of observations is evaluated and stored in A(I) while the absolute value of each difference is stored in B(I). This is accomplished by the DO-loop ending on statement 2.

(b) By calling the subroutines RANK and TIE, the absolute differences are ranked, adjusted for ties, and the adjusted ranks returned to the calling program in array B.

Flowchart for Program 35

Wilcoxon's signed ranks test.

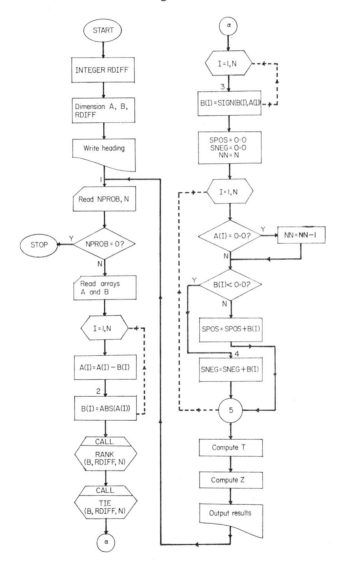

(c) The DO-loop ending on statement 3 then gives to each adjusted rank the sign of the corresponding algebraic difference stored in A. This characteristic feature of the test ('signed ranks') is carried out by a standard Fortran library function SIGN, by which the sign of the second argument is attached to the absolute value of the first argument.

(d) The array of signed ranks, B, is now scanned in the DO-loop ending on 5 so as to produce two results: firstly, if A(I) is equal to zero (i.e. if there was no difference originally between the two members of a pair), we reduce by 1 the effective number of comparisons; this is done by the first logical IF of the loop. Secondly, we sum separately the positively and negatively signed ranks; this is accomplished by the remaining statements in the loop, yielding the totals SNEG and SPOS.

(e) We then compute the test-criterion T as the absolutely smaller of these two sums.

If we have been dealing with 25 or fewer effective comparisons (i.e. NN ≤ 25) we must use the published tables (Siegel, 1956) to assess the probability that the populations have the same median. Note that in using these tables, the *smaller* the value of T the smaller the probability of the null hypothesis being true. If, however, our T is based on more than 25 comparisons its significance may be tested by computing a standardised normal deviate (Z) which is used in the usual way. The output includes the three quantities needed to make these tests of significance.

Program 35

Wilcoxon's signed ranks test. The main program listed here calls Subroutines RANK and TIE (p. 329).

```
C
C............................................................
C
C   THIS PROGRAM COMPUTES WILCOXON'S SIGNED RANK TEST CRITERION.
C   INPUT PARAMETERS ARE AS FOLLOWS –
C       NPROB = PROBLEM REFERENCE NUMBER (ZERO ON LAST DATA CARD)
C       N     = NUMBER OF OBSERVATIONS IN EACH DATA ARRAY
C       A     = VECTOR OF OBSERVATIONS (LENGTH N) IN FIRST ARRAY
C       B     = VECTOR OF OBSERVATIONS (LENGTH N) IN SECOND ARRAY
C   SUBROUTINES RANK AND TIE ARE REQUIRED.
C
C............................................................
C
        INTEGER RDIFF
        DIMENSION A(1000), B(1000), RDIFF(1000)
C
C   WRITE HEADING.
        WRITE (6,100)
  100   FORMAT (1H1, 30X, 27HWILCOXON'S SIGNED RANK TEST)
C
```

Program 35 (continued)

```
C   READ PROBLEM NUMBER AND SAMPLE SIZE.
  1     READ (5,101) NPROB, N
101     FORMAT (2I5)
        IF (NPROB .EQ. 0) STOP
C
C   READ IN DATA ARRAYS.
        READ (5,102) (A(I), I = 1,N)
        READ (5,102) (B(I), I = 1,N)
102     FORMAT (10F7.0)
C
C   COMPUTE TEST CRITERION.
        DO 2 I = 1,N
        A(I) = A(I) - B(I)
  2     B(I) = ABS(A(I))
        CALL RANK (B, RDIFF, N)
        CALL TIE (B, RDIFF, N)
        DO 3 I = 1,N
  3     B(I) = SIGN(B(I), A(I))
        SPOS = 0.0
        SNEG = 0.0
        NN = N
        DO 5 I = 1,N
        IF (A(I) .EQ. 0.0) NN = NN - 1
        IF (B(I) .LT. 0.0) GO TO 4
        SPOS = SPOS + B(I)
        GO TO 5
  4     SNEG = SNEG + B(I)
  5     CONTINUE
        T = AMIN1(SPOS,ABS(SNEG))
C
C   COMPUTE NORMAL DEVIATE FOR T.
        AN = NN
        Z = ABS((T-(AN*(AN+1.0))/4.0)/SQRT((AN*(AN+0.5)*(AN+1.0))/12.0))
C
C   OUTPUT RESULTS.
        WRITE (6,103) NPROB, NN, T, Z
103     FORMAT (/// 8H PROBLEM, I5, 12H (CONTAINING, I5,
     A   13H COMPARISONS) // 11X, 16HTEST CRITERION = ,
     B   F10.2//11X, 16HNORMAL DEVIATE = , F7.2//11X,
     C   91HUSE PUBLISHED SIGNIFICANCE TABLES IF 25 OR FEWER COMPARISONS,
     D   OTHERWISE USE NORMAL DEVIATE)
        GO TO 1
        END
```

Program 36: Mann–Whitney Test for two independent samples

The Mann–Whitney test is intended to examine the possibility that two samples represent populations differing in central tendency or location, again making use of the median as a measure of location. Unlike Wilcoxon's signed ranks test, however, this method does not presuppose matched samples; here they may be independent and may even contain different numbers of individuals. The Mann–Whitney test also differs in that it can, if necessary, be applied to ranks alone, though we shall consider it only in relation to measurements made on an ordinary numerical scale. The procedure may be summarised as follows:

(a) Consider two samples, containing respectively n_1 and n_2 samples, n_2 being the larger if the samples differ in size. $n_1 + n_2 = N$.

(b) Rank the *combined* observations, adjusting for ties in the usual way.

(c) Sum the ranks so obtained for the members of the first sample to give R_1 and for the second sample to give R_2.

(d) Calculate the statistic U, which is the smaller of U_1 and U_2 defined as follows:

$$U_1 = n_1 n_2 + \frac{n_1}{2}(n_1 + 1) - R_1 \tag{6}$$

$$U_2 = n_1 n_2 + \frac{n_2}{2}(n_2 + 1) - R_2. \tag{7}$$

Note that $U_1 + U_2 = n_1 n_2$ so that a less complicated method of calculating U_2 is to obtain U_1 and then apply the relation $U_2 = n_1 n_2 - U_1$.

(e) Depending on the size of n_2, find the probability associated with U from published tables or, if $n_2 > 20$, by calculating a standardised normal deviate z from the formula:

$$z = \frac{U - n_1 n_2/2}{\sqrt{\left(\frac{n_1 n_2}{N(N-1)}\right)\left(\frac{N^3 - N}{12} - \Sigma T\right)}} \tag{8}$$

where

$$T = \frac{t^3 - t}{12}$$

with t equal to the number of observations tied for a given rank.

The program that carries out stages (a) to (d) above is quite straightforward. The two samples are read into arrays A (length NA) and B (length NB). The two arrays are then combined—note how this is done very simply by using a computed subscript J in the DO-loop ending on 3—and the combined array AB is ranked by calling RANK and TIE. The ranks of the individuals originally in sample A are then summed in SUMR and U calculated as the lesser of U and UU.

The computation of the normal deviate from (8) is a slightly more complicated operation and is carried out only if NB > 20. Most of the calculation is carried out by the nested DO-loops ending on 7. This takes each of the adjusted ranks in AB in turn, compares it with all subsequent ranks, counting

Flowchart for Program 36

Mann–Whitney test for two independent samples.

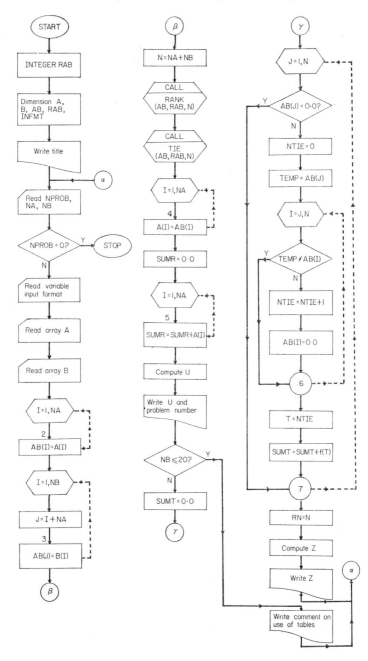

the ties (if any) and marking all tied observations by placing a zero in the location that accommodated them. The ties counted in each pass through the outermost DO-loop are summed in SUMT to give the quantity denoted as ΣT in equation (8). From this and other variables Z is then computed by an arithmetic statement.

A flow-chart for the calling program is given below.

Program 36

Mann–Whitney test for two independent samples. The main program listed here calls Subroutines RANK and TIE (p. 329). Sample data and output are also listed.

```
C
C.............................................................
C
C   THIS PROGRAM COMPUTES THE MANN–WHITNEY TEST CRITERION FOR TWO
C   INDEPENDENT SAMPLES. INPUT PARAMETERS ARE AS FOLLOWS –
C       NPROB = PROBLEM REFERENCE NUMBER (ZERO ON LAST DATA CARD)
C       NA    = NUMBER OF OBSERVATIONS IN FIRST SAMPLE
C       NB    = NUMBER OF OBSERVATIONS IN SECOND SAMPLE
C       A     = VECTOR OF OBSERVATIONS IN FIRST SAMPLE (LENGTH NA)
C       B     = VECTOR OF OBSERVATIONS IN SECOND SAMPLE (LENGTH NB)
C       INFMT = VECTOR HOLDING VARIABLE INPUT FORMAT
C   NOTE THAT NB MUST BE EQUAL TO OR GREATER THAN NA
C
C.............................................................
C
      INTEGER RAB
      DIMENSION A(500), B(500), AB(1000), RAB(1000), INFMT(20)
C
C   WRITE TITLE.
      WRITE (6,100)
  100 FORMAT (1H1, 30X, 19HMANN–WHITNEY U TEST)
C
C   READ PROBLEM SPECIFICATION AND VARIABLE INPUT FORMAT.
    1 READ (5,101) NPROB, NA, NB
  101 FORMAT (3I5)
      IF (NPROB .EQ. 0) STOP
      READ (5,102) (INFMT(I), I = 1,20)
  102 FORMAT (20A4)
C
C   READ OBSERVATIONS.
      READ (5,INFMT) (A(I), I = 1,NA)
      READ (5,INFMT) (B(I), I = 1,NB)
C
C   COMBINE OBSERVATIONS AND RANK.
      DO 2 I = 1,NA
    2 AB(I) = A(I)
      DO 3 I = 1,NB
      J = I + NA
    3 AB(J) = B(I)
      N = NA + NB
      CALL RANK (AB,RAB,N)
      CALL TIE (AB,RAB,N)
C
C   ASSIGN RANKS TO A AND COMPUTE TEST CRITERION (U).
      DO 4 I = 1,NA
    4 A(I) = AB(I)
```

Program 36 (continued)

```
       SUMR = 0.0
       DO 5 I = 1,NA
    5  SUMR = SUMR + A(I)
       RNA = NA
       RNB = NB
       U = RNA * RNB + (RNA * (RNA + 1.0)) / 2.0 – SUMR
       UU = RNA * RNB – U
       U = AMIN1(U,UU)
C
C   PRINT U AND OTHER OUTPUT.
       WRITE (6,103) NPROB, NA, NB, U
  103  FORMAT (///8H PROBLEM, I4, 6H (WITH, I4, 4H AND, I4,
      A 14H OBSERVATIONS)//11X, 16HTEST CRITERION = , F8.1)
C
C   IF NB EXCEEDS 20 COMPUTE NORMAL DEVIATE (Z) WITH CORRECTION FOR TIES.
       IF (NB .LE. 20) GO TO 8
       SUMT = 0.0
       DO 7 J = 1,N
       IF (AB(J) .EQ. 0.0) GO TO 7
       NTIE = 0
       TEMP = AB(J)
       DO 6 I = J,N
       IF (TEMP .NE. AB(I)) GO TO 6
       NTIE = NTIE + 1
       AB(I) = 0.0
    6  CONTINUE
       T = NTIE
       SUMT = SUMT + (T**3 – T) / 12.0
    7  CONTINUE
       RN = N
       Z = ABS (U–RNA*RNB / 2.0) / SQRT ((RNA * RNB / (RN * (RN – 1.0)))
      A * ((RN**3 – RN) / 12.0 – SUMT))
       WRITE (6,104) Z
  104  FORMAT (/11X, 16HNORMAL DEVIATE = , F9.2)
       GO TO 1
    8  WRITE (6,105)
  105  FORMAT (/ 11X, 33HUSE PUBLISHED SIGNIFICANCE TABLES)
       GO TO 1
       END
```

DATA

```
    1    4    5
(10F7.0)
  110.0   70.0   53.0   51.0
   78.0   64.0   75.0   45.0   82.0
    2   16   23
(20F3.0)
 13 12 12 10 10 10 10  9  8  8  7  7  7  7  7  6
 17 16 15 15 15 14 14 14 13 13 13 12 12 12 12 11 11 10 10 10
  8  8  6
  0
```

OUTPUT

MANN–WHITNEY U TEST

PROBLEM   1 (WITH   4 AND   5 OBSERVATIONS)

        TEST CRITERION =   9.0

           USE PUBLISHED SIGNIFICANCE TABLES

Program 36 (*continued*)

PROBLEM 2 (WITH 16 AND 23 OBSERVATIONS)

TEST CRITERION = 64.0

NORMAL DEVIATE = 3.45

Problem 37: Kruskal–Wallis test for several independent samples

Just as a one-way analysis of variance provides a parametric test of significance for the differences among several independent samples, so an analogous nonparametric test is available. This method, the Kruskal–Wallis test, is therefore sometimes spoken of as the Kruskal–Wallis analysis of variance by ranks, though it is not, of course, an analysis of variance at all. Like the Mann–Whitney test, it assumes that the populations being sampled differ only in location, as measured by their medians. The data may consist of measurements made on an ordinary numerical scale or of the ranks alone, but our program considers only the former. The test procedure is as follows:

(a) Consider k independent samples, each of size n_i ($i = 1, 2, 3, \ldots k$).

$$\sum_{i=1}^{k} n_i = N.$$

(b) Rank all N observations, correcting for ties as usual.

(c) Evaluate the number of observations in each group of tied observations (t) and calculate

$$\frac{1 - \Sigma T}{N^3 - N}$$

where $T = t^3 - t$ and ΣT denotes summation over all groups of ties.

(d) Sum the ranks for each sample, so obtaining the sums R_i ($i = 1, 2, 3, \ldots k$).

(e) Calculate the test-statistic H from the relation

$$H = \frac{\dfrac{12}{N(N+1)} \sum_{i=1}^{k} \dfrac{R_i^2}{n_i} - 3(N-1)}{\dfrac{1 - T}{N^3 - N}} \tag{9}$$

(f) Assess the significance of H using published tables from Siegel (1956).

Flowchart for Program 37

Kruskal–Wallis test for several independent samples.

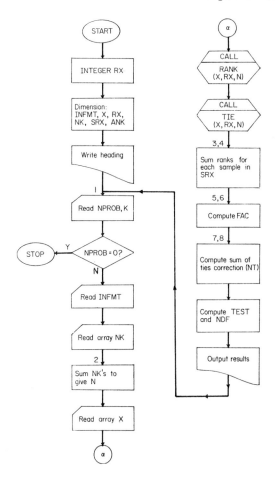

The program based on the above outline is quite easy to follow. The input includes a single array X which contains the observations from all samples arranged in a continuous series, and a vector NK which specifies the number of observations in each sample. The entire array X is then ranked in the usual way by calling RANK and TIE. The ensuing program segment sums the ranks for each sample separately, placing the sums in the vector SRX. The way in which the summation is carried out should be noted, since it depends on recalculating the indexing parameters N1 and N2 on each run through the nested DO-loops ending on 5. The quantity FAC, which corresponds to

$$\sum_{i=1}^{k} \frac{R_i^2}{n_i}$$

of equation (9) is then calculated easily.

The correction for tied observations is computed by the next program segment. Essentially, this is very similar to the correction for ties carried out in Program 36 except that each time a tie is recorded the location containing that observation is marked by being set to ANP1, a number one greater than the highest possible rank originally in the array X. Program 36 used a zero in place of ANP1—either is equally suitable. The quantity ΣT of equation (9) is accumulated in NT and once this is available the test criterion TEST is computed from it and FAC by an arithmetic statement corresponding exactly to relation (9).

The method of assessing the significance of TEST depends on the magnitude of k and the numbers of observations in each sample. If $k = 3$ and the three samples each contain five or fewer observations, Table O of Siegel (1956) should be consulted; otherwise TEST is distributed approximately as χ^2 with $(k - 1)$ degrees of freedom and its probability can be assessed in the usual way. It is worth noting that compared with the parametric variance-ratio test used in an analysis of variance, the Kruskal–Wallis test has asymptotic efficiency of 95.5 per cent.

Program 37

Kruskal–Wallis test for several independent samples. The main program listed here calls Subroutines RANK and TIE (p. 329); data and output are also listed.

```
C................................................................
C................................................................
C
C   THIS PROGRAM EXECUTES A KRUSKAL-WALLIS ONE-WAY ANALYSIS OF VARIANCE
C   ON UP TO 100 INDEPENDENT SAMPLES. ANY NUMBER OF PROBLEMS MAY BE
C   RUN CONSECUTIVELY. INPUT PARAMETERS ARE AS FOLLOWS -
C       NPROB = PROBLEM REFERENCE NUMBER (ZERO ON LAST DATA CARD)
C       K     = NUMBER OF INDEPENDENT SAMPLES
```

Program 37 (continued)

```
C      INFMT = VECTOR HOLDING VARIABLE INPUT FORMAT
C      NK    = VECTOR (LENGTH K) GIVING NUMBER OF OBSERVATIONS
C              IN EACH OF THE K SAMPLES
C      X     = VECTOR CONTAINING OBSERVATIONS FOR ALL SAMPLES
C              PUNCHED IN ONE CONTINUOUS ARRAY
C
C.................................................................
C
       INTEGER RX
       DIMENSION INFMT(20), X(1000), RX(1000), NK(100), SRX(100),
      A ANK(100)
C
C   WRITE HEADING.
       WRITE (6,99)
    99 FORMAT (1H1, 30X, 20HKRUSKAL–WALLIS TESTS)
C
C   READ PROBLEM NUMBER, NUMBER OF SAMPLES AND VARIABLE INPUT FORMAT.
     1 READ (5,100) NPROB, K
   100 FORMAT (2I5)
       IF (NPROB .EQ. 0) STOP
       READ (5,101) (INFMT(I), I = 1,20)
   101 FORMAT (20A4)
C
C   READ NUMBERS IN EACH SAMPLE AND SUM.
       READ (5,102) (NK(I), I = 1,K)
   102 FORMAT (10I7)
       N = 0
       DO 2 I = 1,K
     2 N = N + NK(I)
C
C   READ ALL OBSERVATIONS CONSECUTIVELY IN ORDER OF SAMPLES AND RANK.
       READ (5,INFMT) (X(I), I = 1,N)
       CALL RANK (X, RX, N)
       CALL TIE (X, RX, N)
C
C   SUM RANKS FOR EACH SAMPLE AND COMPUTE FAC.
       DO 3 I = 1,K
     3 SRX(I) = 0.0
       N1 = 1
       DO 5 I = 1,K
       N2 = N1 + NK(I) – 1
       DO 4 J = N1, N2
     4 SRX(I) = SRX(I) + X(J)
     5 N1 = N2 + 1
       FAC = 0.0
       DO 6 I = 1,K
       ANK(I) = NK(I)
     6 FAC = FAC + SRX(I) * SRX(I) / ANK(I)
C
C   COMPUTE CORRECTION FOR TIES
       ANP1 = N + 1
       NT = 0
       DO 8 J = 1,N
       IF (X(J) .EQ. ANP1) GO TO 8
       NTIE = 0
       TEMP = X(J)
       DO 7 I = J, N
       IF (TEMP .NE. X(I)) GO TO 7
       NTIE = NTIE + 1
       X(I) = ANP1
```

Program 37 (continued)

```
      7 CONTINUE
        NT = NT + NTIE ** 3 - NTIE
      8 CONTINUE
C
C  COMPUTE TEST CRITERION.
        AN = N
        ANT = NT
        ANT = 1.0 - ANT / (AN**3 - AN)
        TEST = (12.0 / (AN*AN+AN) * FAC - 3.0 * (AN+1.0)) / ANT
        NDF = K - 1
C
C  OUTPUT RESULTS.
        WRITE (6,103) NPROB, K, TEST
    103 FORMAT (///8H PROBLEM, I4, 6H (WITH, I4, 9H SAMPLES)/
       A   11X, 16HTEST CRITERION = , F10.3)
        IF (K .GT. 3) GO TO 10
        DO 9 I = 1,K
      9 IF (NK(I) .GT. 5) GO TO 10
        WRITE (6,104) (NK(I), I = 1,K)
    104 FORMAT (/11X, 14HSAMPLE SIZES = , 3I5)
        GO TO 1
     10 WRITE (6,105) NDF
    105 FORMAT (/11X, 20HDEGREES OF FREEDOM = , I5)
C
        GO TO 1
        END
```

DATA

```
    1    3
(10F7.0)
    3    4    5
   2.1  2.0  3.0  5.2  5.0  6.0  5.5  8.0  8.0  8.0
   8.8  9.2
    2    8
(10F7.0)
   10    8   10    8    6    4    6    4
   2.0  2.8  3.3  3.2  4.4  3.6  1.9  3.3  2.8  1.1
   3.5  2.8  3.2  3.5  2.3  2.4  2.0  1.6  3.3  3.6
   2.6  3.1  3.2  3.3  2.9  3.4  3.2  3.2  3.2  3.3
   3.2  2.9  3.3  2.5  2.6  2.8  2.6  2.6  2.9  2.0
   2.0  2.1  3.1  2.9  3.1  2.5  2.6  2.2  2.2  2.5
   1.2  1.2  2.5  2.4  3.0  1.5
    0
```

OUTPUT

    KRUSKAL–WALLIS TESTS

PROBLEM   1 (WITH   3 SAMPLES)

        TEST CRITERION =   9.830

        SAMPLE SIZES =   3   4   5

PROBLEM   2 (WITH   8 SAMPLES)

        TEST CRITERION =   18.565

        DEGREES OF FREEDOM =   7

Chapter 12

Fitting Theoretical Distributions to Data

Program 6 enables one to express data in the form of a frequency table, from which the distribution can be plotted as a histogram. Empirical frequency distributions of this kind are informative, but they may be more useful if it can be shown that they correspond to one of the commonly used theoretical distributions of mathematical statistics, such as the normal distribution (for continuous variables) or the Poisson or binomial distribution for discrete variables. If such a correspondence can be established, then the theoretical distribution—whose mathematical properties have been extensively examined—can serve as a model and the data can be effectively summarised in terms of the parameters used to define the theoretical distribution. Further, in many branches of statistics the theory of the subject is based on the assumption that the data conform to the normal, or, less commonly, to some other distribution.

It is therefore desirable to develop means of comparing the empirical distribution with the best-fitting form of one of the theoretical distributions. The method used to do this is typically to tabulate the data as a frequency array, then to estimate from the data the required parameters of the theoretical distribution being fitted, and from these parameters to compute the theoretical frequencies in the same classes as those first observed. The agreement between observed and theoretical frequencies can then be tested by a χ^2-test. The situation differs in detail for each type of distribution being fitted and in this chapter we shall deal with the following:

Program 38: Fitting a normal distribution and testing goodness of fit.
Program 39: Fitting a Poisson distribution and testing goodness of fit.
Program 40: Fitting a binomial distribution and testing goodness of fit.
Program 41: Fitting a negative binomial distribution and testing goodness of fit.

In the discussion of each program some indications are given of the biological uses and significance of the distribution concerned. It is assumed in all cases that the observations are ungrouped when read into central memory.

Program 38: Fitting a normal distribution and testing goodness of fit

Many kinds of biological data represent continuous variables whose frequency distributions yield histograms like that of Fig. 31. Examples include the length or breadth of many plant and animal structures, the yields of crop-plants or measurements of environmental factors such as rainfall, average temperatures and so forth. The histograms are symmetrical, with large central frequencies decreasing towards each extreme. Such data can often be fitted, at least to a good approximation, by the normal frequency curve, a symmetrical bell-shaped curve which tails off to infinity in each direction and is represented by the equation

$$f(x) = \frac{1}{\sigma\sqrt{2\pi}} e^{\{-\frac{1}{2}[(x-\mu)/\sigma]^2\}} \tag{1}$$

where π is the usual constant (approx. 3.1416) and e is the base of natural logarithms (approx. 2.71828) while μ and σ are the mean and standard deviation, the two parameters needed to determine the distribution.

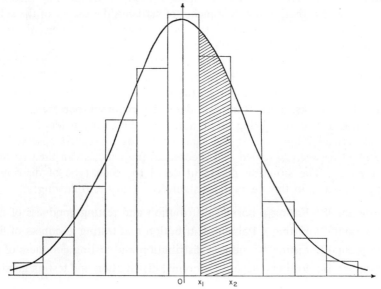

FIG. 31. Histogram and fitted normal curve. The shaded area represents the proportion of observations expected to lie between x_1 and x_2.

In fitting such a curve the mean and standard deviation can be estimated from the data in the usual way, the estimates being denoted as $\bar{x}$ and s. We also allow for the number of observations (N) and the class interval (h) used in constructing the histogram, so arriving at the following formula for the normal curve we must fit to the data:

$$Y = \frac{Nh}{s\sqrt{2\pi}} e^{\{-\frac{1}{2}[(x-\bar{x})/s]^2\}}. \tag{2}$$

In order to test the fit between the data and the curve, however, we need not the value of Y above, but the proportion of individuals lying between each pair of successive values of x, i.e. the areas under the various sections of the curve such as that depicted in Fig. 31. To do such a calculation by hand we would consult tabulated values of the definite integral of (1) as given in any set of statistical tables. This integral has the form

$$F(s) = \frac{1}{\sigma\sqrt{2\pi}} \int_{-\infty}^{s} e^{\{-\frac{1}{2}[(x-\mu)/\sigma]^2\}} dx \tag{3}$$

and by finding the tabulated values of $F(x_1)$ and $F(x_2)$ and subtracting the first from the second we would obtain the proportion of observations expected to lie between x_1 and x_2. We use tabulated values because the integral cannot be solved analytically. In a computer program we must replace the use of tables by a numerical approximation, the exact nature of which is discussed later. With this preliminary explanation we can examine the program given below.

The program assumes that we shall form a frequency array of 20 classes and the DIMENSION statement allows for this in assigning arrays for the class-marks (CLMK) and frequencies (F) of the observed distribution and for the expected frequencies to be calculated from the normal curve (FN). We also assume that the ungrouped data are quite extensive (as they usually must be if non-normality is to be detected) and that the total number of observations has not been counted. The input procedure therefore reads the data, punched according to a variable input format that uses NCHK to identify the last card of each problem: this is left unpunched on all previous cards of that problem. Using the statement IF (NCHK) 4,2,4 to loop back to statement 2 enables an inner DO-loop ending on 3 to count the observations and transfer them successively to a vector X. This is dimensioned at only 1000 in the program below, and might well need to be enlarged. An alternative form of input for a grouped frequency array would be quite easy to devise but is unsuitable if, as in the present program, we wish to examine the possibility of transforming the data.

Tests of normality are commonly made in order to confirm that some subsequent method of statistical analysis may be used. Non-normal data can often be analysed by such methods after bringing the data to normality by a suitable transformation. Transformations may also be applied to equalise the variances of a number of samples (i.e. to promote homoscedasticity). While there are some general principles governing the choice of a transformation (Bartlett, 1947) it is desirable to test the transformed observations for normality. Our program allows this kind of investigation by calling SUBROUTINE TRANS, which operates on the vector of observations X, transforming them in accordance with an option selected by the user through the variable NOPT, which will have been read in as part of the program specification. The options available in TRANS have been selected as among the most useful; others can very easily be added by writing additional DO-loops incorporating the necessary arithmetic statements and altering the computed GO TO statement based on NOPT. As it stands, however, TRANS will return any one of the following transformed values of X:

1. No transformation. This possibility must obviously be included, otherwise the main program could never test the raw data.

2. Logarithmic transformation, i.e. the transformed variable x' is given by $x' = \log x$. Logarithmic transformations are often successfully applied to moderately skew distributions or to equalise variances when the standard deviations are proportional to the sample means. A variable whose logarithm is normally distributed is said to have a *lognormal distribution* (Aitchison and Brown, 1957). This transformation cannot be applied if $x \leqslant 0$ and most computers will end execution if instructed to take the logarithm of zero or of a negative number.

3. The transformation $x' = \log(x + 1)$ is a convenient alternative to 2 when the data include zeros.

4. The inverse sine transformation $x' = \sin^{-1}\sqrt{x}$. This and other inverse sine transformations are often used to equalise variances when the primary data are in the form of proportions or percentages. Note that the inverse sine (or arc sine) is a standard Fortran library function but that some compilers require it to be spelt ASIN instead of the ARSIN used here. The value returned is always in radians and the argument must lie between -1 and $+1$. This option should only be used, therefore, if the primary data are proportions between these limits. Percentage data would have to be expressed as proportions to produce a correct result; the additional statement $A(I) = A(I)/100.0$ inserted between statements 6 and 7 of TRANS will be sufficient, provided that no percentage exceeds 100 in absolute value.

5. The square-root transformation $x' = \sqrt{x}$ is appropriate when the variances of samples are approximately proportional to their means or when it is known that the data conform to a Poisson distribution. Obviously it cannot be applied if the data include negative numbers.

6. The reciprocal transformation $x' = 1/x$ is used to equalise variances when these are approximately proportional to, say, the fourth or some higher power of the mean. Again, it cannot be used when the data include zeros.

7. Finally, the inverse hyperbolic sine transformation $x' = \sinh^{-1}(\sqrt{x})$ is one of a group of transformations which are similar to but sometimes more effective than the logarithmic transformation of option 2. Since the inverse hyperbolic sine is not a standard Fortran library function we must associate with SUBROUTINE TRANS a FUNCTION subprogram ASINH which will compute the necessary quantity. This utilises the relation $\sinh^{-1} x = \log_e(x + \sqrt{(x^2 + 1)})$ for all x; the values returned are in radians. The example illustrates how TRANS may be extended with the aid of specially written FUNCTION subprograms.

Having executed whatever transformation option was originally chosen, the program tabulates the frequencies of the values in array X (transformed if necessary). This is done by calling SUBROUTINE FREQ which uses a

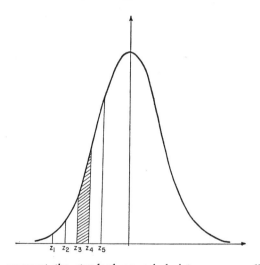

FIG. 32. $z_1 - z_5$ represent the standard normal deviates corresponding to the upper boundaries of the first five classes of a tabulated frequency array. The expected frequency in the fourth class is based on the cross-hatched area beneath the curve and bounded by the ordinates at z_3 and z_4. To compute this area we determine (from tables) the total area below the curve to the left of the ordinate at z_4 and subtract from it the total area to the left of the ordinate at z_3.

tabulating algorithm essentially similar to that of Program 6. In the latter program the user had to specify the number of classes, the class interval and the lower limit of the lowest interval. In SUBROUTINE FREQ this work is done automatically. We call FREQ with the parameter NCL set at 20 (a convenient number of classes for work of this kind). The class interval CLINT is then set at 1/19 of the range of X while the lowest classmark (CLMK) is set at the value of the smallest observation in the data. These are arbitrary choices but they are rather like the decisions one might normally make before beginning to construct a frequency table by hand. Most of the statements in FREQ are concerned with these preliminaries and it is only the last DO-loop (ending on 5) which contains the tabulating algorithm. The main program then uses the array of observations X to calculate their mean and standard deviation, which are needed to compute the areas beneath the normal curve that will give us the expected frequencies (Fig. 32).

The computation of expected frequencies is carried out in three main steps, each represented by a single statement in the DO-loop ending on 6. The first step is to convert the upper boundary of each class into a standardised normal deviate (Z). This is done by subtracting the mean (XBAR) from it and dividing by the standard deviation (SDEV). The second step is to compute that proportion of the area beneath the curve lying to the left of the normal deviate Z (Fig. 32). To do this we use the FUNCTION subprogram PNORM, which provides the answer by a numerical approximation. This is based on other approximations given by Hastings (1955) and may be expressed in the form given by the IBM Manual (IBM, 1968):

$$P = 1.0 - .3989423\, e^{-x^2/2}(1.330274\, w^5 - 1.821256\, w^4 + 1.781478\, w^3 - .3565638\, w^2 + .3193815\, w)$$

where $w = \dfrac{1}{1 + .2316419\, |x|}$ for $x \geqslant 0.0$.

For $x < 0.0$ the polynomial is equal to $(1 - P)$.

In the FUNCTION subprogram PNORM, the polynomial is evaluated in nested form for efficiency (see p.442). The third step is to subtract from this area the area to the left of the corresponding normal deviate for the class below, and to multiply the difference by the total number of individuals in the sample, thus giving the expected frequency FN. It will be seen from Fig. 32 and the Fortran program that the two tails of the distribution are dealt with separately: FN(1) is obtained by taking the entire area to the left of the upper boundary of the first class and FN(20) by finding the entire area to the right of the upper boundary of the last

Outline flowchart for Program 38

Fitting a normal distribution and testing goodness of fit.

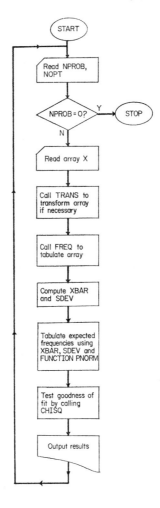

class. Reference to Fig. 32 should make clear what is achieved by the function PNORM and the DO-loop of the main program ending on 6.

Having obtained the theoretical frequencies it is a simple matter to calculate χ^2 using the same subroutine CHISQ that we employed in Program 33. Since we have estimated from our data the two parameters of the normal curve, however, we must reduce by 2 the number of degrees of freedom for the χ^2 returned by CHISQ. This is done by the main program which then prints the output in tabulated form and returns to statement 1 to read the specification card for the next problem.

The flow-chart for this program does not give all the details but outlines the broader features only.

Program 38

Fitting a normal distribution and testing goodness of fit. The main program is listed here, together with Subroutines TRANS, FREQ and CHISQ and Function Sub-programs PNORM and ASINH. Sample data and output are also listed.

```
C
C............................................................
C
C   THIS PROGRAM FITS A NORMAL FREQUENCY DISTRIBUTION TO AN ARRAY
C   OF UNGROUPED OBSERVATIONS, TRANSFORMING THEM IF NECESSARY, AND
C   MEASURING GOODNESS OF FIT BY A CHI-SQUARED TEST. INPUT PARAMETERS
C   ARE AS FOLLOWS -
C       NPROB = PROBLEM REFERENCE NUMBER (ZERO ON LAST DATA CARD)
C       NOPT  = OPTION FOR TRANSFORMATION (SEE SUBROUTINE TRANS)
C       NAME  = TITLE OF PROBLEM
C       INFMT = VECTOR HOLDING VARIABLE INPUT FORMAT
C       NCHK  = NON-ZERO NUMBER ON LAST DATA CARD OF EACH PROBLEM
C       K     = NUMBER OF OBSERVATIONS ON CARD
C       XR    = VECTOR OF OBSERVATIONS ON CARD (LENGTH K)
C   FUNCTION SUBPROGRAMS PNORM AND ASINH AND SUBROUTINES TRANS, FREQ,
C   AND CHISQ ARE CALLED.
C
C............................................................
C
        DIMENSION NAME(20), INFMT(20), XR(50), X(1000), CLMK(20),
     A  F(20), FN(20)
C
C   READ PROBLEM SPECIFICATION AND DATA PUNCHED K OBSERVATIONS PER CARD.
      1 READ (5,100) NPROB, NOPT
    100 FORMAT (2I5)
        IF (NPROB .EQ. 0) STOP
        READ (5,101) (NAME(I), I = 1,20), (INFMT(I), I = 1,20)
    101 FORMAT (20A4)
        N = 0
      2 READ (5,INFMT) NCHK, K, (XR(I), I = 1,K)
        DO 3 I = 1,K
        N = N + 1
      3 X(N) = XR(I)
        IF (NCHK) 4,2,4
C
```

Program 38 (*continued*)

```
C     TRANSFORM DATA IF NECESSARY, FORM FREQUENCY DISTRIBUTION AND
C     COMPUTE MEAN AND STANDARD DEVIATION.
    4 CALL TRANS(X, N, NOPT)
      CALL FREQ (X,N,20,CLINT,CLMK,F)
      SX = 0.0
      SX2 = 0.0
      DO 5 I = 1,N
      SX = SX + X(I)
    5 SX2 = SX2 + X(I) * X(I)
      AN = N
      XBAR = SX / AN
      SDEV = SQRT((SX2 - SX*SX/AN) / (AN - 1.0))
C
C     COMPUTE NORMAL FREQUENCY DISTRIBUTION.
      P1 = 0.0
      DO 6 I = 1,19
      Z = (CLMK(I) + 0.5 * CLINT - XBAR) / SDEV
      P2 = PNORM(Z)
      FN(I) = (P2 - P1) * AN
    6 P1 = P2
      FN(20) = (1.0 - P1) * AN
C
C     TEST GOODNESS OF FIT.
      CALL CHISQ (F, FN, 20, CHI2, NDF, N1, N5)
      NDF = NDF - 2
C
C     OUTPUT RESULTS.
      WRITE (6,102) NPROB, (NAME(I), I = 1,20)
  102 FORMAT (20H1NORMAL DISTRIBUTION//8H PROBLEM, I4// 1X, 20A4///
     A   13X, 5HCLASS, 11X, 5HCLASS, 9X, 8HOBSERVED, 8X,
     B   11HTHEORETICAL/ 13X, 6HNUMBER, 10X, 4HMARK, 10X,
     C   9HFREQUENCY, 7X, 9HFREQUENCY//)
      DO 7 I = 1,20
    7 WRITE (6,103) I, CLMK(I), F(I), FN(I)
  103 FORMAT (I16, F18.3, F16.1, F18.3)
      WRITE (6,104) XBAR, SDEV, CHI2, NDF, N1, N5
  104 FORMAT (//7H MEAN = ,F11.4,4X,21H STANDARD DEVIATION = , F11.4//
     A   14H CHI-SQUARED = , F10.2, 5H WITH, I6,
     B   19H DEGREES OF FREEDOM// 51H NO. OF CELLS WITH EXPECTED FREQUENC
     C   Y LESS THAN 1 = ,I3//51H NO. OF CELLS WITH EXPECTED FREQUENCY LESS
     D   THAN 5 = , I3)
C
      GO TO 1
      END


      SUBROUTINE TRANS(A, N, NOPT)
C
C.............................................................
C
C     THIS SUBROUTINE TRANSFORMS VECTOR A OF LENGTH N ACCORDING TO
C     OPTION SPECIFIED BY NOPT AS FOLLOWS -
C        1 = NO TRANSFORMATION
C        2 = LOG A
C        3 = LOG (A+1)
C        4 = INVERSE SINE
C        5 = SQUARE ROOT
C        6 = RECIPROCAL
C        7 = INVERSE HYPERBOLIC SINE
C
C.............................................................
```

Program 38 (*continued*)

```
C
      DIMENSION A(N)
      GO TO (1,2,4,6,8,10,12), NOPT
    1 RETURN
    2 DO 3 I = 1,N
    3 A(I) = ALOG10(A(I))
      RETURN
    4 DO 5 I = 1,N
    5 A(I) = ALOG10(A(I)+1.0)
      RETURN
    6 DO 7 I = 1,N
    7 A(I) = ARSIN(SQRT(A(I)))
      RETURN
    8 DO 9 I = 1,N
    9 A(I) = SQRT(A(I))
      RETURN
   10 DO 11 I = 1,N
   11 A(I) = 1.0 / A(I)
      RETURN
   12 DO 13 I = 1,N
   13 A(I) = ASINH(SQRT(A(I)))
      RETURN
      END


      SUBROUTINE FREQ (A,N,NCL,CLINT,CLMK,F)
C
C.............................................................
C
C   THIS SUBROUTINE TABULATES A FREQUENCY DISTRIBUTION WITH NCL CLASSES
C   FROM DATA IN VECTOR A OF LENGTH N. ON RETURN CLINT CONTAINS THE
C   CLASS-INTERVAL, CLMK THE CLASS-MARKS AND F THE OBSERVED
C   FREQUENCIES (IN FLOATING-POINT MODE).
C
C.............................................................
C
      DIMENSION A(N), CLMK(NCL), F(NCL)
      AMAX = A(1)
      AMIN = A(1)
      DO 2 I = 2,N
      IF (AMAX .GE. A(I)) GO TO 1
      AMAX = A(I)
    1 IF (AMIN .LE. A(I)) GO TO 2
      AMIN = A(I)
    2 CONTINUE
      CL = NCL
      CLINT = (AMAX - AMIN) / (CL-1.0)
      DO 3 I = 1,NCL
      AIM = I - 1
    3 CLMK(I) = AMIN + AIM*CLINT
      AMIN = AMIN - 0.5*CLINT
      DO 4 I = 1,NCL
    4 F(I) = 0.0
      DO 5 I = 1,N
      K = (A(I) - AMIN) / CLINT + 1.0
    5 F(K) = F(K) + 1.0
      RETURN
      END
```

Program 38 (*continued*)

```
      SUBROUTINE CHISQ (O, E, N, CHI2, NDF, N1, N5)
C
C.........................................................
C
C   THIS SUBROUTINE COMPUTES CHI-SQUARED FOR VECTORS OF OBSERVED AND
C   EXPECTED FREQUENCIES, O AND E, BOTH OF LENGTH N. ALSO RETURNED
C   ARE DEGREES OF FREEDOM (NDF) AND NUMBER OF CELLS WITH EXPECTED
C   FREQUENCIES LESS THAN 1 (N1) AND 5 (N5).
C
C.........................................................
C
      DIMENSION O(N), E(N)
      CHI2 = 0.0
      DO 1 I = 1,N
    1 CHI2 = CHI2 + (O(I) - E(I)) * (O(I) - E(I)) / E(I)
      NDF = N - 1
      N1 = 0
      N5 = 0
      DO 2 I = 1,N
      IF (E(I) .GE. 5.0) GO TO 2
      N5 = N5 + 1
      IF (E(I) .GE. 1.0) GO TO 2
      N1 = N1 + 1
    2 CONTINUE
      RETURN
      END


      FUNCTION PNORM(X)
C
C.........................................................
C
C   THIS FUNCTION SUBPROGRAM COMPUTES THE AREA UNDER THE NORMAL
C   PROBABILITY CURVE FROM MINUS INFINITY TO X, USING AN
C   APPROXIMATION DUE TO HASTINGS (APPROXIMATIONS FOR
C   DIGITAL COMPUTERS).
C
C.........................................................
C
      FX = EXP(-X*X/2.0) * 0.3989423
      W = 1.0 / (1.0+ABS(X)*0.2316419)
      PNORM = 1.0 - FX * W * ((((1.330274 * W - 1.821256) * W +
     A  1.781478) * W - 0.3565638) * W + 0.3193815)
      IF (X) 1,2,2
    1 PNORM = 1.0 - PNORM
    2 RETURN
      END


      FUNCTION ASINH(X)
C
C.........................................................
C
C   THIS FUNCTION SUBPROGRAM COMPUTES THE INVERSE HYPERBOLIC SINE OF X
C
C.........................................................
C
      AX = ABS(X)
      ASINH = ALOG(AX + SQRT(AX*AX+1.0))
      IF (X) 1,2,2
```

FITTING THEORETICAL DISTRIBUTIONS TO DATA

Program 38 (continued)

```
  1  ASINH = - ASINH
  2  RETURN
     END
```

DATA

```
    1    1
TEST 1 OF NORMAL DISTRIBUTION PROGRAM.
(I3,I2,10F6.0)
   10   1.2   1.2   1.4   1.4   1.4   1.6   1.6   1.6   1.6   1.8
   10   1.8   1.8   1.8   1.8   2.0   2.0   2.0   2.0   2.0   2.0
   10   2.2   2.2   2.2   2.2   2.2   2.2   2.2   2.5   2.5   2.5
   10   2.5   2.5   2.5   2.5   2.5   2.7   2.7   2.7   2.7   2.7
   10   2.7   2.7   2.7   2.7   2.7   2.9   2.9   2.9   2.9   2.9
   10   2.9   2.9   2.9   2.9   3.1   3.1   3.1   3.1   3.1   3.1
   10   3.1   3.3   3.3   3.3   3.3   3.3   3.3   3.5   3.5   3.5
   10   3.5   3.5   3.7   3.7   3.7   3.7   3.9   3.9   3.9   3.9
   10   4.1   4.1   4.3   4.3   4.3   4.3   4.5   4.5   4.8   4.8
  999 3  4.8   5.0   5.0
    0
```

OUTPUT

NORMAL DISTRIBUTION

PROBLEM   1

TEST 1 OF NORMAL DISTRIBUTION PROGRAM.

CLASS NUMBER	CLASS MARK	OBSERVED FREQUENCY	THEORETICAL FREQUENCY
1	1.200	2.0	3.946
2	1.400	3.0	2.157
3	1.600	4.0	2.975
4	1.800	5.0	3.922
5	2.000	6.0	4.938
6	2.200	7.0	5.941
7	2.400	0.	6.828
8	2.600	8.0	7.499
9	2.800	10.0	7.868
10	3.000	9.0	7.887
11	3.200	13.0	7.554
12	3.400	0.	6.912
13	3.600	5.0	6.043
14	3.800	8.0	5.048
15	4.000	2.0	4.029
16	4.200	4.0	3.072
17	4.400	0.	2.238
18	4.600	2.0	1.558
19	4.800	3.0	1.036
20	5.000	2.0	1.550

MEAN =   2.9108   STANDARD DEVIATION =   0.9348

CHI–SQUARED =   30.22 WITH   17 DEGREES OF FREEDOM

NO. OF CELLS WITH EXPECTED FREQUENCY LESS THAN 1 = 0

NO. OF CELLS WITH **EXPECTED FREQUENCY** LESS THAN 5 = 11

Program 39: Fitting a Poisson distribution and testing goodness of fit.

A frequent task, arising in many fields of quantitative biology, consists of taking a relatively large number of samples and noting the numbers of samples containing 0, 1, 2, 3, ... individuals. If the individuals are comparatively rare, then the relative frequencies of samples with 0, 1, 2, 3 ... individuals will often approximate to a Poisson series, represented by

$$e^{-m}\left(1, m, \frac{m^2}{2!}, \frac{m^2}{3!} \dots \right) \quad (4)$$

where m is the mean number of individuals per sample. We may write the general term of this series as:

$$f(x) = e^{-m} m^x / x! \quad (5)$$

where x is the number of individuals per sample ($x = 0, 1, 2, 3, \dots$). If the total number of samples taken is N, then the frequencies will, of course, be obtained by multiplying each term of (4) by N, and we shall have an expected frequency table of the form:

x	expected frequency
0	Ne^{-m}
1	Nme^{-m}
2	$Nm^2 e^{-m}/2$
3	$Nm^3 e^{-m}/6$
⋮	⋮

Examining successive frequencies the reader will see that the number in each class after the first is derived from the preceding class by multiplying the latter by m/x. Since the value of x in the i-th class is $(i-1)$ we have a so-called *recursive relationship* for the relative frequencies in a Poisson distribution:

$$f(x)_i = \frac{m f(x)_{i-1}}{(i-1)} \quad (6)$$

enabling us, once we know the frequency in the first class ($x = 0$), to calculate the others in succession.

Frequency tables conforming closely to such a distribution occur, for example, when one records the number of some kind of cell per microscope field (as in a haemocytometer used for blood-counts or for work with other cell-suspensions). They may also arise when a plant ecologist records the number of specimens of a particular species in a large number of quadrats, or in counts of bacterial and fungal colonies on plate-cultures, or in the distribution of insects on their host-plants and so on. The number of disintegrations per unit time in a given amount of radioactive material also forms a Poisson distribution. All these examples are the consequence of a random distribution of individuals in the space or time being sampled. It will be appreciated that the observed distribution is a discrete one, in the sense that the variable x (the number of individuals in a sample) must be zero or a positive integer $-0, 1, 2, 3 \ldots -$ and cannot take intermediate values. It is also clear from (4) or (5) that the only parameter needed to determine the Poisson series is m, the mean number of individuals per sample. It can also be shown, though we do not use the fact in this program, that the mean of a Poisson distribution is numerically equal to its variance.

The general structure of this program is very much like that for the normal distribution (Program 38). The first segment causes the values of X (the number of individuals in each sample) to be read in the form of K values of XR on each data card and the total number of samples (N) counted. The array X can then be arranged in the form of a vector of observed frequencies (F). To produce this we need first to know the largest value of X (XMAX). The vector F will then contain (XMAX + 1) elements and this number (stored as NCL) forms the upper indexing parameter of the DO-loop which ends on 7 and embodies the tabulating algorithm (cf. Program 38, p.350). In this case the computed subscript is IP and it is merely one greater than the value of X assigned to the class, i.e. we are merely counting all the zeros in F(1), all the 1's in F(2), the 2's in F(3) and so on.

The next segment computes XBAR, the mean value of X in the usual way and calls SUBROUTINE POISSN to provide the expected frequencies for a Poisson distribution. The input parameters of this subroutine—N, XBAR and NCL—have already been defined in the main program and the subroutine is intended to return FP, the vector of expected frequencies for the NCL classes of a Poisson distribution with mean sample-size XBAR. The frequency in the first class, FP(1), is computed simply as EXP(−XBAR)∗ AN where EXP is the exponential library function and AN the real equivalent of N. The second to penultimate frequencies are then computed by the recursive relationship (6) given above in the DO-loop ending on 1. This DO-loop also sums the expected frequencies in all classes up to and including the penultimate class, storing this sum as SFP. The reason for this operation is that the expected frequency in the last of the NCL classes is not derived as the next term of a Poisson series, but is the difference between the observed total number of samples N and the sum of all expected frequencies in the (NCL − 1) previous classes. An analogous situation arose in obtaining the expected normal frequency for the last class

Outline flowchart for Program 39

Fitting a Poisson distribution and testing goodness of fit.

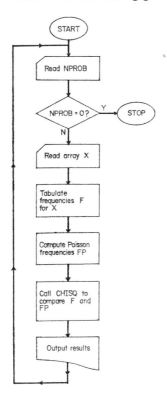

in Program 38. The last arithmetic statement of SUBROUTINE POISSN therefore defines FP(NCL) as AN − SFP.

Once POISSN has returned FP, the array of expected frequencies, to the main program it only remains to call CHISQ to compute χ^2 for this and the array of observed frequencies F. The number of degrees of freedom (NDF) returned by CHISQ must, however, be reduced by 1 since the one required parameter of the Poisson frequency function was estimated from the data as XBAR. Having arranged for output, control then returns to statement 1 to begin the next problem. As before the flowchart gives only the main outlines of the program.

Program 39

Fitting a Poisson distribution and testing goodness of fit. The main program calls Subroutines POISSN and CHISQ (the latter listed on p. 353). Sample data and output are also listed.

```
C
C............................................................
C
C   THIS PROGRAM FITS A POISSON FREQUENCY DISTRIBUTION TO AN ARRAY OF
C   UNGROUPED OBSERVATIONS, TESTING GOODNESS OF FIT BY A CHI-SQUARED
C   TEST. INPUT PARAMETERS ARE AS FOLLOWS -
C       NPROB = PROBLEM REFERENCE NUMBER (ZERO ON LAST DATA CARD)
C       NAME  = TITLE OF PROBLEM
C       INFMT = VECTOR HOLDING VARIABLE INPUT FORMAT
C       NCHK  = NON-ZERO NUMBER ON LAST DATA CARD OF EACH PROBLEM
C       K     = NUMBER OF OBSERVATIONS ON CARD
C       XR    = VECTOR OF OBSERVATIONS ON CARD (LENGTH K)
C   SUBROUTINES POISSN AND CHISQ ARE REQUIRED.
C
C............................................................
C
        DIMENSION NAME(20), INFMT(20), XR(50), X(1000), F(50), FP(50)
C
C   READ PROBLEM SPECIFICATION AND DATA PUNCHED K OBSERVATIONS PER CARD.
      1 READ (5,100) NPROB
    100 FORMAT (I5)
        IF (NPROB .EQ. 0) STOP
        READ (5,101) (NAME(I), I = 1,20), (INFMT(I), I = 1,20)
    101 FORMAT (20A4)
        N = 0
      2 READ (5,INFMT) NCHK, K, (XR(I), I = 1,K)
        DO 3 I = 1,K
        N = N + 1
      3 X(N) = XR(I)
        IF (NCHK) 4,2,4
C
C   TABULATE OBSERVED FREQUENCY DISTRIBUTION.
      4 XMAX = X(1)
        DO 5 I = 2,N
```

Program 39 (continued)

```
          IF (XMAX .GE. X(I)) GO TO 5
          XMAX = X(I)
        5 CONTINUE
          NCL = XMAX + 1.0
          DO 6 I = 1,NCL
        6 F(I) = 0.0
          DO 7 I = 1,N
          IP = X(I) + 1.0
        7 F(IP) = F(IP) + 1.0
C
C   COMPUTE THEORETICAL FREQUENCY DISTRIBUTION AND TEST GOODNESS OF FIT.
          XBAR = 0.0
          DO 8 I = 1,N
        8 XBAR = XBAR + X(I)
          AN = N
          XBAR = XBAR / AN
          CALL POISSN (N, XBAR, NCL, FP)
          CALL CHISQ (F, FP, NCL, CHI2, NDF, N1, N5)
          NDF = NDF - 1
C
C   OUTPUT RESULTS.
          WRITE (6,102) NPROB, (NAME(I), I = 1,20)
      102 FORMAT (21H1POISSON DISTRIBUTION//8H PROBLEM, I4// 1X, 20A4///
         A    13X, 5HCLASS, 11X, 5HCLASS, 9X, 8HOBSERVED, 8X,
         B    11HTHEORETICAL/13X, 6HNUMBER, 10X, 4HSIZE, 10X, 9HFREQUENCY,
         C    7X, 9HFREQUENCY//)
          DO 9 I = 1,NCL
          NZ = I - 1
        9 WRITE (6,103) I, NZ, F(I), FP(I)
      103 FORMAT (2I16, F18.1, F18.3)
          WRITE (6,104) XBAR, CHI2, NDF, N1, N5
      104 FORMAT (//7H MEAN = , F12.5//14H CHI-SQUARED = , F10.2, 5H WITH, I6,
         A    19H DEGREES OF FREEDOM// 51H NO. OF CELLS WITH EXPECTED FREQUENC
         B    Y LESS THAN 1 = ,I3// 51H NO. OF CELLS WITH EXPECTED FREQUENCY LESS
         C    THAN 5 = , I3)
C
          GO TO 1
          END


          SUBROUTINE POISSN (N, XBAR, NCL, FP)
C
C.............................................................
C
C   THIS SUBROUTINE CONSTRUCTS A POISSON FREQUENCY DISTRIBUTION FP
C   CONTAINING N INDIVIDUALS, WITH MEAN XBAR, ARRANGED IN NCL CLASSES.
C
C.............................................................
C
          DIMENSION FP(NCL)
          NCLM = NCL - 1
          AN = N
          FP(1) = EXP(-XBAR) * AN
          SFP = FP(1)
          DO 1 I = 2,NCLM
          AIM = I - 1
          FP(I) = FP(I - 1) * XBAR / AIM
        1 SFP = SFP + FP(I)
          FP(NCL) = AN - SFP
          RETURN
          END
```

Program 39 (continued)

DATA

```
    1
TEST DATA - SERIES 1.
(I3, I2, 10F2.0)
    10  0  1  2  0  1  3  0  1  1  2
    10  2  1  4  3  0  1  1  3  1  0
   999  3  0  4  1
     0
```

OUTPUT

POISSON DISTRIBUTION

PROBLEM   1

TEST DATA - SERIES 1.

CLASS NUMBER	CLASS SIZE	OBSERVED FREQUENCY	THEORETICAL FREQUENCY
1	-0	6.0	5.721
2	1	9.0	7.960
3	2	3.0	5.537
4	3	3.0	2.568
5	4	2.0	1.213

MEAN =   1.39130

CHI-SQUARED =   1.90 WITH   3 DEGREES OF FREEDOM

NO. OF CELLS WITH EXPECTED FREQUENCY LESS THAN 1 = 0

NO. OF CELLS WITH EXPECTED FREQUENCY LESS THAN 5 = 2

Program 40: Fitting a binomial distribution and testing goodness of fit.

The binomial distribution is another discrete frequency distribution, arising when we have repetitive observations, in each of which an event may either occur or not occur. For example, each animal in a litter may be a male or not a male (i.e. a female); or each individual from a Mendelian cross may either have a certain phenotype or have some other phenotype. Under such circumstances, if the proportion of individuals in the population with the condition is p and the proportion without it is $q = (1 - p)$, then the proportions in which we shall find the condition occurring $0, 1, 2, 3 \ldots n$ times in randomly selected groups of n individuals are given by the terms in the expansion of $(q + p)^n$. By the binomial theorem this expansion is

$$(q + p)^n = q^n + nq^{n-1}p + \frac{n(n-1)}{2}q^{n-2}p^2 + \ldots + p^n. \tag{7}$$

The expected relative frequency of the condition occurring x times ($x = 0, 1, 2, 3 \ldots n$) is therefore given by the general term of (7), i.e.

$$f(x) = \frac{n!}{x!(n-x)!} p^x q^{n-x}. \tag{8}$$

This is often written more compactly as

$$f(x) = \binom{n}{x} p^x q^{n-x} \tag{9}$$

where $\binom{n}{x}$ denotes the number of combinations of n things taken x at a time. If, then, we take N samples, each of size n, the expected absolute frequencies with which x takes the values $0, 1, 2, 3, \ldots n$ will be $Nf(x)$.

As a very simple example, supposing that the proportions of boys and girls in the population are equal, what are the expected frequencies with which families of 5 children will contain 0, 1, 2, 3, 4 or 5 boys? Clearly $p = q = \frac{1}{2}$ and $n = 5$. Expanding $(\frac{1}{2} + \frac{1}{2})^5$ by the binomial theorem we get the relative frequencies $f(x)$ denoted below. Further, if our total sample consisted of 1000 such families of 5 children the expected absolute frequencies for the 6 classes of x would be the numbers given in the third column of the table.

x	$f(x)$	$Nf(x)$
0	1/32 = 0.03125	31.25
1	5/32 = 0.15625	156.25
2	10/32 = 0.31250	312.50
3	10/32 = 0.31250	312.50
4	5/32 = 0.15625	156.25
5	10/32 = 0.03125	31.25
	$\Sigma f(x) = 1.00000$	$N = 1000.00$

Expected frequencies like this could be compared with observed data by a chi-squared test of goodness of fit as in Programs 38 and 39. Instead, however, we shall employ a *dispersion test* as described by Maxwell (1968). We consider only the case where n is constant and p is estimated from the data, this being the situation most often encountered. The test may be employed even when the number of observations is too small for goodness of fit to be assessed in the usual way. It requires us to compute

$$\chi^2 = \frac{\sum_{x=0}^{n} f_x (x - np)^2}{npq} \tag{10}$$

in which f_x is the observed frequency for $x = 0, 1, 2, 3 \ldots n$ and the χ^2 has $(N - 1)$ degrees of freedom. np is estimated as $\bar{x}$, the mean of the observed x's. It is to be noted that (10) may be calculated without knowing the expected frequencies, though our program will calculate them and they are included in the output for their own sake.

The form of the program can readily be understood from the above account and the first two program-segments are similar to those of Program 39. The first results in the observed values of x being stored in vector X and their number counted to give N (note that this Fortran variable corresponds to N of the above theoretical discussion, not to n). As before, the second segment tabulates the frequencies of X in the vector F, the number of classes, NCL, having already been read as data. NCL corresponds, of course, to $(n + 1)$ in terms of the above discussion, so that by specifying NCL the user is defining one parameter of the binomial frequency function he intends to fit to the data. To compute the expected frequencies, however, we need also to estimate from the data the proportion p of (7). This is done by the Fortran instruction P = SX/AN/ANCLM where SX is the sum of X while AN and ANCLM are the real equivalents of N and (NCL − 1). This arithmetic statement is the Fortran equivalent of

$$p = \sum_{i=1}^{N} x_i/Nn = \bar{x}/n. \tag{11}$$

Having estimated P and knowing N and NCL we can compute the expected binomial frequencies by calling SUBROUTINE BINOM. This begins by computing the expected frequency in the first class (i.e. the quantity Nq^n of our theoretical account) using the arithmetic statement

FB(1) = Q ∗∗ (NCL − 1)∗AN.

The expected frequencies of the second, third, ... classes are then obtained by the recursive relationship

$$f(x)_i = f(x)_{i-1} \cdot \frac{p}{q} \cdot \frac{(n + i - 2)}{(i - 1)} \tag{12}$$

where $f(x)_i$ is the expected frequency in the i-th class $(i \geqslant 2)$. The necessary operations are accomplished by the DO-loop ending on statement 1 which, if it is followed carefully, will be seen to correspond to (12).

Finally the χ^2 for the dispersion test is computed from formula (10) above. Again, the Fortran seems superficially different but will be found to correspond if examined closely. Output follows the pattern for the previous program and control returns to statement 1 for the next problem to be dealt with.

Outline flowchart for Program 40

Fitting a binomial distribution and testing goodness of fit.

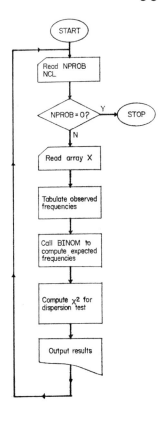

Program 40

Fitting a binomial distribution and testing goodness of fit. The main program is listed here together with SUBROUTINE BINOM. Sample data and output are also listed.

```
C
C.................................................................
C
C   THIS PROGRAM FITS A BINOMIAL FREQUENCY DISTRIBUTION TO AN ARRAY OF
C   UNGROUPED OBSERVATIONS, TESTING GOODNESS OF FIT BY A DISPERSION TEST.
C   INPUT PARAMETERS ARE AS FOLLOWS –
C      NPROB = PROBLEM REFERENCE NUMBER (ZERO ON LAST DATA CARD)
C      NCL   = NUMBER OF CLASSES REQUIRED IN FREQUENCY ARRAY
C      NAME  = TITLE OF PROBLEM
C      INFMT = VECTOR HOLDING VARIABLE INPUT FORMAT
C      NCHK  = NON–ZERO NUMBER ON LAST DATA CARD OF EACH PROBLEM
C      K     = NUMBER OF OBSERVATIONS ON CARD
C      XR    = VECTOR OF OBSERVATIONS ON CARD (LENGTH K)
C   FOR COMPUTATIONAL DETAILS SEE MAXWELL, A.E., ANALYSING QUALITATIVE
C   DATA, 1967, P.103.
C   SUBROUTINE BINOM IS REQUIRED.
C
C.................................................................
C
      DIMENSION NAME(20), INFMT(20), XR(50), X(1000), F(50), FB(50)
C
C   READ PROBLEM SPECIFICATION AND DATA PUNCHED K OBSERVATIONS PER CARD.
    1 READ (5,100) NPROB, NCL
  100 FORMAT (2I5)
      IF (NPROB .EQ. 0) STOP
      READ (5,101) (NAME(I), I = 1,20), (INFMT(I), I = 1,20)
  101 FORMAT (20A4)
      N = 0
    2 READ (5,INFMT) NCHK, K, (XR(I), I = 1,K)
      DO 3 I = 1,K
      N = N + 1
    3 X(N) = XR(I)
      IF (NCHK) 4,2,4
C
C   TABULATE OBSERVED FREQUENCY DISTRIBUTION.
    4 DO 5 I = 1,NCL
    5 F(I) = 0.0
      DO 6 I = 1,N
      IP = X(I) + 1.0
    6 F(IP) = F(IP) + 1.0
C
C   COMPUTE THEORETICAL FREQUENCY DISTRIBUTION.
      SX = 0.0
      DO 7 I = 1,N
    7 SX = SX + X(I)
      AN = N
      ANCLM = NCL – 1
      P = SX / AN / ANCLM
      CALL BINOM (P, NCL, N, FB)
C
C   EXECUTE DISPERSION TEST.
      CHI2 = 0.0
      DO 8 I = 1,NCL
      AIM = I – 1
```

Program 40 (continued)

```
    8 CHI2 = CHI2 + F(I) * (AIM – P * ANCLM) ** 2
      CHI2 = CHI2 / ANCLM / P /(1.0 – P)
      NDF = N – 1
C
C   OUTPUT RESULTS.
      WRITE (6,102) NPROB, (NAME(I), I = 1,20)
  102 FORMAT (22H1BINOMIAL DISTRIBUTION//8H PROBLEM, I4//1X, 20A4///
     A  13X, 5HCLASS, 11X, 5HCLASS, 9X, 8HOBSERVED, 8X, 11HTHEORETICAL/
     B  13X, 6HNUMBER, 10X, 4HSIZE, 10X, 9HFREQUENCY, 7X, 9HFREQUENCY//)
      DO 9 I = 1,NCL
      NZ = I – 1
    9 WRITE (6,103) I, NZ, F(I), FB(I)
  103 FORMAT (2I16, F18.1, F18.3)
      WRITE (6,104) P, CHI2, NDF
  104 FORMAT (//51H ESTIMATE OF P IN BINOMIAL EXPANSION (Q+P)**N. P = ,
     A  F8.4///32H DISPERSION TEST. CHI–SQUARED = , F10.2, 5H WITH, I6,
     B  19H DEGREES OF FREEDOM)
C
      GO TO 1
      END


      SUBROUTINE BINOM (P,NCL, N, FB)
C
C.........................................................
C
C   THIS SUBROUTINE CONSTRUCTS A BINOMIAL FREQUENCY DISTRIBUTION FB
C   CONTAINING N OBSERVATIONS IN NCL CLASSES AND HAVING A GIVEN VALUE
C   OF P IN EXPANSION OF (Q + P) ** (NCL – 1)
C
C.........................................................
C
      DIMENSION FB(NCL)
      Q = 1.0 – P
      AN = N
      FAC = P/Q
      FB(1) = Q ** (NCL – 1) * AN
      DO 1 I = 2,NCL
      A = NCL – I + 1
      B = I – 1
    1 FB(I) = FB(I–1) * A/B * FAC
      RETURN
      END

DATA

    1    7
TEST PROBLEM NUMBER ONE.
(I3, I2, 10F2.0)
    10  0  1  2  0  1  3  0  1  1  2
    10  2  1  4  3  0  1  1  3  1  0
   999  3  0  4  1
      0

OUTPUT

BINOMIAL DISTRIBUTION

PROBLEM    1

TEST PROBLEM NUMBER ONE.
```

Program 40 (*continued*)

CLASS NUMBER	CLASS SIZE	OBSERVED FREQUENCY	THEORETICAL FREQUENCY
1	−0	6.0	4.724
2	1	9.0	8.556
3	2	3.0	6.458
4	3	3.0	2.599
5	4	2.0	0.589
6	5	0.	0.071
7	6	0.	0.004

ESTIMATE OF P IN BINOMIAL EXPANSION (Q+P)**N. P =   0.2319

DISPERSION TEST. CHI–SQUARED =   33.20 WITH   22 DEGREES OF FREEDOM

Program 41: Fitting a negative binomial distribution and testing goodness of fit.

In the discussion of Program 39 we considered the Poisson distribution, which summarises the numbers of sampling units containing 0, 1, 2, 3, ... x individuals. The expected variance of a Poisson distribution is equal to its mean, but in practice it is often found when sampling in this way that the variance, as estimated from the data, is significantly greater than the mean. This situation is referred to as *over-dispersion* and indicates that the assumptions underlying the Poisson distribution are not fulfilled in the data. Many theoretical distributions have been devised to deal with over-dispersion, one of the best known being the negative binomial distribution, discussed for example by Bliss (1953) and Evans (1953). Instances of the kind of ecological problem in which the negative binomial distribution arises include quadrat-counts in plant ecology, plankton-sampling, the distribution of insects on their host-plants and of tapeworm cysts or ticks in relation to the animals they parasitise.

The expected relative frequencies of sampling units containing 0, 1, 2, 3, ... x individuals are given by:

$$f(x) = \frac{(k + x - 1)!}{x!(k - 1)!} \frac{p^x}{(1 + p)^{k+x}} \tag{13}$$

where $p = \bar{x}/k$. If the total number of sampling units is N, the frequencies are, of course, given by $Nf(x)$. The distribution is evidently determined by two parameters $\bar{x}$ and k, which must be estimated from the data. The mean $\bar{x}$ is, as usual, estimated as $\sum_{i=1}^{N} x_i/N$ but the estimation of k presents greater difficulty. Several approximate methods are available, but in this program we shall use the maximum likelihood method described by Fisher (1953), which is outlined below in connection with the program. Having estimated k it is quite easy to calculate the expected frequencies and thence to obtain a χ^2 for goodness of fit to the original data. Other methods of testing the agreement between observed and expected frequencies are discussed by Bliss (1953) but will not be included in the present program.

The first program-segment arranges, as before, for the observed values of x to be read into a singly-subscripted array X and counted. They are then tabulated as a frequency array by the second program-segment using essentially the same method as in Program 39 for the Poisson distribution. The only difference here is that we have tabulated the data with one more class than those needed to accommodate the maximum recorded value in X. From the tabulated frequencies in vector F we can easily calculate the accumulated frequencies; these are computed by the next program-segment and arranged in the vector FA.

Returning to the vector X, we use it to compute the mean (XBAR) and variance (VAR) in the usual way and test for over-diepersion by the relationship:

$$\chi^2 = \frac{s^2(n-1)}{\bar{x}} \qquad (14)$$

with $(n-1)$ degrees of freedom, allowing the user to assess significance from tabulated percentage points of χ^2 in the usual way. The results required for this test are stored as CHI2 and NDF and are printed out. It is anticipated that the test will show the variance to be significantly greater than the mean so that the fitting of a negative binomial distribution is justified.

The next program segment is concerned with finding two trial values of k which we denote in Fortran as K0 and K1 and which will later be used to obtain the maximum likelihood estimate, K. The first trial estimate K0 is easily derived from the expression

$$k_0 = \frac{\bar{x}^2}{s^2 - \bar{x}} \qquad (15)$$

but for reasons of numerical accuracy we must check that the value of the denominator in (15) is sufficiently great to avoid undue loss of significant digits. Our criterion is that $(s^2 - \bar{x})$ should not be less than $10^{-4}s^2$ and if our data fail to satisfy this test the program terminates execution after a diagnostic message, indicated in statement 14. A further check on K0 is also needed before we can use it. Fisher's maximum likelihood method requires that we eventually find a value of k that reduces z_i in the expression below to zero:

$$z_i = \sum\left(\frac{A_x}{k_i + x}\right) - N \log_e\left(1 + \frac{\bar{x}}{k_i}\right) \qquad (16)$$

where A_x are the accumulated frequencies and k_i are the successive values of k in the iterative procedure to be followed. Clearly, the second term on the right can only be evaluated if $\bar{x}/k_i > -1.0$, so that we test K0 for this condition, again ending execution if it cannot be fulfilled. The reader will realise that in the ordinary way K0 will satisfy both this and the previous condition. They should nevertheless be tested for, since if some unusual data were to yield quite atypical values of K0 the program would fail or would produce quite inaccurate results. Having obtained K0 we compute the corresponding value of z from equation (16), calling it Z0. This and all later values of z are computed using the FUNCTION subprogram ZF. If by sheer good fortune Z0 turned out to be zero we could immediately accept K0 as the maximum likelihood estimate of k and proceed to statement 19. It is much more likely, however, that Z0 will have some positive or negative value. Our object is, therefore, to obtain a second estimate of k which we call K1 and which, when substituted in (16) will produce a value of z (Z1) that is of opposite sign to Z0 (or, by good fortune, is equal to zero). The Fortran statements from

K1 = K0

AK0 = ABS(K0)

as far as statement 13 are intended to produce the required value of K1. Essentially, we obtain a provisional value of K1 by increasing or decreasing K0 by 0.2 of its value, then testing the resulting Z1 to see if its sign differs from Z0. This procedure is repeated using a counter NCYC to record the number of cycles through the loop between statements 12 and 13. We set a maximum of 100 cycles and end execution with a diagnostic warning if a suitable value of K1 has not been found at the end of that time. All being well, however, we should emerge from the loop via the statement

IF (Z0 * Z1) 15, 20, 13

by passing to statement 20 (in the very improbable event of Z1 being zero and hence K1 our required maximum likelihood estimate of k) or to statement 15 in the more likely event that Z0 and Z1 are now of opposite sign. The logic of this section of the program is quite straightforward once the reader appreciates what we have been trying to do, namely to choose values of K0 and K1 using as our criterion of suitability the values of Z0 and Z1 given by equation (16).

Assuming, therefore, that the two estimates K0 and K1 give us values of Z0 and Z1 that straddle the desirable condition $z_i = 0$, we must now use some method of interpolation to obtain the final estimate of K. Our procedure can be understood more clearly if we represent the position diagrammatically (Fig. 33). In terms of this diagram, we require the value

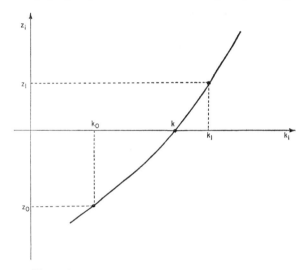

FIG. 33. Diagram illustrating method of estimating k by interpolation. k_0 and k_1 are the first two trial values. For explanation see text.

of k at which the curve joining (k_0, z_0) and (k_1, z_1) intersects the abcissa. We do not, of course, know the nature of this curve, but we can find the required value of k by an approximation known as the method of false position, discussed in most text-books of numerical analysis. In the present context this entails the following steps:

1. Choose two approximations k_0 and k_1 such that $z_0 . z_1 < 0$. This we have already done.

2. Find a new approximation k from the formula

$$k = \frac{k_0 z_1 - k_1 z_0}{z_1 - z_0}. \tag{17}$$

3. Compute from k a new value of z.

4. If z lies close to zero—say within the range ± 0.0001—accept k as the maximum likelihood estimate required. Otherwise proceed to step 5.

Outline flowchart for Program 41

Fitting a negative binomial distribution and testing goodness of fit.

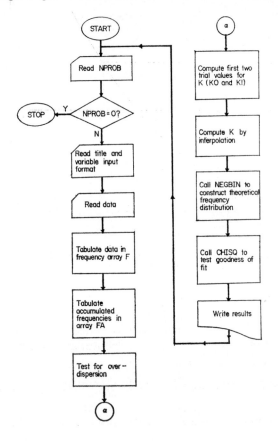

5. If z and z_0 are of like sign, put $k_1 = k$ and $z_1 = z$ and return to step 2.
If z and z_0 are of opposite sign put $k_0 = k$ and $z_0 = z$ and return to step 2.

In the program segment carrying out this interpolation we also allow for certain potential sources of inaccuracy, for the possibility that $\bar{x}/k < -1$ (see above) and for the possibility that a suitable value of k has not materialised after 100 cycles of iteration. All these possibilities, which are unlikely to cause trouble in any real example, lead back to the error message written by statement 14.

Normally, however, interpolation will produce the required estimate of k and we can proceed to statement 21 where the subroutine NEGBIN is called to compute the expected frequencies in a negative binomial distribution having the estimated values of $\bar{x}$ and k. The subroutine first computes the frequency in the first class of the distribution—FN(1)—from the relationship

$$f(x)_1 = \frac{N}{(1 + \bar{x}/k)^k}. \tag{18}$$

The frequencies in the second to the penultimate class are then derived from the relationship

$$f(x)_i = f(x)_{i-1} \cdot \frac{\bar{x}}{(k + \bar{x})} \cdot \frac{(k + i - 2)}{(i - 1)}. \tag{19}$$

Finally the frequency in the last class is obtained by subtracting the sum of the previous frequencies from N. Once the expected frequencies have been returned by NEGBIN to the main program we can compute χ^2 in the usual way and print out the results.

Program 41

Fitting a negative binomial distribution and testing goodness of fit. The main program calls Function Sub-program ZF and Subroutines NEGBIN and CHISQ (the latter listed on p. 353). Sample data and output are also listed.

```
C
C............................................................
C
C   THIS PROGRAM FITS A NEGATIVE BINOMIAL FREQUENCY DISTRIBUTION TO AN
C   ARRAY OF UNGROUPED OBSERVATIONS, USING THE MAXIMUM LIKELIHOOD
C   METHOD OF BLISS AND FISHER (1953) BIOMETRICS, VOL. 9, PP. 176-200.
C   GOODNESS OF FIT IS TESTED BY A CHI-SQUARED TEST.
C   INPUT PARAMETERS ARE AS FOLLOWS -
C       NPROB = PROBLEM REFERENCE NUMBER (ZERO ON LAST DATA CARD)
C       NAME  = TITLE OF PROBLEM
C       INFMT = VECTOR HOLDING VARIABLE INPUT FORMAT
C       NCHK  = NON-ZERO NUMBER ON LAST DATA CARD OF EACH PROBLEM
```

Program 41 (*continued*)

```
C      M     = NUMBER OF OBSERVATIONS ON CARD
C      XR    = VECTOR OF OBSERVATIONS ON CARD (LENGTH K)
C   SUBROUTINES NEGBIN AND CHISQ AND FUNCTION SUBPROGRAM ZF ARE NEEDED.
C
C...............................................................
C
       REAL K0, K1, K
       DIMENSION NAME(20), INFMT(20), XR(50), X(1000), F(200),
      A FA(200), FN(200)
C
C   READ PROBLEM SPECIFICATION AND DATA PUNCHED M OBSERVATIONS PER CARD.
     1 READ (5,100) NPROB
   100 FORMAT (I5)
       IF (NPROB .EQ. 0) STOP
       READ (5,101) (NAME(I), I = 1,20), (INFMT(I), I = 1,20)
   101 FORMAT (20A4)
       N = 0
     2 READ (5,INFMT) NCHK, M, (XR(I), I = 1,M)
       DO 3 I = 1,M
       N = N + 1
     3 X(N) = XR(I)
       IF (NCHK) 4,2,4
C
C   TABULATE OBSERVED FREQUENCY DISTRIBUTION.
     4 XMAX = X(1)
       DO 5 I = 2,N
       IF (XMAX .GE. X(I)) GO TO 5
       XMAX = X(I)
     5 CONTINUE
       NCL = XMAX + 2.0
       DO 6 I = 1,NCL
     6 F(I) = 0.0
       DO 7 I = 1,N
       IP = X(I) + 1.0
     7 F(IP) = F(IP) + 1.0
C
C   TABULATE ACCUMULATED FREQUENCIES.
       AN = N
       S = 0.0
       DO 8 I = 1,NCL
       S = S + F(I)
     8 FA(I) = AN - S
C
C   COMPUTE MEAN AND VARIANCE AND TEST FOR OVER-DISPERSION.
       SX = 0.0
       SX2 = 0.0
       DO 9 I = 1,N
       SX = SX + X(I)
     9 SX2 = SX2 + X(I) * X(I)
       XBAR = SX / AN
       VAR = (SX2 - SX*SX/AN) / (AN - 1.0)
       CHI2 = (AN - 1.0) * VAR / XBAR
       NDF = N - 1
C
       WRITE (6,103) NPROB, (NAME(I), I = 1,20), CHI2, NDF
   103 FORMAT (31H1NEGATIVE BINOMIAL DISTRIBUTION///8H PROBLEM, I4///
      A 1X, 20A4/// 24H TEST OF OVER-DISPERSION// 11X,
      B 14H CHI-SQUARED =, 1PE12.4, 5H WITH, I4, 19H DEGREES OF FREEDOM)
C
```

Program 41 (*continued*)

```
C   COMPUTE FIRST TWO TRIAL VALUES OF K.
        DEN = VAR - XBAR
        IF (ABS(DEN) .LT. 0.0001 * ABS(VAR)) GO TO 14
        K0 = XBAR * XBAR / DEN
        IF (XBAR/K0 .LE. (-1.0)) GO TO 14
        Z0 = ZF(K0, NCL, FA XBAR, AN)
        K1 = K0
        AK0 = ABS(K0)
        NCYC = 0
        IF(Z0) 10,19,11
    10  DIFF = -0.2 * AK0
        GO TO 12
    11  DIFF = 0.2 * AK0
    12  K1 = K1 + DIFF
        IF (XBAR/K1 .LE. (-1.0)) GO TO 14
        Z1 = ZF(K1, NCL, FA, XBAR, AN)
        NCYC = NCYC + 1
        IF (Z0 * Z1) 15,20,13
    13  IF (NCYC .LT. 100) GO TO 12
    14  WRITE (6,104)
   104  FORMAT (//50H NO SUITABLE ESTIMATES OF K. EXECUTION TERMINATED.)
        GO TO 1
C
C   ESTIMATE K BY INTERPOLATION (METHOD OF FALSE POSITION).
    15  NCYC = 0
    16  DEN = Z1 - Z0
        IF (ABS(DEN) .LT. 0.0001 * ABS(Z1)) GO TO 14
        K = (K0 * Z1 - K1 * Z0) / DEN
        IF (XBAR/K .LE. (-1.0)) GO TO 14
        Z = ZF(K, NCL, FA, XBAR, AN)
        NCYC = NCYC + 1
        IF (ABS(Z) .LT. 0.0001) GO TO 21
        IF (NCYC .GE. 100) GO TO 14
        IF (Z * Z0) 17,21, 18
    17  K1 = K
        Z1 = Z
        GO TO 16
    18  K0 = K
        Z0 = Z
        GO TO 16
C
C   CONSTRUCT THEORETICAL FREQUENCY DISTRIBUTION AND TEST
C   GOODNESS OF FIT.
    19  K = K0
        GO TO 21
    20  K = K1
    21  CALL NEGBIN (NCL, N, XBAR, K, FN)
        CALL CHISQ (F, FN, NCL, CHI2, NDF, N1, N5)
        NDF = NDF - 2
C
C   OUTPUT RESULTS.
        WRITE (6,105)
   105  FORMAT (///13X, 5HCLASS, 11X, 5HCLASS, 9X, 8HOBSERVED,
       A  8X, 11HTHEORETICAL/ 13X, 6HNUMBER, 10X, 4HSIZE, 10X,
       B  9HFREQUENCY, 7X, 9HFREQUENCY//)
        DO 22 I = 1,NCL
        NX = I - 1
    22  WRITE (6,106) I, NX, F(I), FN(I)
   106  FORMAT (2I16, F18.1, F18.3)
        WRITE (6,107) XBAR, K, CHI2, NDF, N1, N5
```

Program 41 (*continued*)

```
  107 FORMAT (//7H MEAN = , F10.4, 10X, 3HK = , F11.5//14H CHI–SQUARED = ,
     A  1PE12.4, 5H WITH, I4, 19H DEGREES OF FREEDOM//
     B  54H NUMBER OF CELLS WITH EXPECTED FREQUENCY LESS THAN 1 = , I4//
     C  54H NUMBER OF CELLS WITH EXPECTED FREQUENCY LESS THAN 5 = , I4)
      GO TO 1
      END



      SUBROUTINE NEGBIN (NCL, N, XBAR, K, FN)
C
C.............................................................................
C
C   THIS SUBROUTINE CONSTRUCTS A NEGATIVE BINOMIAL FREQUENCY
C   DISTRIBUTION FN CONTAINING N OBSERVATIONS IN NCL CLASSES AND HAVING
C   GIVEN VALUES OF MEAN (XBAR) AND K.
C
C.............................................................................
C
      REAL K
      DIMENSION FN(NCL)
      AN = N
      NCLM = NCL – 1
      FN(1) = AN / ((1.0 + XBAR/K) ** K)
      SFN = FN(1)
      DO 1 I = 2,NCLM
      AIM = I – 1
      FN(I) = FN(I – 1) * (K + AIM – 1.0) * XBAR / (K + XBAR) / AIM
    1 SFN = SFN + FN(I)
      FN(NCL) = AN – SFN
      RETURN
      END


      FUNCTION ZF(K,NCL, FA, XBAR, AN)
C
C.............................................................................
C
C   THIS FUNCTION SUBPROGRAM COMPUTES Z FOR FISHER'S MAXIMUM LIKELIHOOD
C   SOLUTION OF A NEGATIVE BINOMIAL DISTRIBUTION WITH PARAMETER K,
C   CONTAINING AN OBSERVATIONS, MEAN XBAR, AND ACCUMULATED FREQUENCIES
C   FA, ARRANGED IN NCL CLASSES.
C
C.............................................................................
C
      REAL K
      DIMENSION FA(NCL)
      ZF = 0.0
      DO 1 I = 1,NCL
      XN = I – 1
    1 ZF = ZF + FA(I) / (K + XN)
      ZF = ZF – AN * ALOG(1.0 + XBAR/K)
      RETURN
      END

DATA

    1
DATA FROM BLISS AND FISHER PAGE 178, TABLE 1.
(I3, I2, 30F2.0)
      30 0 0 0 0 0 0 0 0 0 0 0 0 0 0 0 0 0 0 0 0 0 0 0 0 0 0 0 0 0 0
      30 0 0 0 0 0 0 0 0 0 0 0 0 0 0 0 0 0 0 0 0 0 0 0 0 0 0 0 0 0 0
```

Program 41 (*continued*)

```
30 0 0 0 0 0 0 0 0 0 0 1 1 1 1 1 1 1 1 1 1 1 1 1 1 1 1 1 1 1 1
30 1 1 1 1 1 1 1 1 1 1 1 1 1 1 1 1 1 1 1 2 2 2 2 2 2 2 2 2 2 2
99930 2 2 2 2 2 3 3 3 3 3 3 3 3 3 3 4 4 4 4 4 4 4 4 4 5 5 5 6 6 7
    0
```

OUTPUT

NEGATIVE BINOMIAL DISTRIBUTION

PROBLEM   1

DATA FROM BLISS AND FISHER PAGE 178, TABLE 1.

TEST OF OVER-DISPERSION

CHI-SQUARED = 2.9544E 02 WITH 149 DEGREES OF FREEDOM

CLASS NUMBER	CLASS SIZE	OBSERVED FREQUENCY	THEORETICAL FREQUENCY
1	-0	70.0	69.488
2	1	38.0	37.600
3	2	17.0	20.101
4	3	10.0	10.703
5	4	9.0	5.687
6	5	3.0	3.018
7	6	2.0	1.600
8	7	1.0	0.848
9	8	0.	0.955

MEAN =   1.1467   K =   1.02460

CHI-SQUARED =   3.5445E 00 WITH   6 DEGREES OF FREEDOM

NUMBER OF CELLS WITH EXPECTED FREQUENCY LESS THAN 1 =   2

NUMBER OF CELLS WITH EXPECTED FREQUENCY LESS THAN 5 =   4

Chapter 13

Models and Simulation

The digital computer's capacity for storing large bodies of numerical information and manipulating the numbers very rapidly has led to its widespread use in mathematical modelling. Indeed, this activity is hardly to be contemplated on a serious scale without access to a computer. For the most part, however, the construction of a model depends not on the use of a computer but on the modeller's understanding of the biological and physical characteristics of the problem and on his ability to express it in mathematically tractable form without resorting to undue simplification. Fortran contributes little of importance to this aspect of the problem. Once a model has been constructed, however, its numerical consequences or behaviour can be examined quickly by computer and the effects of systematic changes in the different variables and parameters can be assessed with relative ease.

To understand some of the possibilities open to the investigator we need first to distinguish between *deterministic* and *stochastic* models. A deterministic model is one in which the variables do not involve any random element, so that given the same initial conditions and parameters the same values of the dependent variables will result, no matter how often the model situation is repeated. On the other hand, a stochastic model is one in which the effects of sampling are operative. Thus if we assume a fixed value for, say, the fecundity of a species under given conditions the model is, to that extent anyway, a deterministic one. We could, however, regard fecundity as a random variable taking values obtained by sampling a particular frequency distribution characterised by, say, a certain mean and variance. We would then have incorporated a stochastic element in our model and would find that no two repetitions yielded identical results. How important the random element might be in determining some future state of a system would, of course, depend on various properties of the model and the system it was intended to describe. One might, for example, find that in describing the behaviour of large animal populations a deterministic ecological model was perfectly satisfactory, whereas in dealing with small fluctuating populations, where chance events play a larger role, a stochastic model might be necessary. In all biological investigations it is, of course, necessary to validate the model by comparing predictions derived from it with empirical data.

Deterministic models are not necessarily simple, especially when they involve large numbers of interacting variables, and when, as is nearly always true in biology, the whole system is subject to change or is in a state of dynamic equilibrium. Stochastic models, however, almost invariably contain very difficult mathematics, and as their consequences cannot always be evaluated analytically it may be necessary to resort to simulation techniques, in which the stochastic features are reproduced 'experimentally'. The most primitive kind of simulation is that where numbers are selected randomly by hand or by some simple machine in such a way as to imitate the stochastic processes that occur naturally in a biological system. With the advent of computers simulation can be practised very much more rapidly and on a far larger scale.

Since the formulation of adequate biological models—deterministic or stochastic—depends so much on the field of study, no attempt will be made here to discuss any particular situation in detail or to develop programs for handling specific problems. Those interested will find a stimulating elementary discussion in Maynard–Smith (1968) and a more elaborate introduction by Bailey (1967). The fields of application are many and diverse and there is little doubt that mathematical approaches of this kind will become increasingly important in most branches of biology. As an indication of their scope one might mention, almost at random, the books by Watt (1968) on ecology and resource-management, Bailey (1957) on the theory of epidemics, Bartlett (1960) on ecology and epidemiology, Morton (1969) on genetics, Grodins (1963) on control mechanisms in respiratory and cardiovascular physiology or Goodwin (1963) on cellular metabolism. Cybernetics has generalised the significance of control systems in biology (Wiener, 1961; Ashby, 1956; Milsum, 1966) and information theory provides other widely applicable concepts (e.g. Yockey, 1958; Margalef, 1969). Population genetics and evolution theory has long had its special literature, notable for the early contributions of Fisher (1930), Haldane (1932) and Wright (1931). These latter form only a part of the older tradition of biomathematics, exemplified also by the work of Lotka (1924), Rashevsky (1960) and others, as well as in the more static and descriptive approach initiated by D'Arcy Thompson in 1917.

Few biologists have anything approaching the mathematical competence needed to elaborate the subject theoretically, but there are many simple aspects of modelling and simulation which are useful adjuncts in a variety of biological investigations and are well within the capacity of anyone with a taste for numerical methods. This chapter is intended to provide a few programming techniques of wide applicability. Since the methods are likely to be employed in a variety of quite different biological contexts, the programs are usually presented as subroutines and function subprograms or as short program-segments rather than as self-contained programs. For a general introduction to work of this kind Martin (1968) and Naylor et al. (1966) are useful, though neither deals with biology. The subjects touched on here include parameter studies, numerical integration, the solution of differential equations, the generation of random variables and the evaluation of probabilities from statistical distributions.

Parameter studies: These include some of the simplest examples of deterministic models in biology. It may happen that a relationship in which one is interested can be expressed by a relatively simple formula, whose implications are nevertheless not obvious at sight. One way to explore them is to tabulate the function, varying each parameter or independent variable systematically over an appropriate range. The tabulated values may then be inspected or plotted in various ways—by hand or through various on- and off-line graph-plotting methods that we shall not deal with here. To take a very simple example from human genetics, the frequency of a rare recessive allele can be estimated from data on first-cousin marriages using the equation:

$$q = \frac{c(1-k)}{16k - 15c - ck} \tag{1}$$

where q is the frequency of the recessive allele, c the frequency of first-cousin marriages in the general population and k the frequency of first-cousin marriages among the parents of affected individuals (Dahlberg, 1947). Though simple, with the right-hand side involving only multiplication and division with two variables (c and k), it is by no means obvious at sight how q changes with variations in c and k. We can, however, very readily tabulate q for all values of c between, say, 0.001 and 0.1 at intervals of .001 and for all values of k between 0.001 and 0.2 at the same intervals. To do this we require two nested DO-loops, enclosing equation (1) within the inner loop and using the loop indices to compute appropriate values of c and k. For example:

```
REAL K
DIMENSION Q (200), K(200)
DO 2 I = 1,100
C = I
C = C * 0.001
WRITE (6,100) C
```

```
100  FORMAT (//F6.3)
     DO 1 J = 1,200
     K(J) = J
     K(J) = K(J) * 0.001
  1  Q(J) = C * (1.0-K)/(16.0* K-15.0 * C-C * K)
  2  WRITE (6,101) (K(J), Q(J), J = 1,200)
101  FORMAT (5(F10.3, F10.6))
     STOP
     END
```

A few things may be noted about this program. Firstly it is run without any data cards, the values of C and K being generated within the program itself. If one wished to generalise it, initial values and step-sizes could be read in, but there is not usually much point in generalising such programs. Secondly, the indices I and J are used to compute successive values of C and K by simple arithmetic statements. Thirdly, the WRITE statements must be suitably placed to record changes in C and K. Fourthly, although a very short and simple set of instructions, it will generate over 4000 lines of output, a common characteristic of such programs. The general principles of successively nested DO-loops and the use of their indices to compute parameter and variable values will apply to all programs of this kind. If, however, tabulation occurs over certain ranges or at unusual intervals this may require a little thought. Bearing in mind that DO-loop indices cannot assume zero or negative values or run in descending order we offer a few fairly obvious suggestions:

(i) *Tabulating with integers in descending order.*

Here we require loops of the form:

$$DO\ 1\ I = 1, N$$
$$J = N - I + 1$$
$$1\ A(J) = ...$$

when the values of J will run from N to 1.

(ii) *Tabulating with integers from zero*, e.g.

$$DO\ 1\ I = 1, N$$
$$J = I - 1$$
$$1\ LAMBDA\ (I) = M * J + K$$

when J will run 0, 1, 2, 3, ... (N − 1)

(iii) *Tabulating in fractions from a given lower limit*, e.g.

$$DO\ 1\ I = 1, N$$
$$XI = I - 1$$
$$1\ A(I) = 3.5 + 0.1 * XI$$

when A will run upwards from 3.5 at intervals of 0.1 in the independent variable XI.

(iv) *Tabulating on a logarithmic scale*, e.g.

```
    XI = 0.5
    DO 1 I = 1, N
    XI = XI * 2.0
  1 Y(I) = XI/C − D
```

when XI will assume the successive values 1, 2, 4, ... 2^{N-1}.

Very little ingenuity is needed to devise other variations on these themes and to produce a neatly-organised lay-out on the printed page with all variables fully identified.

Numerical integration: The more complicated deterministic models will probably include a variety of mathematical functions and in developing the model it may be required to integrate some expression. When the integral can be evaluated analytically (i.e., by the normal procedures of integral calculus) this will be done by the modeller and the computer will still only deal as a rule with functions of the kind that are generally available in Fortran (e.g. exponentiation, extraction of roots, logarithmic and trigonometric functions). Less commonly one may encounter expressions involving the hyperbolic functions, inverse hyperbolic and trigonometric functions or gamma-functions. The larger scientific computing systems often include these as built-in functions or they can readily be incorporated into a program as FUNCTION subprograms (e.g. the inverse hyperbolic sine ASINH on p. 353). Very many expressions—even quite simple ones—cannot, however, be integrated analytically and if they feature in a model it may be necessary to determine numerically the value of the integral between certain specified limits. This is the problem of numerical quadrature, discussed at length in works on numerical analysis. Of the many available techniques we shall consider only that based on Simpson's rule.

Program 42: Numerical integration by Simpson's rule

As is well known, the definite integral of a function may be represented by the area beneath the curve of the function. For example, in Fig. 34 the numerical value of the integral $\int_a^b f(x)\,dx$ is the area APSD bounded by the curve $y = f(x)$, the x-axis and the ordinates at $x = a$ and $x = b$.

Clearly, an approximation to the area can be obtained by dividing it into a number of narrow strips (such as the shaded area in Fig. 34), calculating the approximate area of each strip and summing these areas. This is the principle underlying numerical integration by several methods. Simpson's rule takes three neighbouring points on the curve, such as P, Q and R in Fig. 34 and, in effect, draws through them a parabola with centre B so as to approximate to the original curve joining P, Q and R. The area APRC composed of a pair of strips then lies beneath the parabola and can be calculated. If the original area APSD is divided into an even number of strips by equally spaced ordinates one can sum the approximate areas of the pairs of strips and so evaluate the integral. It can be shown that the following approximation holds:

$$\int_a^b f(x)\,dx \approx \frac{h}{3}\left(f(a) + f(b) + 4\text{ (sum of even ordinates)}\right.$$
$$\left. + 2\text{ (sum of odd ordinates)}\right) \qquad (2)$$

where h is the interval between successive ordinates. Other things being equal, the greater the number of strips into which the area is divided the greater will be the accuracy of the approximation, a matter discussed below.

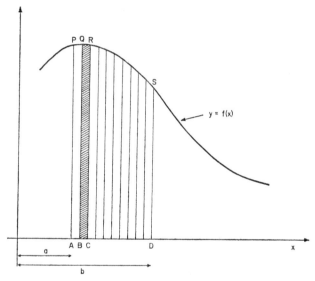

FIG. 34. Diagram illustrating numerical integration using Simpson's rule. The area APSD corresponds to $\int_a^b f(x)\,dx$. By dividing it into a number of vertical strips and summing their approximate areas the integral can be evaluated numerically.

The program that carries out the numerical integration is written as SUBROUTINE SIMP and is intended to be used in conjunction with a main program and a FUNCTION subprogram that defines F, the function to be integrated. The need for the FUNCTION subprogram requires a little explanation. In order to use the subroutine for numerical integration it is necessary to transmit, as one of its arguments, the function F. This function cannot be defined in the main program by an arithmetic function statement since the latter only acts *within* a single program. We could, of course, include the arithmetic function statement in SUBROUTINE SIMP but this would limit one's capacity to a single function per run or would require a subroutine dealing separately with a series of separately defined functions. A more convenient method, enabling one to integrate any number of functions in succession is to associate a specially written FUNCTION subprogram with SUBROUTINE SIMP and the main program. According to the value of an index (here denoted as NF) any of the functions in FUNCTION F can then be passed to the subroutine any number of times for integration. It is also possible using this method to integrate functions that cannot be defined simply by an arithmetic statement function; the steps necessary to compute such functions can form part of the FUNCTION subprogram. The present example contains three functions to be integrated, each defined by an arithmetic statement. The first two are simple functions whose integral is known analytically ($y = \sin x$ and $y = \sin^2 x$); the third is the normal distribution function which cannot be integrated analytically but which has, of course, been tabulated in detail by statisticians.

In the form given here the main program allows any number of problems to be executed. For each one a problem specification card is read giving the reference number of the problem (NPROB); the option number NF used to select from FUNCTION F the expression to be integrated; the number of intervals to be used in applying Simpson's rule (N, which for reasons given above must be an even number); and the upper and lower limits of integration (A and B). Supplied with these parameters the main program can call SUBROUTINE SIMP to carry out the integration and will then print out the results.

The arguments of SUBROUTINE SIMP are F, NF, A, B, N and S. Of these F is the function that will be selected by the option NF. The mechanism for doing this is the computed GO TO instruction in the subprogram FUNCTION F (X, NF), whose construction presents no difficulties. One must bear in mind, however, that F is a function defined outside the main program and must therefore be declared in the latter by the statement EXTERNAL F. The arguments A, B and N of SIMP are the same as those read by the main program under these variable-names. S is the numerical value of the integral which SIMP will return to the main program.

The first two instructions in SUBROUTINE SIMP compute H, the interval size. S1 is an intermediate variable used to accumulate the sums of the even ordinates of equation (2) using the DO-loop ending on 1. S2 and the ensuing DO-loop fulfill the same purpose for the odd ordinates. With S1 and S2 computed we need only write

$$S = H/3.0 * (F(A, NF) + F(B, NF) + 4.0 * S1 + 2.0 * S2)$$

Flowchart for Program 42

Numerical Integration using Simpson's Rule.

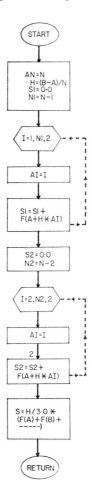

which is the Fortran equivalent of (2) above and completes the process of numerical integration

Program 42

Numerical integration using Simpson's Rule. The computation is carried out by SUBROUTINE SIMP. The user must supply a Function Sub-program F and an appropriate calling program; those listed here are merely given as examples. Sample data and output are also listed.

```
C
C   THIS PROGRAM CALLS SUBROUTINE SIMP. USER MUST SUPPLY APPROPRIATE
C   FUNCTION SUBPROGRAM TO DEFINE F (THE FUNCTION TO BE INTEGRATED).
C
        EXTERNAL F
C
        WRITE (6,100)
  100   FORMAT (1H1,3X,7HPROBLEM,11X,1HA,14X,1HB,14X,1HN,10X,8HINTEGRAL)
    1   READ (5,101) NPROB, NF, N, A, B
  101   FORMAT (3I5, 2F10.0)
        IF (NPROB .EQ. 0) STOP
        CALL SIMP (F,NF,A,B,N,S)
        WRITE (6,102) NPROB, A, B, N, S
  102   FORMAT (/I8, F18.3, F15.3, I12, F18.6)
        GO TO 1
        END


C
        SUBROUTINE SIMP (F,NF,A,B,N,S)
C
C.........................................................
C
C   THIS SUBROUTINE COMPUTES THE NUMERICAL VALUE OF AN INTEGRAL BY
C   SIMPSON'S RULE. PARAMETERS ARE AS FOLLOWS —
C       F  = SUPPLIED FUNCTION TO BE INTEGRATED. THIS MUST BE
C            DEFINED IN A SEPARATE FUNCTION SUBPROGRAM AND DECLARED
C            AS EXTERNAL IN CALLING PROGRAM.
C       NF = OPTION USED TO DEFINE F IN FUNCTION SUBPROGRAM F
C       A  = LOWER LIMIT OF INTEGRAL
C       B  = UPPER LIMIT OF INTEGRAL
C       N  = NUMBER OF INTERVALS (N MUST BE EVEN, N = 40 MAY PROVE
C            TO BE A SUITABLE VALUE)
C       S  = NUMERICAL VALUE OF INTEGRAL
C
C.........................................................
C
        AN = N
        H = (B−A) / AN
        S1 = 0.0
        N1 = N−1
        DO 1 I = 1,N1,2
        AI = I
    1   S1 = S1 + F(A+H*AI,NF)
        S2 = 0.0
        N2 = N−2
        DO 2 I = 2,N2,2
        AI = I
```

Program 42 (continued)

```
    2 S2 = S2 + F(A+H*AI,NF)
      S = H/3.0 * (F(A,NF) + F(B,NF) + 4.0*S1 + 2.0*S2)
      RETURN
      END


C
      FUNCTION F(X,NF)
C
C.........................................................
C
C     THIS FUNCTION SUBPROGRAM DEFINES THE FUNCTION TO BE INTEGRATED
C     AND IS USED IN CONJUNCTION WITH SUBROUTINE SIMP AND A MAIN PROGRAM.
C     FUNCTION F IS SELECTED ACCORDING TO OPTION SPECIFIED BY NF
C
C.........................................................
C
      GO TO (1,2,3), NF
    1 F = SIN(X)
      RETURN
    2 F = SIN(X) * SIN(X)
      RETURN
    3 F = 0.39894228 * EXP(-X*X/2.0)
      RETURN
      END

DATA

    1  1  40      0.01  .57079633
    2  2  40      0.01  .57079633
    3  3  40      0.01   0.5
    4  3  40      0.01   1.0
    5  3  40      0.01   1.5
    6  3  40      0.01   2.0
    0  3  40      0.01

OUTPUT
PROBLEM  A       B      N     INTEGRAL

    1    0.    1.571   40    1.000000
    2    0.    1.571   40    0.785398
    3    0.    0.500   40    0.191462
    4    0.    1.000   40    0.341345
    5    0.    1.500   40    0.433193
    6    0.    2.000   40    0.477250
```

The accuracy of this program depends on a number of factors that must be discussed very briefly. One such factor is clearly the size of N, the number of intervals. Other things being equal, the larger this number the more accurate the answer will be. As an illustration the program was tested by using it to evaluate $\int_0^{\pi/2} \sin x \, dx$ with various values of N. The right answer is 1 and was computed correct to 4 decimal places with $N = 6$, to 5 decimal places with $N = 10$ and to 6 or more decimal places with $N = 20, 40$ or 80. Using $N = 40$ the test-program also produced results for $\int_0^{\pi/2} \sin^2 x \, dx$ correct to 6 decimal places (the answer is known to be $\pi/4$, approximately 0.785398). The program was also used to calculate the normal distribution function:

$$\Phi(b) = \int_0^b \frac{1}{\sqrt{2\pi}} e^{-x^2/2} \, dx \tag{3}$$

with b set successively at 0.5, 1.0, 1.5 and 2.0. Again these were computed correct to 6 decimal places with $N = 40$. It can in fact be shown that the inherent error in Simpson's rule is such that the number of intervals required for a given maximum error ΔI is

$$N \geqslant \sqrt[4]{\frac{M_4(b-a)^5}{90 \, \Delta I}} \tag{4}$$

where M_4 is the maximum value attained between the limits of integration by the fourth derivative of $f(x)$. See, for example, Pennington (1970) for further details. The required number of intervals will therefore depend on the nature of the function and the limits between which integration is carried out. One must also remember, however, that as N is increased there will be a greater accumulation of round-off errors (p. 460) and these may actually result in a decrease of accuracy! This can be demonstrated by comparing the results obtained for various values of N, using first the single-precision facilities of Fortran and then specifying DOUBLE PRECISION, which reduces round-off error. Simpson's rule has a great advantage over another simple method of numerical integration—the trapezoidal formula—in that it requires far fewer intervals to yield results of similar accuracy. On the other hand the rather more complicated Romberg method (discussed with Fortran IV programs by Conte, 1965) is preferable to Simpson's rule for maximum accuracy in single-precision mode since it works with coarser intervals so that round-off error is less.

Comparison of single- and double-precision results or those obtained by varying the value of N is often used as a method of checking the accuracy of numerical integration. This may not help, however, if the function has an unusual form. For example a very abrupt maximum or minimum could lie entirely within one of the intervals and would not be detected when the adjacent ordinates were computed. A further complication in numerical integration occurs when the integrand (function being integrated) becomes infinite at some point within the required limits or when one or both of the limits of integration are infinite. Special methods are available for some such 'improper integrals' and are given in works on numerical analysis. The point to notice is that one should not apply any method of numerical integration blindly, though for any reasonably well-behaved functions Simpson's rule is perfectly satisfactory.

Differential equations. Most biological systems are in a state of change and to describe them in a mathematical model it may be necessary to use differential equations. These are equations involving one or more derivatives of the variables and some simple examples of their use in biological models are given by Maynard-Smith (1968). Suppose, for example, that a population of micro-organisms is provided with unlimited food and has no predators or competitors. The rate at which the population increases might well be proportional to x, the number of individuals present at any one time. That is,

$$\frac{dx}{dt} = kx \tag{5}$$

To solve this differential equation we proceed as follows:

$$\frac{dt}{dx} = \frac{1}{kx}$$

$$dt = \frac{dx}{kx}$$

Integrating,

$$\int dt = \frac{1}{k} \int \frac{dx}{x}$$

$$\therefore \log_e x = kt + C$$

i.e. $x = Ae^{kt}$ \hfill (6)

where e is the base of natural logarithms (2.71828...) and A and k are constants. Equation (6) is the logarithmic growth equation and is the solution of the original differential equation (5). The solution is thus seen to be another equation, from which the derivative dx/dt has been eliminated and which we can use directly to predict the size of the population at some future time if we can assign numerical values to the parameters A and k. The simple solution is due to the fact that the right-hand side of the re-arranged version of (5) can be integrated so easily. In many cases this will not be so: quite innocuous-looking differential equations may have solutions of alarming complexity, involving integrals that cannot be evaluated analytically. It may therefore be necessary to solve the differential equation numerically. Such a solution will not be an equation like (6) but would consist, in the above example, of a set of different numerical values of t, each accompanied by the corresponding value of x. From these we could see how x changes with t by plotting values of one variable against the other in the usual way. The resulting graph would correspond to equation (6) which we would not know explicitly but which we would have tabulated. Many methods are available for the numerical solution of differential equations, but we shall consider only the Runge–Kutta technique.

Program 43: Solution of first-order ordinary differential equations by a Runge–Kutta method

Differential equations are of various kinds and the present program deals only with the solution of a first-order ordinary differential equation, i.e. one involving only first derivatives such as dy/dx and having only one dependent and one independent variable. Equation (5) above is an example, the independent variable being the time t and the only derivative dx/dt. An equation of this kind may be written in its most general form as

$$\frac{dy}{dx} = f(x, y) \qquad (7)$$

As part of a mathematical model we shall have decided on the form of the function $f(x, y)$—which, incidentally, may be a function of x or y or both—and it is our object to obtain a set of x and y values, so as to tabulate the unknown function whose derivative is $f(x, y)$.

The principle underlying numerical methods of solving such equations may be understood from Fig. 35.

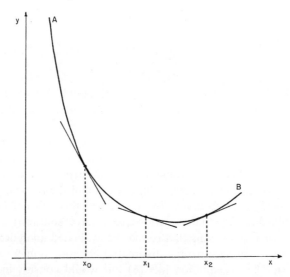

FIG. 35. Diagram illustrating how the slope of the curve AB at x_0, x_1 and x_2 indicates the direction taken by the curve.

The curve AB represents the unknown function we are to tabulate. We know $dy/dx = f(x, y)$ and this gives us the slope of the unknown function at various points corresponding to x_0, x_1, x_2 etc. Knowing these slopes we can tell which direction the curve AB takes and we can hope to approximate to the curve by moving that way in a series of steps. We pass from a first point, say $x_0 y_0$ a short distance along the slope to a second point, where we determine a new slope, passing from this to a third point, and so on. The result is an approximation which tends to wander away from the true curve AB as shown in an exaggerated fashion in Fig. 36.

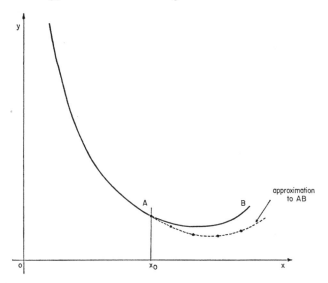

FIG. 36. Diagram illustrating how the course of a curve AB may be approximated by moving from an initial point x_0 along the tangents at successive points.

The object of a satisfactory numerical technique is to keep the error as small as possible. The Runge–Kutta method requires initial values for x and y—we may call them x_0, y_0 as above—and then proceeds to alter x by a series of steps, each of size h, until it reaches the final value of x that we require. At each step we generate an approximation by the recursion formula:

$$y_{n+1} = y_n + \tfrac{1}{6}(k_1 + 2k_2 + 2k_3 + k_4) \tag{8}$$

where $k_1 = hf(x_n, y_n)$

$$\begin{aligned} k_2 &= hf(x_n + h/2, y_n + k_1/2) \\ k_3 &= hf(x_n + h/2, y_n + k_2/2) \\ k_4 &= hf(x_n + h, y_n + k_3) \end{aligned} \tag{9}$$

The error of each step is of the order of h^5 so that by choosing a suitably small value of h we may hope to obtain a tolerably accurate solution. The process of calculation from (8) is quite simple but may not be very fast if a complicated function is involved since this must be evaluated four times in each step.

As in Program 42 we make use of a special FUNCTION sub-program F which enables us to select any number of functions $f(x, y)$ successively. This subprogram must, of course, be provided by the user and declared as EXTERNAL in the calling program. An example is given, relating to three simple equations that can in fact be solved analytically and were used for checking the program. As before we use NF to select the particular function dealt with on any one occasion.

The solution of the equation is carried out by SUBROUTINE RUNKUT, in which the following arguments occur. F is the supplied function $f(x, y)$ of equation (7) and NF is the variable used to select F from several which may be given in the FUNCTION subprogram. XI and YI are the initial values which we need as the starting point. Unlike some other numerical methods of solving differential equations, the Runge–Kutta technique requires only these two initial values. XF is the final value of X which must also be specified. XF may be greater or less than XI and tabulation will be carried out from XI to XF. If necessary we can run two problems with the same initial value, one ending with an XF larger than XI and the other with an XF smaller than XI. We can thus tabulate the solution on both sides of the initial value. The last input variable is N, which specifies the number of values of X and Y in the tabulated solution. These include the initial and final values so that the interval between XI and XF will be divided into $(N - 1)$ steps and h of formula (9) is then fixed automatically. Thus if XI and XF were respectively 0 and 1 and N were set at 51, the interval would be divided into 50 steps, each of size 0.02. Finally we note the two output vectors X and Y which on return will contain the tabulated values we require; X(1) and Y(1) will contain the supplied initial values XI and YI while X(N) will contain XF. The subroutine RUNKUT begins, in fact, by assigning XI and YI to X and Y and computing the step-size H. There remains only the single DO-loop in which the recursion formula (8) is applied to obtain the successive elements of Y, while those of X are derived by simply adding H on each pass through the loop. Note that by using XIM and YIM we improve efficiency by avoiding the repeated use of subscripts in computing K_1, K_2, K_3 and K_4.

The program is very simple and straightforward but nevertheless requires a little thought on the part of the user. In specifying initial and final values of x one must be sure that $f(x, y)$ does not become infinite at some point within the interval. As presented here the Runge–Kutta method is applied to only a single equation. It is, however, possible to extend it quite easily to a set of first-order equations and in this way to use Runge–Kutta methods for solving higher-order differential equations. The process may, though, show instability when applied to a system of equations. The solution of differential equations is a major topic of numerical analysis and a variety of techniques are available for use with digital computers (quite apart from those tech-

Flowchart for Program 43

Solution of first-order ordinary differential equations by a Runge–Kutta method.

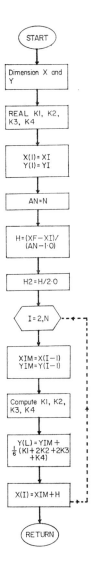

niques that employ analog machines). The present program, though entirely adequate for its purpose, is only one of several which might be required for the numerical solution of models employing differential equations.

Program 43

Solution of first-order ordinary differential equations by a Runge–Kutta method. The computation is carried out by SUBROUTINE RUNKUT. The user must supply a Function Sub-program F and an appropriate calling program; those listed are merely given as examples. Sample data and output are also listed.

```
C
C  THIS PROGRAM CALLS SUBROUTINE RUNKUT AND TABULATES RESULTS. USER
C  MUST SUPPLY FUNCTION SUBPROGRAM F TO DEFINE DERIVATIVE.
C
      DIMENSION X(1001), Y(1001)
      EXTERNAL F
C
    1 READ (5,100) NPROB, NF, XI, YI, XF, N
  100 FORMAT (2I5, 3F10.0, I5)
      IF (NPROB .EQ. 0) STOP
      CALL RUNKUT (F,NF,XI,YI,XF,N,X,Y)
      WRITE (6,101) NPROB
  101 FORMAT (32H1RUNGE-KUTTA SOLUTION OF PROBLEM, I3//
     A  13H NO. OF STEPS, 10X, 1HX, 16X, 1HY/)
      DO 2 I = 1,N,5
      NSTEP = I - 1
    2 WRITE (6,102) NSTEP, X(I), Y(I)
  102 FORMAT (I7, F22.5, F17.5)
      GO TO 1
      END


      SUBROUTINE RUNKUT (F,NF,XI,YI,XF,N,X,Y)
C
C.............................................
C
C  THIS SUBROUTINE SOLVES FIRST-ORDER ORDINARY DIFFERENTIAL EQUATIONS
C  BY THE RUNGE-KUTTA METHOD OF ORDER 4. PARAMETERS ARE AS FOLLOWS -
C    F  = SUPPLIED FUNCTION FOR DERIVATIVE. THIS MUST BE DEFINED
C           IN A SEPARATE FUNCTION SUBPROGRAM AND DECLARED AS EXTERNAL
C           IN CALLING PROGRAM
C    NF = OPTION USED TO DEFINE F IN FUNCTION SUBPROGRAM
C    XI = INITIAL VALUE OF X
C    YI = INITIAL VALUE OF Y
C    XF = FINAL VALUE OF X
C    N  = NUMBER OF VALUES OF X AND Y (INCLUDING INITIAL AND
C           FINAL VALUES)
C    X  = OUTPUT VECTOR OF X-VALUES (LENGTH N)
C    Y  = OUTPUT VECTOR OF Y-VALUES (LENGTH N)
C
C.............................................
C
      DIMENSION X(N), Y(N)
      REAL K1, K2, K3, K4
C
      X(1) = XI
      Y(1) = YI
```

Program 43 (*continued*)

```
      AN = N
      H = (XF-XI) / (AN-1.0)
      H2 = H/2.0
      DO 1 I = 2,N
      XIM = X(I-1)
      YIM = Y(I-1)
      K1 = H * F(XIM,YIM,NF)
      K2 = H * F(XIM+H2,YIM+K1/2.0,NF)
      K3 = H * F(XIM+H2,YIM+K2/2.0,NF)
      K4 = H * F(XIM+H,YIM+K3,NF)
      Y(I) = YIM + (K1+2.0*(K2+K3)+K4)/6.0
    1 X(I) = XIM + H
      RETURN
      END


      FUNCTION F (X,Y,NF)
C
C.............................................................
C
C     THIS FUNCTION SUBPROGRAM DEFINES THE USER-SUPPLIED DERIVATIVE FOR THE
C     RUNGE-KUTTA METHOD OF SOLVING DIFFERENTIAL EQUATIONS. IT IS USED
C     IN CONJUNCTION WITH SUBROUTINE RUNKUT AND A MAIN PROGRAM.
C     FUNCTION F IS SELECTED ACCORDING TO OPTION NF.
C
C.............................................................
C
      GO TO (1,2,3), NF
    1 F = 1.0/X**2 - Y/X - Y*Y
      RETURN
    2 F = X**4 + SIN(X)
      RETURN
    3 F = (1.0+X*Y) / (1.0-X*X)
      RETURN
      END
```

DATA

```
    2    1       1.0      -1.0     2.0    51
    0
```

OUTPUT

RUNGE-KUTTA SOLUTION OF PROBLEM   2

NO. OF STEPS	X	Y
-0	1.00000	-1.00000
5	1.10000	-0.90909
10	1.20000	-0.83333
15	1.30000	-0.76923
20	1.40000	-0.71429
25	1.50000	-0.66667
30	1.60000	-0.62500
35	1.70000	-0.58824
40	1.80000	-0.55556
45	1.90000	-0.52632
50	2.00000	-0.50000

Generation of Random Variables: Simulation studies, in which a stochastic model is made to work by using the computer to sample statistical distributions, require a supply of random numbers and random variables. In principle these could be obtained from a supply stored in the central memory or auxiliary store, rather as one would use published tables of random numbers when doing a sampling experiment by hand. In practice it is usually better to use the arithmetic unit of the computer to generate the required numbers as and when they are needed in the course of a calculation. The methods used to generate them are, of course, normal arithmetic procedures which, starting from a given quantity, will always produce the same answer. In this sense, therefore, the numbers generated are not really random at all. They do, however, obey many of the accepted statistical tests of randomness (for which reasons they are often referred to as pseudorandom) and they can be used in stochastic models. As we shall see, the numbers are generated in a sequence which will start repeating itself only after very many numbers have been produced. The fact that a sequence can if necessary be repeated deliberately by the user is in itself a considerable advantage in testing programs and even in certain simulation experiments. Works like those by Tocher (1963), Martin (1968), Naylor et al. (1966), Gurland (1964) and Moshman (1967) discuss at some length the mathematical aspects and uses of random variables in computer simulation. The techniques employed are widely known as Monte Carlo methods (Hammersley and Handscomb, 1964). In this section we shall describe subroutines for generating (a) random real numbers distributed uniformly (i.e. with a rectangular distribution) between 0 and 1; (b) random integers uniformly distributed between 1 and N (N less than, say, 1000); (c) random normal variates; (d) random binomial variates; (e) random variates from the Poisson distribution and from the (f) Chi-squared and (g) exponential distributions. Of these the first (random real numbers between 0 and 1) is the most important since the other subroutines either use it directly or employ the method involved in it. We shall consider the subroutines in sequence, bearing in mind that the Fortran coding is very short and simple but the principles underlying it require some explanation.

(a) *Random uniform real numbers, 0 to 1 (SUBROUTINE RAND1).*

The generator described here depends on the power residue or multiplicative congruential method, whereby pseudorandom integers are formed by the recursive relationship

$$r_{n+1} = kr_n \,(\text{mod}\, p) \tag{10}$$

where r_n and r_{n+1} are successive numbers, while k and p are integer constants. Equation (10) means that r_{n+1} is the remainder obtained after multiplying r_n by k and dividing by p. Like all pseudorandom sequences this one will eventually repeat and the object is to choose values of k and p which will minimise both the arithmetic involved and the frequency of repetition. One finds, firstly, that by making p equal to one more than the maximum fixed point number capable of being stored in a given computer one can eliminate any explicit arithmetical steps for taking the modulus. Thus if the computer word contains b bits (excluding the sign bit) p should have the value 2^b. When any number is divided by this, the remainder will be the b least significant digits of the $2b$-bit word formed by multiplying k and r_n. Formula (10) can therefore be simplified to:

$$r_{n+1} = kr_n$$

and the modulus will be taken automatically. Secondly, we find that a suitable value for k is about $2^{b/2}$ provided that $k = 8i - 3$, where i is any integer. For a computer with a 36-bit word structure, b will be 35 and a suitable value of k is therefore 131075 ($= 2^{17} + 3$). A starting value r_0—which may be any odd positive integer—has also to be supplied by the user. This is denoted in SUBROUTINE RAND1 as NR1 and the first line of the subroutine multiples NR1 by 131075. This will produce a new value of NR1 and the recursion having started will continue with a period of 2^{b-2} ($= 8,591,507,456$) terms. Eight and a half (American) billion pseudorandom numbers should satisfy the most ardent simulator.

So far, however, we have only devised a method for obtaining a sequence of random integers. To convert these to real numbers in the range 0 to 1 we need, in effect, to move the decimal point to the extreme left of the computer word containing the integer. A shift of this kind can be accomplished by simple division but we must bear in mind that numbers are represented in the computer in binary form. We therefore need to divide the current value of NR1 by 2^{35}. This is equivalent to multiplying by $.2910383 \times 10^{-10}$ or, using the Fortran E-specification we can convert NR1 to a real equivalent X and then write

$$X = X * .2910383E-10$$

to give us the required random real number X which is returned by RAND1. The new value of NR1 generated in the process will, of course, ensure that the subroutine works without further input whenever it is called again in the same program.

It is now easy to see how we derive the rather mysterious-looking constants in the subroutine. Clearly, the program on p. 396 will only work in a binary computer with 35 bits per word (excluding the sign bit). Users of, say, an IBM System 360 machine, which has a word-length of 31 bits (excluding the sign bit) will find, if they care to apply the principles indicated above, that a suitable value of k is 65539 (i.e. $2^{16} + 3$) and that the multiplying factor becomes .4656613E–9. All large computer installations contain an efficient subroutine or function subprogram for producing pseudorandom real numbers and any user can safely rely on it in place of RAND1. The rather detailed discussion is justified by the importance of the subroutine in the generation of other random variables in (b) to (g) below. If a system subroutine or function is used in place of RAND1, these other subroutines will need slight amendment.

Subroutines RAND1 to RAND7

A short calling program is given only for Subroutine RAND1. Similar calling programs can easily be written for the other subroutines if the pseudorandom numbers or variates are to be placed in an array as they are generated.

```
C
C   THIS PROGRAM CALLS SUBROUTINE RAND1.
C
      DIMENSION XX(1000)
C
      READ (5,100) NR1, K
  100 FORMAT (2I5)
      DO 1 I = 1,K
      CALL RAND1 (NR1, X)
    1 XX(I) = X
      WRITE (6,101) (XX(I), I = 1,K)
  101 FORMAT (/10F12.6)
      STOP
      END


C
      SUBROUTINE RAND1 (NR1,X)
C
C.............................................................
C
C   THIS SUBROUTINE GENERATES A SUPPLY OF PSEUDORANDOM REAL NUMBERS (X)
C   UNIFORMLY DISTRIBUTED BETWEEN 0.0 AND 1.0. A STARTING VALUE FOR
C   NR1 MUST BE SUPPLIED (ANY ODD POSITIVE INTEGER). THIS SUBROUTINE
C   IS SUITABLE FOR A COMPUTER WITH A 36–BIT WORD–STRUCTURE E.G., IBM 7094
C
C.............................................................
C
      NR1 = NR1 * 131075
      X = NR1
      X = X * .2910383E–10
      RETURN
      END
```

Apart from its use in the generation of other random variables, RAND1 has various functions in simulation programs. Suppose, for example, that a model postulates an event E to occur with probability p. We can construct a 'probability switch' to tell us whether, on a particular occasion in the simulation E will occur or not. To do this we generate an uniform random variable x between 0 and 1 and compare it with p. If $p \geqslant x$ then E occurs, otherwise it does not. In Fortran:

```
      ⋮
      CALL RAND1 (NR1, X)
      IF (P − X) 10, 20, 20
   10 ...
      ⋮
   20 ...
```

where 10 and 20 are statement numbers allowing control to pass either to 10 (E has not occurred) or to 20 (E has occurred).

(b) *Random uniform Integers from 1 to N (SUBROUTINE RAND2)*

Some kinds of simulation program involve sampling a finite population or selecting a member of an array at random. For this purpose we need a supply of random integers varying between 1 and some previously defined maximum number N. There are several ways in which such numbers can be obtained. It will be recalled that in SUBROUTINE RAND1 we first obtained a random integer NR1 and subsequently converted this to a real number X. We can, however, take NR1 and derive from it another integer fulfilling our present requirements. One way would be to reject the current value of NR1 if it exceeded N and to continue rejecting until we obtained a value of NR1 between 1 and N. This would obviously be very wasteful and our first step should be to derive a number not greater than the greatest value which we allow N to assume in our new subroutine RAND2. We take this to be 999 and we can therefore start with the three most significant digits of NR1 (i.e. the three *leftmost* digits). We take the most significant digits because it is a property of the recursion formula (10) that the period of repetition is greatest for the leftmost digit and decreases for each successive digit to the right of this. Once again we could if we wished reject any of the three-digit numbers that exceeded N, but this would still be a very wasteful procedure. If N were 20, for instance, about 98% of the pseudorandom numbers we generated would still be wasted. We therefore adopt a principle similar to that used when working with a table of random numbers. In effect we divide our 3-digit random number by N and use the remainder, bearing in mind that a few of our random numbers will still have to be rejected, i.e. those equal to or greater than the largest multiple of N below 999.

For example, suppose our subroutine generates numbers between 1 and 999 (both inclusive) and that on a particular occasion we need a supply of numbers between 1 and 20. Let us further suppose that a random number of 278 were generated by the first few statements in the subroutine. We first check that the number generated does not exceed 979. This is the current value of M, computed in the first line of the subroutine. We then obtain the remainder when 278 is divided by 20 (i.e. 18) and add 1 to it because our remainders will have possible values of 0 to 19 whereas we require values from 1 to 20. The result, 19, is the random integer we require. A subroutine to carry out these operations could call RAND1 to supply a value of X and then work from this, but it is simpler and quicker to incorporate all the calculations in SUBROUTINE RAND2. In all subsequent subroutines we have followed the practice of calling RAND1 explicitly as the user will then find it easier to adapt the calling subroutine to an alternative system routine in place of RAND1.

```
C
      SUBROUTINE RAND2 (NR1,N,NR2)
C
C.................................................................
C
C  THIS SUBROUTINE GENERATES A SUPPLY OF PSEUDORANDOM NUMBERS (NR2) FROM
C  1 TO N (BOTH INCLUSIVE, N NOT EXCEEDING 999). A STARTING VALUE FOR NR1
C  MUST BE SUPPLIED (ANY ODD POSITIVE INTEGER). THIS SUBROUTINE IS
C  SUITABLE FOR A COMPUTER WITH A 36-BIT WORD-STRUCTURE, E.G., IBM 7094.
C
C.................................................................
C
      M = 998 - MOD(999,N)
    1 NR1 = NR1 * 131075
      R2 = NR1
      NR2 = R2 * .2910383E-7
      IF (NR2 .GT. M) GO TO 1
      NR2 = MOD(NR2,N) + 1
      RETURN
      END
```

Subroutine RAND2 can easily be used in a simulation study to select items at random from a singly subscripted array A containing, say, NP values. All we do is to make a call of the form

CALL RAND2 (NR1, NP, NR2)

and then use NR2 as the subscript of A. Similarly if we wished to select at random an element from the doubly-subscripted array B, composed of NP rows and NQ columns, we could do so as follows:

CALL RAND2 (NR1, NP, NR2)
IR = NR2
CALL RAND2 (NR1, NQ, NR2)
JR = NR2

and then use B(IR, JR) as the randomly selected element of the array B. In this and similar ways we can use the computer as a device for sampling experiments based on a finite population of observations previously stored in central memory.

(c) *Random normal variates* (*SUBROUTINE RAND3*)

In many simulation experiments we require a supply of random variables from a normal distribution with a specified mean and standard deviation. One way of obtaining these would be to employ (i) a supply of random real numbers between 0 and 1 with (ii) a numerical approximation for the inverse normal distribution function. This is an application of a general technique for obtaining random variables from any specified distribution, but in practice it is more convenient to use for normal variates an algorithm which may be stated as follows:

$$R = \frac{\sum_{i=1}^{n} u_i - n/2}{(n/12)^{\frac{1}{2}}} \qquad (11)$$

where R is the required random normal variate and u_i a random real number uniformly distributed between 0 and 1. The algorithm is based on the fact that the sum of n random numbers u_i tends as n tends to infinity to be normally distributed with mean $n/2$ and variance $n/12$. In fact reasonable approximations, reliable to within ± 3 standard deviations, can be obtained with values of n as low as 10, though such approximations might be unsatisfactory if the problem in which they were used depended critically on values near the tails of the normal distribution. A value of $n = 12$ is particularly convenient, since the denominator of (11) then becomes 1. Formula (11) carries out the standardising operation that ensures the random variates are, in effect, selected from a normal distribution with zero mean and unit standard deviation. The Fortran program that forms SUBROUTINE RAND3 enables one to obtain random variables from a normal distribution in which the mean XBAR and the standard deviation S are specified by the user. This is easily accomplished by taking the value of R from formula (11), multiplying by S and subtracting XBAR.

```
C
      SUBROUTINE RAND3 (XBAR,S,NR1,R)
C
C.................................................................
C
C   THIS SUBROUTINE GENERATES A SUPPLY OF RANDOM VARIATES (R),
C   NORMALLY DISTRIBUTED WITH MEAN XBAR AND STANDARD DEVIATION S.
C   SUBROUTINE RAND1 IS REQUIRED. SUITABLE FOR IBM 7094.
C
C.................................................................
C
      R = 0.0
      DO 1 I = 1,12
      CALL RAND1 (NR1,X)
    1 R = R + X
      R = R - 6.0
      R = R * S + XBAR
      RETURN
      END
```

This subroutine can be used in various ways. The calling program reproduced with it shows how we can use it to fill a singly subscripted array with random normal deviates which might subsequently be used to represent, say, the dimensions of individual animals in a sample or the time they take to complete some phase of their development or any similar continuous variable likely to be normally distributed. Similar sequences in a calling program can, of course, be used to generate and store an array of values from any of the other subroutines discussed in this section.

(d) *Random binomial variates (SUBROUTINE RAND4)*

Many simulation programs require variates from a binomial distribution, whose properties and applications have been briefly indicated on p. 360 and which, it will be recalled, may be defined as the expansion of $(q + p)^n$ where $q = (1 - p)$. Values of n and p must therefore be supplied by the user and generation of the random variables proceeds in either of two ways.

(i) If n is reasonably small (say ≤ 10) and p reasonably large (say ≥ 0.1) we use the following algorithm. Generate n random numbers uniformly distributed between 0 and 1 (using SUBROUTINE RAND1 or a comparable system subroutine) and count how many of these are equal to or less than p. The number so counted is a random binomial variate.

(ii) If n is large (say > 10) and p small (say < 0.1) we use a slightly more complicated method, counting the number of iterations needed to satisfy the inequality

$$u \leq \sum_{i=0}^{N} r_i \qquad (12)$$

where u is a uniform random number between 0 and 1,

$$r_0 = (1 - p)^n$$

and

$$r_{i+1} = r_i \left(\frac{n-1}{i+1}\right) \left(\frac{p}{1-p}\right)$$

SUBROUTINE RAND4 incorporates both methods, using an IF statement based on the magnitudes of n and p.

```
C
      SUBROUTINE RAND4 (NR1,N,P,NB)
C
C.............................................................
C
C   THIS SUBROUTINE GENERATES RANDOM VARIATES (NB) FROM A BINOMIAL
C   DISTRIBUTION WITH GIVEN PARAMETERS N AND P. SUBROUTINE RAND1
C   IS REQUIRED. SUITABLE FOR IBM 7094.
C
C.............................................................
C
      NB = 0
      IF (N .GT. 10 .AND. P .LT. 0.1) GO TO 2
```

```
      DO 1 I = 1,N
      CALL RAND1 (NR1,X)
      IF (X .LE. P) NB = NB+1
    1 CONTINUE
      RETURN
    2 CALL RAND1 (NR1,X)
      AN = N
      S = 0.0
      PP = P / (1.0-P)
      R = (1.0-P) ** N
    3 S = S + R
      IF (S .GE. X) RETURN
      ANB = NB
      R = R * PP * ((AN-ANB)/(ANB+1.0))
      NB = NB + 1
      GO TO 3
      END
```

(e) *Random Poisson variates (SUBROUTINE RAND5)*

A convenient method of generating random variates from a Poisson distribution with mean $\bar{x}$ is to form the continued product of N uniform real numbers (derived from SUBROUTINE RAND1) subject to the inequality

$$\prod_{i=1}^{N} u_i < e^{-\bar{x}} \tag{13}$$

The random Poisson variate is then given by $(N - 1)$. If the inequality is satisfied by the first uniform random number generated, then the random Poisson variate is zero. The subroutine RAND5 based on algorithm (13) is self-explanatory and if called successively will produce a vector of Poisson variates with mean XBAR. Such a vector would, perhaps, be used to denote the number of randomly distributed organisms found in successive samples or the number of events in successive time-intervals or any of the other numerical quantities which may be represented in a model by a Poisson distribution.

```
C
      SUBROUTINE RAND5 (XBAR,NR1,NP)
C
C.................................................................
C
C THIS SUBROUTINE GENERATES A SUPPLY OF RANDOM VARIATES (NP) FROM A
C POISSON DISTRIBUTION WITH MEAN XBAR. SUBROUTINE RAND1 IS REQUIRED.
C SUITABLE FOR IBM 7094.
C
C.................................................................
C
      TEST = EXP(-XBAR)
      PI = 1.0
      NP = -1
    1 CALL RAND1(NR1,X)
      PI = PI * X
      NP = NP + 1
      IF (PI .GE. TEST) GO TO 1
      RETURN
      END
```

(f) *Random variates from the Chi-squared distribution* (*SUBROUTINE RAND6*)

This is a continuous distribution whose frequency function forms an unimodal bell-shaped curve of positive values with unlimited range. This frequency function is a complicated mathematical expression but random chi-squared variates can be generated very simply. For any specified number of degrees of freedom (denoted in SUBROUTINE RAND6 by the Fortran variable NDF) we generate NDF random normal variates (zero mean, unit standard deviation) by the appropriate calls to SUBROUTINE RAND3. These variates are then squared and summed to yield the required random chi-squared variate denoted by CHI2 (Tocher, 1963).

```
C
      SUBROUTINE RAND6 (NR1,NDF,CHI2)
C
C............................................................
C
C  THIS SUBROUTINE GENERATES A SUPPLY OF RANDOM VARIATES (CHI2)
C  FROM A CHI-SQUARED DISTRIBUTION ON NDF DEGREES OF FREEDOM.
C  SUBROUTINES RAND1 AND RAND3 ARE REQUIRED. SUITABLE FOR IBM 7094.
C
C............................................................
C
      CHI2 = 0.0
      DO 1 I = 1,NDF
      CALL RAND3 (0.0, 1.0, NR1, R)
    1 CHI2 = CHI2 + R * R
      RETURN
      END
```

(g) *Random exponential variates* (*SUBROUTINE RAND7*)

The time-intervals between events are often found to be distributed exponentially. Given the mean XBAR of an exponential distribution we can generate random exponential variates very simply. We merely multiply the negative of the natural logarithm of a random uniform variate by XBAR, as indicated in SUBROUTINE RAND7.

```
C
      SUBROUTINE RAND7(NR1,XBAR,EX)
C
C............................................................
C
C  THIS SUBROUTINE GENERATES A SUPPLY OF RANDOM VARIATES (EX) FROM AN
C  EXPONENTIAL DISTRIBUTION OF MEAN XBAR. SUBROUTINE RAND1 IS REQUIRED.
C  SUITABLE FOR IBM 7094.
C
C............................................................
C
      CALL RAND1 (NR1,X)
      EX = -ALOG(X) * XBAR
      RETURN
      END
```

(h) *Tests for randomness*

Much attention has been paid to the construction of suitable statistical tests of randomness to be applied to the variates produced by procedures like the above. The variates from statistical distributions such as the normal, Poisson, binomial, etc. can if necessary be tested using goodness-of-fit methods such as are given in Programs 38 to 40. The results of these generator programs will, however, depend on the randomness of the numbers produced by the procedures involved in RAND1. This and its equivalent RAND2 therefore need rather careful scrutiny and Tocher (1963), for example, discusses a number of tests which may be applied to numbers generated in this way. For most purposes it is enough that a set of pseudorandom numbers should (i) be uniformly distributed over a set of equally spaced sub-intervals as shown by a chi-squared test, and (ii) that there should be no first-order serial correlation within the sequence. The latter provision can also be made the subject of a chi-squared test. An array of pseudorandom numbers is scanned and a count made of the number of times a number in the i-th class of the frequency table is followed by one in the j-th class. This count is kept in the cell F_{ij} of a two-dimensional array containing n^2 cells (where n is the number of classes in the frequency table). In the absence of first-order serial correlation the F_{ij} should be equal and this null hypothesis can be tested by computing chi-squared with $(n-1)^2$ degrees of freedom. Further details are given in Golden (1965), from whose account a simple program to test randomness can be written. When a set of 500 random integers generated by RAND2 was subjected to these tests it gave χ^2 values of 24.24 with 19 d.f. for uniformity of distribution and 401.20 with 361 d.f. for absence of serial correlation. Both results indicate that the numbers may be regarded as randomly distributed.

Calculation of Probabilities from Statistical Distributions

The significance tests used in statistical analyses end with the computation of a test-statistic such as χ^2, t, a variance-ratio (F) or a standardised normal deviate z,† together with the degrees of freedom where appropriate. In the ordinary way, the user of a computer program would examine the test-statistics in the output and determine the probabilities associated with them by consulting the values tabulated in works such as Fisher and Yates (1963). It may happen however, more particularly in simulation experiments, that a test-statistic is produced during a computation and that the course of further computation depends on the associated probability. Obviously one can hardly allow a run to stop while tables are consulted and the next branch of the

---

† It is unfortunate that z is the symbol used by some writers to denote a standardised normal deviate ($z = (x - \bar{x})/s$) while others use it to denote a function related to the variance ratio ($F = e^{2z}$). We use it here only in the former sense.

program decided. Equally it is not very convenient to store a large set of statistical tables in the central memory or auxiliary store of a computer and to consult them as required. There is, in fact, a need for a short function subprogram or subroutine which can be called when needed to compute the probability associated with a given test-statistic. The frequency functions of the distributions involved are mathematically complicated and need to be integrated numerically if exact probabilities are required. A practicable program can, however, be developed using a polynomial approximation for the normal distribution function and relying on the relationships that exist between the four test-statistics. These may conveniently be expressed by indicating the equivalence between t, χ^2 and z in terms of F. The following table summarises the position.

Test Statistic	d.f.	Equivalent in terms of F	d.f. n_1	d.f. n_2
χ^2	n	χ^2/n	1	n
t	n	t^2	1	n
z	—	z^2	1	∞

This means, for example, that if a test yields a χ^2 of 10.0 with 4 degrees of freedom, the associated probability would be the same as that for $F = 10.0/4 = 2.5$ with 1 and 4 degrees of freedom. The probabilities involved require rather careful definition (see below) but clearly if we could compute the probability corresponding to any variance ratio we could use the result to compute probabilities for χ^2, t and z. Jaspen (1965) summarises work on this problem and points out that a variance ratio may be normalised by the transformation:

$$z = \frac{\left(1 - \frac{2}{9n_2}\right) F^{1/3} - \left(1 - \frac{2}{9n_1}\right)}{\left(\frac{2}{9n_2} F^{2/3} + \frac{2}{9n_1}\right)^{\frac{1}{2}}} \quad (14)$$

where $F \geqslant 1$. For $F < 1$ we replace F by $1/F$, interchange n_1 and n_2 and subtract the probability eventually computed from 1. Knowing z as calculated from (14), or as given by a test yielding a normal deviate, we can compute the probability by using the same polynomial approximation as was used to write the FUNCTION subprogram PNORM in Program 38 (p. 348).

It is necessary to be clear, however, whether the probabilities concerned relate to one or both tails of the frequency distribution. Fig. 37 shows the frequency functions for a χ^2 and F distribution, the shaded area below the right-hand part of the curve being the probability of obtaining a value of the test-statistic exceeding that shown on the x-axis. If the shaded area is less than 0.05 or 0.01 or whatever level one sets before-hand, then the observations being tested differ significantly from the situation postulated by the null hypothesis in the test. This constitutes a so-called one-tailed test, which is the usual sort involving χ^2 and F. The normal and t-distributions, however, have frequency functions that are symmetrical about zero and when employing a t or normal deviate test we are usually interested in the probability of a given deviation in either direction. In other words, the usual critical probability of 0.05 or 0.01 is here the sum of the two equal areas at opposite ends of the curve. This corresponds to a two-tailed test. The equivalent probabilities indicated in the table on p. 404 (and computed by the program below) can now be defined more precisely: they are the probabilities represented by the right-hand tail of the χ^2 and F distributions or by both tails of the t and normal distribution. For most tests, therefore, the computed probability can be compared with the usual critical values of 0.05 or 0.01. Less often, however, it might be necessary to carry out, say, a one-tailed normal deviate test, in which one were concerned with the probability of a given positive deviation only.

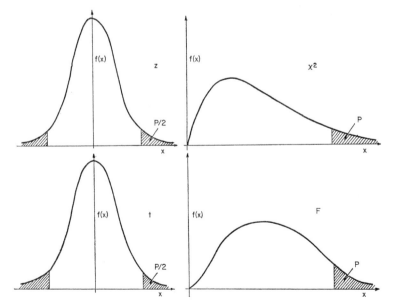

FIG. 37. Frequency functions $f(x)$ for the four important sampling distributions—normal, Student's t, chi-squared and F. The shaded areas indicate the values of P computed by Program 43.

The critical probability level would then be fixed at twice the value used in the two-tailed normal deviate test (e.g. at 0.1 in place of the more usual 0.05). With these explanations it is possible to describe the program very briefly.

Program 44: Computation of probabilities corresponding to the test-statistics χ^2 t, z and F

This will return only a single value, the probability, and may therefore conveniently be written as a FUNCTION subprogram, to which we give the name FUNCTION PROB with the four formal parameters NOPT, X, N1 and N2. These have the following significance:

NOPT   is an option number used to specify the test-statistic ($1 = \chi^2$, $2 = t$, $3 = z$ and $4 = F$).

X   is the value of the test-statistic computed by the main program or some subroutine.

N1 and N2 are the degrees of freedom. When F has been specified N1 corresponds to the numerator and N2 to the denominator. If χ^2 or t are specified the degrees of freedom are given by N1 alone and N2 can be set at zero (it cannot be omitted since the actual parameters listed in a call must correspond in number to those given in the FUNCTION subprogram). For z, both N1 and N2 are set at zero.

The program starts by selecting the option. The values for χ^2 and t are then converted to corresponding values of F, using the relationships tabulated on p. 404. If z has been specified the absolute value is taken and control passes immediately to the polynomial approximation beginning with statement 7. This is a slightly modified version of the form in Program 38 and yields the probability corresponding to the right-hand shaded area of Fig. 37. This value is subsequently doubled to give the probability of such a deviation occurring in either direction from zero. A specified F value and the F equivalent or χ^2 and t are first normalised by formula (13) and the resulting normal deviate used to compute a probability.

F values smaller than unity are treated as described above and the z values obtained from (13) with N2 $\leqslant$ 3 are adjusted for greater accuracy by the formula

$$Z = Z * (1.0 + 0.08 * Z ** 4/AN2 ** 3)$$

The polynomial approximation yields a probability correct to six decimal places but formula (13) is appreciably less accurate, especially for very large values of F with very few degrees of freedom in the denominator. The results are sufficiently accurate for any of the usual significance tests but should not be trusted in work where the exact probability is critical.

Flowchart for Program 44

Computation of probabilities corresponding to the test-statistics χ^2, t, z and F.

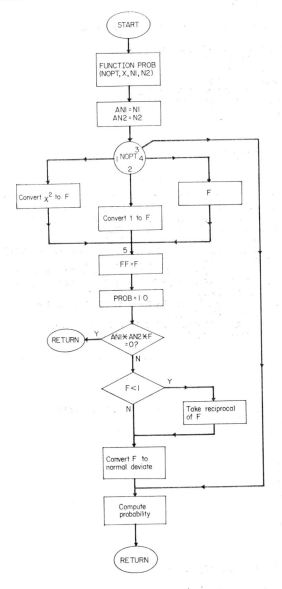

Program 44

Computation of probabilities corresponding to the test-statistics χ^2, t, z and F. The computation is carried out by the Function Subprogram PROB.

```
C
C   THIS PROGRAM CALLS FUNCTION SUBPROGRAM PROB.
C
      1 READ (5,100) NPROB, NOPT, X, N1, N2
    100 FORMAT (2I5, F10.0, 2I5)
        IF (NPROB .EQ. 0) STOP
        P = PROB (NOPT,X, N1,N2)
        WRITE (6,101) NPROB, NOPT, X, N1, N2, P
    101 FORMAT (/2I10, F12.3, 2I10, F12.5)
        GO TO 1
        END


        FUNCTION PROB(NOPT,X,N1,N2)
C
C.............................................................
C
C   THIS FUNCTION SUBPROGRAM COMPUTES THE PROBABILITY CORRESPONDING TO
C   GIVEN VALUE OF A VARIANCE-RATIO, CHI-SQUARED, STUDENT'S T, OR
C   STANDARDISED NORMAL DEVIATE. PARAMETERS ARE AS FOLLOWS -
C       NOPT = 1 FOR CHI-SQUARED (ONE-TAILED TEST)
C              2 FOR STUDENT'S T (TWO-TAILED TEST)
C              3 FOR STANDARDISED NORMAL DEVIATE (TWO-TAILED TEST)
C              4 FOR VARIANCE RATIO (ONE-TAILED TEST)
C       X    = NUMERICAL VALUE OF TEST-STATISTIC SPECIFIED BY NOPT
C       N1   = DEGREES OF FREEDOM (FOR NUMERATOR IF NOPT = 4. SPECIFY
C              ZERO IF NOPT = 3)
C       N2   = DEGREES OF FREEDOM FOR DENOMINATOR IF NOPT = 4,
C              OTHERWISE SPECIFY ZERO
C   NOTE - FOR ACCURACY SEE GOLDEN, WEISS AND DAWIS (1968)
C   EDUC. PSYCHOL. MEASUREMENT, VOL.28, PP.163-165
C
C.............................................................
C
        AN1 = N1
        AN2 = N2
C
C   CONVERT TEST STATISTIC TO VARIANCE RATIO IF NECESSARY.
        GO TO (1,2,3,4), NOPT
      1 F = X / AN1
        AN2 = 1.0E10
        GO TO 5
      2 F = X * X
        AN1 = 1.0
        AN2 = N1
        GO TO 5
      3 Z = ABS(X)
        F = 10.0
        GO TO 7
      4 F = X
      5 FF = F
        PROB = 1.0
        IF (AN1 * AN2 * F .EQ. 0.0) RETURN
C
C   TAKE RECIPROCAL IF F LESS THAN 1.
        IF (F .GE. 1.0) GO TO 6
```

Program 44 (*continued*)

```
            FF = 1.0 / F
            TEMP = AN1
            AN1 = AN2
            AN2 = TEMP
C
C     NORMALISE VARIANCE RATIO.
         6  A1 = 2.0/AN1/9.0
            A2 = 2.0/AN2/9.0
            Z = ABS(((1.0-A2) *FF**0.3333333 - 1.0 + A1) / SQRT(A2 *FF**
          A    0.6666667 + A1))
            IF (AN2 .LE. 3.0) Z = Z * (1.0 + 0.08*Z**4/AN2**3)
C
C     COMPUTE PROBABILITY.
         7  FZ = EXP(-Z*Z/2.0) * 0.3989423
            W = 1.0 / (1.0 + Z * 0.2316419)
            PROB = FZ * W * ((((1.330274 * W - 1.821256) * W +
          A    1.781478) * W - 0.3565638) * W + 0.3193815)
            IF (NOPT .EQ. 3) PROB = 2.0*PROB
            IF (F .LT. 1.0) PROB = 1.0 - PROB
            RETURN
            END
```

Chapter 14

Three Special-Purpose Programs

The programs given in earlier chapters of this book may be used in a great variety of investigations and many of them are equally applicable outside the biological field. Other programs are, however, developed to deal with more specifically biological problems and this chapter considers three such specialised programs. They cover fields where numerical methods are now of considerable interest and importance. Probit analysis (Program 45) is employed in biological assay (including the bioassay of insecticides and fungicides as well as in pharmacology and toxicology). Cluster analysis (Program 46) is increasingly employed in taxonomy and in other areas where classificatory problems are important, such as biogeography or vegetation analysis. The estimation of animal populations by marking, release and recapture (Program 47) is widely used in fisheries biology and in studies of the population dynamics of terrestrial animals, especially insects. In each case the program is only one of many which might have been chosen from the field concerned. The programs are presented more briefly than has been the case in previous chapters and emphasis centres on the principles they embody rather than the technicalities of Fortran coding. The reader should study them carefully if he wishes to develop some facility in the art of translating a set of numerical procedures into the form of a Fortran program.

Program 45: Probit analysis.

This program deals with a technique of statistical analysis used in biological assays based on a quantal response, i.e. a response which has an all-or-none character. Examples of quantal responses are, say, the death of the test-organism in a toxicological or insecticidal assay, the failure of a fungal spore to germinate within the test period, or the onset of a physiological condition like oestrus or paralysis. If one plots the proportion of test-organisms responding against the dose applied, one commonly finds an asymmetrical sigmoid curve like that shown in the upper diagram of Fig. 38. The asymmetry can often be eliminated by transforming the dose logarithmically (lower diagram of Fig. 38). Since the slope of the resulting symmetrical sigmoid curve is greatest when 50% of the test-organisms respond, one can compare different preparations most accurately in terms of the dose at this level of response. This is the so-called *median effective dose* or *ED*50. If the response is the death of the test-organism one usually speaks of the *median lethal dose* or *LD*50. In order to determine this and its variability with some accuracy it is necessary to transform the symmetrical sigmoid curve of Fig. 38 into a straight line.

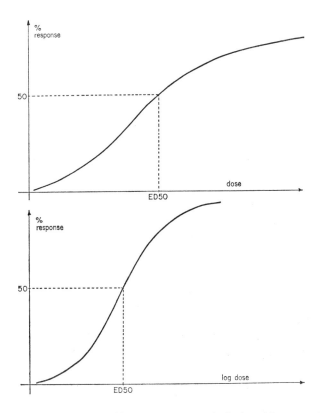

FIG. 38. Dosage-response curves Above, an asymmetrical sigmoid curve obtained by plotting the percentage response against the dose. Below, the symmetrical sigmoid curve obtained after the doses have been transformed to logarithms.

To understand the transformation used it is necessary to appreciate how the sigmoid curve arises. If one considers the individual test-organisms it is probable that they will differ in their susceptibility to the substance being assayed. It would not be surprising if this individual susceptibility (as measured by the logarithm of the dose) were distributed normally. One can therefore readily imagine a situation in which the proportions of individuals just susceptible to a given set of doses were as depicted in Fig. 39. In

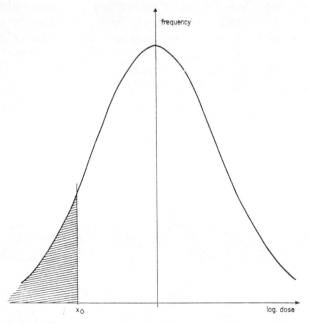

FIG. 39. Frequency distribution of individual susceptibility in a sample of test-organisms. The abcissa gives the logarithm of the dose required to produce a given response.

practice one could not determine such a frequency curve directly, for the dose represented by x_0 in Fig. 39 would produce a response not only in the test-organisms with exactly that level of susceptibility, but also in all those that were more susceptible. The proportion responding to x_0, therefore, is indicated by the shaded area beneath the frequency curve of individual responses. For the normal curve this area is given by the integral

$$P = \int_{-\infty}^{x_0} \frac{1}{\sigma\sqrt{2\pi}} e^{-[(x-\mu)^2/2\sigma^2]} dx. \tag{1}$$

Expression (10) is related to the normal probability integral tabulated in all statistical text-books and a plot of P against x yields the sigmoid curve referred to above. The problem of transforming this curve to give a straight line is therefore solved if we can find the value of x_0 in (1) which produces a given response P. To bring about the transformation we plot against the log dose a function of the mortality known as the *probit*. The probit Y corresponding to a proportion P is given by

$$P = \frac{1}{\sqrt{2\pi}} \int_{-\infty}^{Y-5} e^{-\frac{1}{2}u^2} \, du \tag{2}$$

the constant 5 in this expression having been introduced to avoid negative probits. The relationship between probits and percentage responses is shown graphically in the lower diagram of Fig. 40 and the effect of transforming

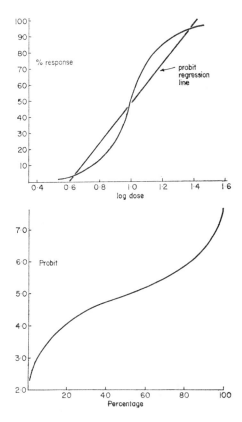

FIG. 40. The probit transformation. The lower diagram indicates the relationship between percentage response and probit. The upper diagram shows the effect of the probit transformation in converting a symmetrical sigmoid curve (log. dose against percentage response) into a straight line.

the log-response relationship in the upper diagram. Having transformed the data to a linear pattern in this way, we compute the regression of probit-values on log. dose. The best ways of doing this are discussed in detail by Finney (1952), who describes a method of computing the maximum-likelihood solution. The stages in this relatively complicated calculation are set out in Appendix 1 of Finney's book and will not be discussed at length here; they have been followed closely in writing the program and only some of the latter's less obvious features will be touched on.

Having read in the concentrations of chemical being assayed, the number of test-organisms used at each concentration and the number responding positively in each case, we transform the concentrations logarithmically and correct the responses for those occurring in the untreated control organisms by Abbott's formula:

$$P = \frac{(P' - C)}{(1 - C)} \tag{3}$$

where P is the corrected response, P' the response observed and C the proportion of untreated controls responding. The vector of corrected responses is then converted to probits. If the calculations were being done by hand, this would entail consulting a published table of probits. In the program, however, we call FUNCTION PROBIT which employs a polynomial approximation to the inverse normal distribution function from which the probit is derived (Formula (2) above). Since a response of 0 or 100% corresponds theoretically to probits of $-\infty$ and $+\infty$ respectively, we include in the FUNCTION subprogram an arrangement for setting these limiting values at $\pm 1 \times 10^{25}$. We then compute the provisional probit/log dose regression line in the usual way (cf. Program 2), omitting from it, however, any points based on 0 or 100% response. This program-segment corresponds to the graphical method of obtaining a provisional probit regression line when doing a hand-calculation.

The maximum likelihood solution entails an iterative procedure beginning with statement 5. From the provisional probit regression line we first compute the expected probit Y(I), corresponding to the I-th dose, and then Z, the ordinate to the normal curve from the relation

$$Z = \frac{1}{\sqrt{2\pi}} e^{-\frac{1}{2}(Y-5)^2}. \tag{4}$$

The mortality P (variable P1 in the program) corresponding to an expected probit Y can then be calculated as the area beneath the normal curve from $-\infty$ to $(Y - 5)$. This is computed using another polynomial approximation in the FUNCTION subprogram PNORM (which we have previously encountered in Program 38, p.348). Armed with Z and P we can then

Flowchart for Program 45

Probit analysis.

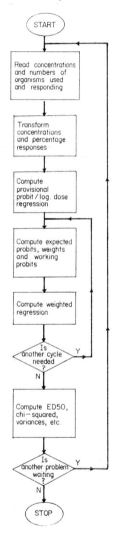

compute the *weighting coefficients* that are needed from the formula (Finney, 1952):

$$w = \frac{Z^2}{Q\left(P + \frac{C}{1-C}\right)} \tag{5}$$

where $Q = (1 - P)$ and C, as before, is the proportion of untreated controls responding. Multiplying each weighting coefficient by the number of test-organisms used at that concentration gives us the *weights* while the *working probits* we also need are given by a simple arithmetic statement (number 6):

$$Y(I) = Y(I) + (P(I) - P1)/Z$$

The weighted regression is then computed in a manner analogous to that used for a simple linear regression earlier in the program, but now applying the appropriate weight to each variable or product of variables being summed. The result is the computation of the slope (BW) and intercept (AW) of the weighted regression line.

When the weighted probit regression line is computed on a desk calculator one commonly accepts the results of the first regression. Using a digital computer, however, we can continue the iterative cycle to whatever limits of accuracy we care to set. In the program we compare BW with B, returning to statement 5 for a maximum of 10 cycles if successive estimates of the regression coefficient yield a relative error greater than 0.005. Assuming that a sufficiently accurate solution is obtained we then calculate the remaining quantities by a few straightforward arithmetic statements. The ED50 is the statistic to which users will normally turn first, but the output includes also the slope and intercept of the regression line, the variance of its slope, and a chi-squared allowing one to test for heterogeneity of response, as well as the log ED50 and its variance. The user who requires fiducial limits for the probit regression line and for the ED50 can use the statistics just mentioned to compute them himself; the limits will depend on whether or not allowance must be made for heterogeneity (as measured by the χ^2) and on the level of probability to be used.

Program 45

Probit analysis. The main program calls the Function Sub-programs PROBIT and PNORM. Sample data and output are also listed.

```
C
C............................................................
C
C   THIS PROGRAM COMPUTES THE PROBIT/LOG.CONCENTRATION REGRESSIONS
C   FOR ANY NUMBER OF CONSECUTIVE BIOASSAYS USING THE MAXIMUM LIKELIHOOD
```

Program 45 (*continued*)

```
C     METHOD GIVEN BY FINNEY (1952), PROBIT ANALYSIS, APPENDIX 1.
C     INPUT PARAMETERS ARE AS FOLLOWS –
C        NTEST = REFERENCE NUMBER OF BIOASSAY
C        NCONC = NUMBER OF CONCENTRATIONS
C        C     = PROPORTION OF CONTROLS RESPONDING
C        X     = VECTOR OF CONCENTRATIONS (LENGTH NCONC)
C        AN    = VECTOR OF NUMBERS OF TEST–ORGANISMS USED
C        AR    = VECTOR OF NUMBERS OF TEST–ORGANISMS RESPONDING
C     FUNCTION SUBPROGRAMS PROBIT AND PNORM ARE REQUIRED.
C
C.......................................................................
C
      DIMENSION X(20), AN(20), AR(20), P(20), Y(20), WN(20)
C
C     WRITE HEADINGS, READ TEST SPECIFICATION AND DATA.
      WRITE (6,100)
  100 FORMAT (16H1PROBIT ANALYSIS////3X, 4HTEST, 7X, 5HSLOPE, 8X,
     A  9HINTERCEPT, 6X, 8HVARIANCE, 7X, 10HCHI-SQUARE, 5X, 4HD.F.,
     B  4X, 9HLOG. ED50, 6X, 11HVARIANCE OF, 7X, 4HED50/ 2X, 6HNUMBER,
     C  7X, 3H(B), 12X, 3H(A), 9X, 8HOF SLOPE, 45X, 9HLOG. ED50//)
    1 READ (5,101) NTEST, NCONC, C
  101 FORMAT (2I5, F7.0)
      IF (NTEST .EQ. 0) STOP
      READ (5,102) (X(I), I = 1,NCONC)
      READ (5,102) (AN(I), I = 1,NCONC)
      READ (5,102) (AR(I), I = 1,NCONC)
  102 FORMAT (10F7.0)
C
C     TRANSFORM CONCENTRATIONS AND COMPUTE EMPIRICAL PROBITS, APPLYING
C     ABBOTT'S FORMULA.
      DO 2 I = 1,NCONC
      X(I) = ALOG10(X(I))
      P(I) = (AR(I) / AN(I) – C) / (1.0 – C)
    2 Y(I) = PROBIT(P(I))
C
C     COMPUTE PROVISIONAL REGRESSION LINE, EXCLUDING OBSERVATIONS WITH
C     ZERO OR 100 PER CENT RESPONSE.
      SXY = 0.0
      SX = 0.0
      SY = 0.0
      SX2 = 0.0
      DEN = NCONC
      DO 4 I = 1,NCONC
      IF(Y(I) .GT. –1.0E25 .AND. Y(I) .LT. 1.0E25) GO TO 3
      DEN = DEN – 1.0
      GO TO 4
    3 SXY = SXY + X(I) * Y(I)
      SX = SX + X(I)
      SY = SY + Y(I)
      SX2 = SX2 + X(I) * X(I)
    4 CONTINUE
      SX2 = SX2 – SX*SX/DEN
      SXY = SXY – SX*SY/DEN
      B = SXY / SX2
      A = SY/DEN – B*SX/DEN
      NCYC = 0
C
C     COMPUTE EXPECTED PROBITS, WEIGHTS, AND WORKING PROBITS.
    5 DO 6 I = 1,NCONC
      Y(I) = B * X(I) + A
```

Program 45 (*continued*)

```
          Z = 0.3989423 * EXP(-0.5*(Y(I)-5.0)**2)
          P1 = Y(I) - 5.0
          P1 = PNORM(P1)
          WN(I) = AN(I) * Z * Z / (1.0-P1) / (P1 + C / (1.0-C))
        6 Y(I) = Y(I) + (P(I) - P1) / Z
C
C     COMPUTE WEIGHTED REGRESSION.
          SWN = 0.0
          SWNX = 0.0
          SWNY = 0.0
          SWNXY = 0.0
          SWNX2 = 0.0
          SWNY2 = 0.0
          DO 7 I = 1,NCONC
          SWN = SWN + WN(I)
          SWNX = SWNX + WN(I) * X(I)
          SWNY = SWNY + WN(I) * Y(I)
          SWNXY = SWNXY + WN(I) * X(I) * Y(I)
          SWNX2 = SWNX2 + WN(I) * X(I) * X(I)
        7 SWNY2 = SWNY2 + WN(I) * Y(I) * Y(I)
          SWNX2 = SWNX2 - SWNX * SWNX / SWN
          SWNXY = SWNXY - SWNX * SWNY / SWN
          SWNY2 = SWNY2 - SWNY * SWNY / SWN
          BW = SWNXY / SWNX2
          AW = SWNY / SWN - BW * SWNX / SWN
          NCYC = NCYC + 1
C
C     TEST WHETHER ANOTHER CYCLE IS TO BE COMPUTED.
          IF (ABS((BW-B) / BW) .LE. 0.005) GO TO 8
          IF (NCYC .EQ. 10) GO TO 9
          B = BW
          A = AW
          GO TO 5
C
C     COMPUTE ED50 AND OTHER RESULTS.
        8 CHI2 = SWNY2 - SWNXY**2 / SWNX2
          NDF = NCONC - 2
          VB = 1.0 / SWNX2
          AM = (5.0 - AW) / BW
          VM = (1.0/SWN + (AM-SWNX/SWN)**2 / SWNX2) / (BW * BW)
          ED50 = 10.0 ** AM
C
C     OUTPUT RESULTS.
          WRITE (6,103) NTEST, BW, AW, VB, CHI2, NDF, AM, VM, ED50
      103 FORMAT (/I6, 1PE16.4, 3E15.4, I7, E16.4, 2E15.4)
          GO TO 1
C
        9 WRITE (6,104) NPROB
      104 FORMAT (/I6, 6X, 34HNO CONVERGENCE AFTER 10 ITERATIONS)
          GO TO 1
          END



          FUNCTION PROBIT(P)
C
C.............................................................
C
C     THIS FUNCTION SUBPROGRAM COMPUTES THE PROBIT CORRESPONDING TO A
C     PROPORTION P, USING AN APPROXIMATION TO THE INVERSE NORMAL
```

Program 45 (continued)

```
C    DISTRIBUTION FROM HASTINGS (APPROXIMATIONS FOR DIGITAL COMPUTERS,
C    PRINCETON, 1955). IF P = 0.0, PROBIT IS SET TO -1.0E25. IF P = 1.0,
C    PROBIT IS SET TO +1.0E25.
C
C.............................................................
C
     IF (P .LE. 0.0) GO TO 1
     IF (P .EQ. 1.0) GO TO 2
     PP = P
     IF (P .GT. 0.5) PP = 1.0 - PP
     W2 = ALOG(1.0/PP/PP)
     W = SQRT(W2)
     X = W - (2.515517 + 0.802853*W + 0.010328*W2) / (1.0 + 1.432788
    A  * W + 0.189269 * W2 + 0.001308 * W * W2)
     IF (P .LE. 0.5) X = -X
     PROBIT = 5.0 + X
     RETURN
   1 PROBIT = -1.0E25
     RETURN
   2 PROBIT = +1.0E25
     RETURN
     END


     FUNCTION PNORM(X)
C
C.............................................................
C
C    THIS FUNCTION SUBPROGRAM COMPUTES THE AREA UNDER THE NORMAL
C    PROBABILITY CURVE FROM MINUS INFINITY TO X, USING AN
C    APPROXIMATION DUE TO HASTINGS (APPROXIMATIONS FOR
C    DIGITAL COMPUTERS).
C
C.............................................................
C
     FX = EXP(-X*X/2.0) * 0.3989423
     W = 1.0 / (1.0+ABS(X)*0.2316419)
     PNORM = 1.0 - FX * W * ((((1.330274 * W - 1.821256) * W +
    A  1.781478) * W - 0.3565638) * W + 0.3193815)
     IF (X) 1,2,2
   1 PNORM = 1.0 - PNORM
   2 RETURN
     END
```

DATA

```
1    5    0.0
2.6    3.8    5.1    7.7    10.2
50.0   48.0   46.0   49.0   50.0
6.0    16.0   24.0   42.0   44.0
0
```

OUTPUT

PROBIT ANALYSIS

TEST NUMBER	SLOPE (B)	INTERCEPT (A)	VARIANCE OF SLOPE	CHI-SQUARE	D.F.	LOG. ED50	VARIANCE OF LOG. ED50	ED50
1	5.0245E 00	-6.0784E 00	3.4577E-01	2.5944E 00	3	2.2049E 00	7.1613E-04	1.6028E 02

Program 46: Single-linkage cluster analysis.

Cluster-analysis is the name given to various procedures whereby a set of individuals or units is divided into two or more assemblages or subgroups (clusters) on the basis of a set of attributes which they share. As might be imagined, techniques of cluster analysis can be applied readily in taxonomy and in many other fields of biology where classificatory problems are important. Examples of its use may be found in the numerical analysis of vegetation complexes or animal associations by ecologists, in the treatment of quantitative biogeographical data, in the recognition of various clinical forms of a disease or in the separation of distinctive racial groups by anthropologists. There are many mathematically quite distinct forms of cluster analysis, including some of the techniques of multivariate analysis discussed in Chapter 10. Here we shall deal with only one of the hierarchical clustering methods, employing single-linkage.

The method may be understood by considering data of the form set out below. Each row of this data matrix represents an individual and each column refers to a character, so that each element of the matrix is a numerically coded representation of the state of a particular character in a particular individual.

$$\begin{matrix} 1 & 1 & 1 & 1 & 2 & 2 \\ 2 & 1 & 1 & 1 & 2 & 2 \\ 1 & 2 & 2 & 1 & 1 & 2 \\ 2 & 2 & 2 & 2 & 2 & 2 \\ 2 & 2 & 2 & 2 & 2 & 1 \end{matrix}$$

In this example every character exists in one or other of two alternative conditions, 2 denoting, say, the presence of a certain structure and 1 its absence. Clearly the first and second individuals agree in 5 out of the 6 characters that have been examined, the first and third agree in 3 out of the 6, and so on. It is easy to see in such a small example that the fourth and fifth individuals are very much alike and that the first and second are also very similar, so that we appear to have two clusters with the third, unclustered, individual in a more or less intermediate position. If one were faced with a much larger array of data—say 50 individuals and 100 characters—it would be out of the question to cluster them by simple inspection and some sort of systematic procedure would have to be adopted. The procedure used for single-linkage clustering in our program is as follows:

1. We compare any two individuals by calculating a simple matching

coefficient between the rows concerned. Thus if the i-th and j-th rows of a data matrix agree in 7 out of 10 characters the matching coefficient S_{ij} would be 0.7. Such a matching coefficient can obviously vary from 0 to 1.

2. From the $m \times n$ data matrix we can construct an $m \times m$ *similarity matrix* **S** of matching coefficients. This will, of course, be a symmetric matrix with 1's in the main diagonal.

3. By examining the similarity matrix we locate the largest element, thus identifying the two individuals that form the nucleus of the first cluster. For example if element $s_{4,5}$ is the largest, then the fourth and fifth individuals form the nucleus. In some cases there may be more than one element with the same maximum value thus indicating either that the first cluster contains 3 or more individuals or that there are two or more separate clusters at this stage.

4. We then set an initial *similarity level* somewhat below the maximum and search for all individuals linked with the first cluster, i.e. we seek those individuals whose matching coefficient with any member of the first cluster exceeds the given similarity level. The term 'single linkage' thus implies that a new individual is admitted to a previously existing cluster if it is linked to at least one member of the cluster in this way.

5. The similarity level is then decreased by a pre-arranged amount, say 0.01, and another search made for new clusters or for individuals now admitted to a previously existing cluster. In the latter case this may lead to the fusion of two or more existing clusters.

6. The process is repeated until all the clusters have finally merged into a single cluster containing all the individuals.

The result of the analysis may be depicted in a branching tree-like diagram such as Fig. 41 which is immediately intelligible and provides an intuitively appealing picture of relationships. Detailed discussions of this and other methods of cluster analysis are given by Sokal and Sneath (1963) and by Cole (1969). A more complicated but more efficient technique of single-linkage clustering is discussed by Gower and Ross (1969). There is a rapidly growing literature on the subject and most methods of clustering are susceptible of many variations. Thus the single-linkage method outlined above may operate on other forms of similarity matrix including, where appropriate, a correlation matrix. For this reason our program comprises a calling program and two subroutines. One of these—SUBROUTINE SIMAT—computes the similarity matrix of matching coefficients while the other—SUBROUTINE SILINK—forms single-linkage clusters at any given similarity level and is called repeatedly each time the similarity level is lowered by the main program. Each of these subroutines can readily be replaced by others so that the main structure of the program can be adapted to cater for other clustering techniques and different methods of expressing similarity.

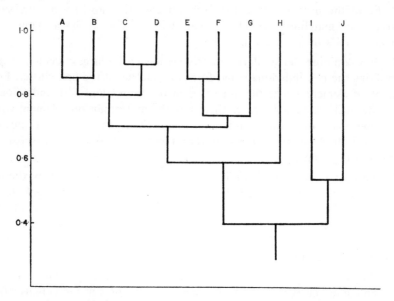

FIG. 41. Diagrammatic representation of clustering of units A–J.

Three aspects of the program deserve attention. Firstly, it is desirable to be able to cluster the data either by rows or by columns, partly because large bodies of data may already be punched with the individuals arranged column-wise or row-wise and partly because the associations between both row- and column-variables may be of interest in their own right. The subroutine SIMAT is therefore provided with a suitable option to bring about whichever form of clustering is required. Secondly, it may happen that the data matrix has an occasional missing element or that a few characters are not strictly comparable between certain individuals. Provision is therefore made in SIMAT for these data to be excluded from the calculation of matching coefficients. The missing or non-comparable elements must be punched as negative values when the data cards are being prepared. Thirdly, the main program is designed to avoid unnecessary output, the cluster pattern being printed only when it changes. Thus the similarity level might fall by steps of 0.01 from 0.95 to 0.53 but there would be little point in reproducing the results of all 43 clustering procedures if only 6 of them represented changes. The method used to check changes in the clustering pattern is to compare the vector assigning individuals to clusters (NA, returned by SIMAT) with its value at the immediately previous similarity level and to print the results only if it has changed.

The most interesting feature of the program is, of course, the logical structure of SUBROUTINE SILINK, which should be followed carefully if it is to be understood, since the functions of the relatively complicated set of nested DO-loops is not obvious at first sight. Essentially, the method used is as follows:

(i) Initialisation includes the elimination of the main diagonal elements of the similarity matrix by setting them to -1 so that the matrix can be scanned more easily for maximum off-diagonal elements (loop ending on statement 1).

(ii) The upper triangle is then scanned row by row until the first matching coefficient above the similarity level is encountered (loop ending on 2, with exit to 3 as soon as such a coefficient is located).

(iii) We then identify and count the individuals (units as they are called in the program) that are directly linked with the unit corresponding to the row in which the large matching coefficient was located by step (ii). This is achieved by the loop ending on statement 4.

(iv) Each unit identified in step (iii) is assigned to cluster 1, and the appropriate element of the vector NA is marked accordingly. Thus if units 3, 4 and 7 were directly linked to 2 the vector NA at this stage would have the form

$$0\ 1\ 1\ 1\ 0\ 0\ 1\ 0\ 0\ ...$$

(v) Each of the similarity matrix columns identified by a 1 as the appropriate element of NA is now scanned to locate subsequent links in the cluster. Any units added to the cluster are again marked in NA.

(vi) If NA changes as a result of (v) that step is repeated until no change occurs. All the individuals in the cluster will then have been identified in NA and control returns to the beginning of the outer DO-loop (DO 9 I = 1, NM) so that steps (ii) to (vi) can be repeated to find a second cluster, the cycle continuing until all clusters have been identified at that particular similarity level. Unclustered individuals will be denoted by the residual zeros in NA, the other elements being the cluster numbers to which each unit has been assigned. Thus, if at the end of the procedure NA had the form

$$0\ 1\ 1\ 2\ 2\ 0\ 1\ 1\ 0\ 2\ 3\ 3\ 3\ 1$$

it would be clear that the 14 units were clustered as follows:

Cluster 1 contains units 2, 3, 7, 8 and 14
Cluster 2 contains units 4, 5 and 10
Cluster 3 contains units 11, 12 and 13

while units 1, 6 and 9 remain unclustered.

Outline flowchart for Program 46

Single-linkage cluster analysis: SUBROUTINE SILINK.

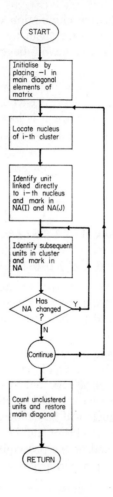

The subroutine then returns the vector of assignments NA, the number of clusters, NCL and the number of unclustered units (NU) to the calling program. From NA it can be decided whether the cluster pattern at that similarity level is to be printed. If so, the output is constructed by the main program, the essential steps being taken by the DO-loops ending on 12 and 13. As a result of these the units in each cluster are identified by the numbers of the rows (or columns) that they occupied in the original data matrix.

Program 46

Single-linkage cluster analysis. The main program calls Subroutines SIMAT and SILINK. Sample data and output are also listed.

```
C
C..............................................................
C
C   THIS PROGRAM CARRIES OUT A SINGLE-LINKAGE CLUSTER ANALYSIS ON A
C   MATRIX OF SIMILARITY COEFFICIENTS. INPUT PARAMETERS ARE AS FOLLOWS -
C      NPROB = PROBLEM REFERENCE NUMBER (ZERO ON LAST DATA CARD)
C      NR    = NUMBER OF ROWS IN DATA MATRIX
C      NC    = NUMBER OF COLUMNS IN DATA MATRIX
C      NOPT1 = 1 FOR BETWEEN-ROWS SIMILARITY MATRIX
C              2 FOR BETWEEN-COLUMNS SIMILARITY MATRIX
C      STEP  = STEP SIZE BETWEEN SUCCESSIVE SIMILARITY LEVELS
C      NAME  = TITLE OF PROBLEM
C      INFMT = VECTOR HOLDING VARIABLE INPUT FORMAT
C      DATA  = PRIMARY DATA MATRIX
C   SUBROUTINES SIMAT AND SILINK ARE REQUIRED
C
C..............................................................
C
      DIMENSION NAME(20), INFMT(20), DATA(50,50), S(50,50), NA(50),
     A NWK(25), NA1(50)
      EQUIVALENCE (DATA(1,1), NWK(1)), (DATA(1,2), NA(1)),
     A (DATA(1,3), NA1(1))
C
C   READ DATA AND WRITE HEADING.
    1 READ (5,100) NPROB, NR, NC, NOPT1, STEP
  100 FORMAT (4I5, F5.0)
      IF (NPROB .EQ. 0) STOP
      READ (5,101) (NAME(I), I = 1,20), (INFMT(I), I = 1,20)
  101 FORMAT (20A4)
      DO 2 I = 1,NR
    2 READ (5,INFMT) (DATA(I,J), J = 1,NC)
      WRITE (6,102) NPROB, (NAME(I), I = 1,20)
  102 FORMAT (8H1PROBLEM, I4//1X,20A4)
C
C   COMPUTE SIMILARITY MATRIX AND MAXIMUM SIMILARITY LEVEL.
      CALL SIMAT (DATA,NR,NC,NOPT1,S,N)
      WRITE (6,103)
  103 FORMAT (//18H SIMILARITY MATRIX)
      DO 3 I = 1,N
      WRITE (6,104) I
  104 FORMAT (/4H ROW, I4)
    3 WRITE (6,105) (J, S(I,J), J = I,N)
  105 FORMAT (10(I6, F6.3))
      SMAX = S(1,2)
      NM = N - 1
```

Program 46 (*continued*)

```
      DO 4 I = 1,NM
      IP = I + 1
      DO 4 J = IP,N
      IF (S(I,J) .LE. SMAX) GO TO 4
      SMAX = S(I,J)
    4 CONTINUE
      MAX = SMAX * 100.0
      IF (MAX) 5,5,6
    5 WRITE (6,106)
  106 FORMAT (/53H MAXIMUM SIMILARITY COEFFICIENT ZERO. EXECUTION ENDED)
      GO TO 1
    6 SL = FLOAT(MAX) / 100.0
C
C     BEGIN CLUSTER ANALYSIS.
      WRITE (6,107)
  107 FORMAT (//17H CLUSTER ANALYSIS)
      DO 7 I = 1,N
    7 NA1(I) = 0.0
    8 CALL SILINK (S,N,SL,NA,NCL,NU)
      DO 9 I = 1,N
      IF (NA(I) .NE. NA1(I)) GO TO 10
    9 CONTINUE
      GO TO 15
C
C     PRINT OUTPUT WHENEVER CLUSTER-PATTERN CHANGES.
   10 WRITE (6,108) SL, NCL
  108 FORMAT (//19H SIMILARITY LEVEL = , F5.2,
     A  24H          NUMBER OF CLUSTERS = , I3)
      DO 12 I = 1,NCL
      K = 0
      DO 11 J = 1,N
      IF (NA(J) .NE. I) GO TO 11
      K = K + 1
      NWK(K) = J
   11 CONTINUE
   12 WRITE (6,109) I, (NWK(J), J = 1,K)
  109 FORMAT (/8H CLUSTER, I3, 4X, (20I5))
      IF (NU .EQ. 0) GO TO 14
      K = 0
      DO 13 I = 1,N
      IF (NA(I) .NE. 0) GO TO 13
      K = K + 1
      NWK(K) = I
   13 CONTINUE
      WRITE (6,110) (NWK(I), I = 1,K)
  110 FORMAT (/12H UNCLUSTERED, 3X, (20I5))
   14 WRITE (6,111)
  111 FORMAT (/121H ...............................................
     A ...............................................................
     B ..)
C
C     TERMINATE CLUSTERING WHEN ALL UNITS ARE IN SAME CLUSTER.
   15 DO 17 I = 1,N
      IF (NA(I) .EQ. 1) GO TO 17
      SL = SL - STEP
      DO 16 J = 1,N
   16 NA1(J) = NA(J)
      GO TO 8
   17 CONTINUE
      GO TO 1
      END
```

Program 46 (*continued*)

```
C
      SUBROUTINE SIMAT (DATA,NR,NC,NOPT1,S,N)
C
C.................................................
C
C   THIS SUBROUTINE COMPUTES A SIMPLE MATCHING COEFFICIENT MATRIX BETWEEN
C   THE ROWS OR COLUMNS OF A PRIMARY DATA MATRIX. PARAMETERS ARE
C   AS FOLLOWS -
C      DATA  = PRIMARY DATA MATRIX (MISSING OR NON-COMPARABLE ELEMENTS
C              SET TO ANY NEGATIVE VALUE, E.G. -1.0)
C      NR    = NUMBER OF ROWS IN DATA MATRIX
C      NC    = NUMBER OF COLUMNS IN DATA MATRIX
C      NOPT1 = 1 FOR BETWEEN-ROWS SIMILARITY MATRIX
C              2 FOR BETWEEN-COLUMNS SIMILARITY MATRIX
C      S     = SIMILARITY MATRIX (OUTPUT)
C      N     = ORDER OF SIMILARITY MATRIX (OUTPUT)
C
C.................................................
C
      DIMENSION DATA(50,50), S(50,50)
C
C   SELECT OPTION.
      IF (NOPT1 - 1) 1,1,2
    1 N = NR
      M = NC
      GO TO 3
    2 N = NC
      M = NR
      GO TO 6
C
C   COMPUTE BETWEEN-ROWS SIMILARITY MATRIX.
    3 DO 5 I = 1,N
      DO 5 J = I,N
      DEN = 0.0
      S(I,J) = 0.0
      DO 4 K = 1,M
      IF (DATA(I,K).LT.0.0 .OR. DATA(J,K).LT.0.0) GO TO 4
      DEN = DEN + 1.0
      IF (DATA(I,K) .NE. DATA(J,K)) GO TO 4
      S(I,J) = S(I,J) + 1.0
    4 CONTINUE
    5 S(I,J) = S(I,J) / DEN
      GO TO 9
C
C   COMPUTE BETWEEN-COLUMNS SIMILARITY MATRIX.
    6 DO 8 I = 1,N
      DO 8 J = I,N
      DEN = 0.0
      S(I,J) = 0.0
      DO 7 K = 1,M
      IF (DATA(K,I).LT.0.0 .OR. DATA(K,J).LT.0.0) GO TO 7
      DEN = DEN + 1.0
      IF (DATA(K,I) .NE. DATA(K,J)) GO TO 7
      S(I,J) = S(I,J) + 1.0
    7 CONTINUE
    8 S(I,J) = S(I,J) / DEN
C
C   COMPLETE LOWER TRIANGLE.
    9 DO 10 I = 1,N
      DO 10 J = I,N
```

Program 46 (*continued*)

```
   10 S(J,I) = S(I,J)
      RETURN
      END


      SUBROUTINE SILINK (S,N,SL,NA,NCL,NU)
C
C.......................................................................
C
C  THIS SUBROUTINE FORMS SINGLE-LINKAGE CLUSTERS AT A SPECIFIED
C  SIMILARITY LEVEL AND ASSIGNS EACH UNIT TO ITS CLUSTER.
C  PARAMETERS ARE AS FOLLOWS -
C     S   = SIMILARITY MATRIX (INPUT)
C     N   = ORDER OF S (INPUT)
C     SL  = SIMILARITY LEVEL (INPUT)
C     NA  = VECTOR (LENGTH N) ASSIGNING UNIT TO CLUSTER. ZERO
C           ELEMENT DENOTES UNCLUSTERED UNIT (OUTPUT)
C     NCL = NUMBER OF CLUSTERS FORMED (OUTPUT)
C     NU  = NUMBER OF UNCLUSTERED UNITS (OUTPUT)
C  IN NORMAL USE THIS SUBROUTINE IS CALLED AT SUCCESSIVE SIMILARITY
C  LEVELS TO PERFORM A CLUSTER ANALYSIS.
C
C.......................................................................
C
      DIMENSION S(50,50), NA(50)
C
C  INITIALISE.
      DO 1 I = 1,N
      S(I,I) = -1.0
    1 NA(I) = 0.0
      NM = N - 1
      NCL = 0
C
C  LOCATE CLUSTER.
      DO 9 I = 1,NM
      IF (NA(I) .NE. 0) GO TO 9
      IP = I + 1
      DO 2 J = IP,N
      IF (S(I,J) .LT. SL) GO TO 2
      NCL = NCL + 1
      NA(I) = NCL
      GO TO 3
    2 CONTINUE
      GO TO 9
C
C  IDENTIFY LINKS WITH I-TH UNIT.
    3 N1 = 1
      DO 4 J = IP,N
      IF (S(I,J) .LT. SL) GO TO 4
      NA(J) = NCL
      N1 = N1 + 1
    4 CONTINUE
C
C  IDENTIFY SUBSEQUENT LINKS IN CLUSTER.
    5 DO 7 K = IP, N
      IF (NA(K) .NE. NCL) GO TO 7
      DO 6 J = I,N
      IF (S(J,K) .LT. SL) GO TO 6
      NA(J) = NCL
    6 CONTINUE
```

Program 46 (*continued*)

```
      7 CONTINUE
        N2 = 0
        DO 8 J = 1,N
        IF (NA(J) .EQ. NCL) N2 = N2 + 1
      8 CONTINUE
        IF (N1 .EQ. N2) GO TO 9
        N1 = N2
        GO TO 5
      9 CONTINUE
C
C     COUNT UNCLUSTERED UNITS AND RESTORE MAIN DIAGONAL.
        NU = 0
        DO 10 I = 1,N
        IF (NA(I) .EQ. 0) NU = NU+1
     10 S(I,I) = 1.0
        RETURN
        END
```

DATA

```
    2   10    8    2 0.01
TEST PROBLEM 2. BETWEEN-COLUMNS SIMILARITY MATRIX.
(20F2.0)
 1 1 1 1 1 1 1 1
 1 2 1 1 1 1 2 1
 1 1 1 1 1 2 2 2
-1 2 2 2 2 2 1 1
 2 2 2 2 2 2 2 2
 2 2 1 1 1 2 2 2
 2 2 2 2 2 1 1 1
 2 2 1 1 1 1 1 1
 2 1 1 1 1 1 1 2
 2 1 1 1 1 1-1 1
    0
```

OUTPUT

PROBLEM   2

TEST PROBLEM 2. BETWEEN-COLUMNS SIMILARITY MATRIX.

SIMILARITY MATRIX

ROW   1
   1 1.000   2 0.667   3 0.556   4 0.556   5 0.556   6 0.444   7 0.375   8 0.556

ROW   2
   2 1.000   3 0.700   4 0.700   5 0.700   6 0.600   7 0.556   8 0.400

ROW   3
   3 1.000   4 1.000   5 1.000   6 0.700   7 0.444   8 0.500

ROW   4
   4 1.000   5 1.000   6 0.700   7 0.444   8 0.500

ROW   5
   5 1.000   6 0.700   7 0.444   8 0.500

ROW   6
   6 1.000   7 0.778   8 0.800

Program 46 (*continued*)

```
ROW    7
    7 1.000     8 0.778

ROW    8
    8 1.000

CLUSTER ANALYSIS

SIMILARITY LEVEL = 1.00    NUMBER OF CLUSTERS = 1
CLUSTER   1         3    4    5
UNCLUSTERED         1    2    6    7    8
..................................................................................

SIMILARITY LEVEL = 0.80    NUMBER OF CLUSTERS = 2
CLUSTER   1         3    4    5
CLUSTER   2         6    8
UNCLUSTERED         1    2    7
..................................................................................

SIMILARITY LEVEL = 0.77    NUMBER OF CLUSTERS = 2
CLUSTER   1         3    4    5
CLUSTER   2         6    7    8
UNCLUSTERED         1    2
..................................................................................

SIMILARITY LEVEL = 0.70    NUMBER OF CLUSTERS = 1
CLUSTER   1         2    3    4    5    6    7    8
UNCLUSTERED         1
..................................................................................

SIMILARITY LEVEL = 0.66    NUMBER OF CLUSTERS = 1
CLUSTER   1         1    2    3    4    5    6    7    8
..................................................................................
```

Program 47: Estimation of population-size by marking and recapture.

Populations of mobile organisms cannot be estimated by the methods employed for plants and sedentary animals. Instead, animal ecologists and biomathematicians have developed a variety of techniques for estimating population size by marking animals, releasing them and, when marked and unmarked individuals have mingled effectively, collecting another sample and recording the numbers of marked and unmarked individuals. The simplest form of this procedure is based on the so-called Lincoln index. Thus, if a animals are marked and released and we subsequently catch a sample of n animals, of which r are found to be marked, then the population size x can be estimated as

$$\hat{x} = an/r. \qquad (6)$$

Various elaborations of this principle allow one not only to estimate the size of the population but also to assess the precision of this estimate and to allow for the effects of births and deaths or other forms of recruitment and depletion. Some methods are deterministic, others allow for stochastic changes. The subject has often been reviewed, e.g. by Bailey (1951, 1952), Jones (1964) and Jolly (1963, 1965). Some methods, including the useful 'triple-catch' technique, involve very little calculation and are hardly worth programming for a digital computer. Others, such as Jackson's 'negative method' involve the iterative solution of a non-linear equation which is tedious by hand-calculation but well suited for a computer. Both the methods just mentioned are based on deterministic models which were introduced into capture-recapture problems in order to simplify the theory. However, Jolly (1965) has shown that a stochastic solution may, in fact, be simpler and his method is programmed here. The computations are not difficult in principle—involving little more than substitution in the formulae given by Jolly—but since these formulae are rather complicated hand-calculation is time-consuming and prone to error. The principles on which the solution are based cannot be discussed here, but the method is well set out in an appendix to Jolly's paper and the present program should be followed in relation to that account and the list of definitions (Jolly, 1965: 226). As far as possible Jolly's notation has been closely followed in the Fortran coding, several variables being declared REAL in order to facilitate this.

The important input variables are as follows:

- L is the number of occasions on which animals are captured. On the first of these there are, of course, no marked animals in the population while on the last occasion there is no further purpose served by marking. Otherwise, marking and release is carried out on each occasion, the marks being different every time.

- P1 is intended to provide a value for the population size at the time of the first sampling. This quantity is ultimately required in order to compute the variance of population size. It cannot be estimated as the other values of N2 are (see below) and if the user does not care to make a guess at it, this field on the data card can remain unpunched. In that case the program will use the estimate of N2(2) in its place. If, however, the program has first been run without it, the results may allow a guess to be made and the program can then be run a second time. In practice the assignment of a value to P1 does not very much affect the results.

N   is a vector giving the number of animals caught on each of the L occasions (corresponding to n_i of Jolly's algebraic notation).

S   is another vector giving the number of animals released from the i-th sample after marking (s_i) and is, of course, recorded only for the first (L − 1) occasions. Note that s_i may be less than n_i owing to injury or death while capturing or marking.

N1 is the number of animals in the i-th sample last captured in the j-th sample (where j varies from 1 to (i − 1)). These numbers are denoted in the algebraic formulae as n_{ij} and they form the lower triangle of the matrix N1.

The first few rows of the matrix given in Jolly's example are as follows:

	$j=1$	2	3	4
$i=2$	10			
3	3	34		
4	5	18	33	
5	2	8	13	30

Thus 10 of the animals captured on the second occasion had previously been captured on the first occasion. Obviously there can be no entry for the first occasion so that the various values of N1 are stored below the main diagonal. The DO-loop indexing parameters arrange for this storage provided that the data cards have been punched appropriately. In the above example the first card would bear only 10, the second 3 and 34, the third 5, 18 and 33 and so on. Each row of the lower triangle begins on a new card. The main diagonal and upper triangle of N1 remain unused throughout the program. This is not a very efficient procedure, but the program is never likely to make great demands on space and the present method of programming allows a more direct comparison with Jolly's account.

Computation begins by summing the columns of the lower triangle of N1 to give a vector R. Note that this summation occurs for the 2nd to the (L − 1)th columns, so that R(1) and R(L) are not computed. The quantity R_i is, of course, the number of the s_i animals released from the i-th sample that were subsequently recaptured. Immediately after this the program computes from N1 the lower triangle of a second matrix A (the numbers a_{ij} of Jolly's algebraic formulae). These are obtained by summing the n_{ij}

from the left. Thus, in the above numerical example a_{31} will be 3 and a_{32} will be 37, while for the fourth row the corresponding numbers will be:

$$a_{41} = 5, \quad a_{42} = 23, \quad a_{43} = 56.$$

From the matrix A we then derive two further vectors: M and Z (in Jolly's terminology m and Z). The values of $Z(I + 1)$ are found by summing all but the uppermost element in the I-th column. In biological terms M(I) is the number of marked animals in the I-th sample while Z(I) is the number marked before time I which are not caught in the I-th sample but are caught subsequently. Given the quantities computed so far it is now possible to estimate the remaining parameters by a set of arithmetic statements enclosed in DO-loops. These parameters are:

(i) M1(I), equivalent to M_i in Jolly's notation and denoting the total number of marked animals in the population at time I.

(ii) ALPHA (I), equivalent to α_i in Jolly's notation and denoting the proportion of marked animals at time I.

(iii) N2(I), equivalent to N_i in Jolly's notation and denoting the total number in the population when the i-th sample is captured. This is the 'population size' at time i—the most important parameter being estimated.

(iv) PHI (I), or ϕ_i in Jolly's notation, which gives the probability that an animal alive when the i-th sample is released will survive until the $(i + 1)$th sample is captured. For the purposes of this analysis death and emigration are synonymous.

(v) B(I), which is the number of new animals joining the population between the i-th and $(i + 1)$th sample and alive at time $(i + 1)$. This is the B_i of Jolly's paper.

There is no point in repeating here the formulae used in computing the above. They are given in full in Jolly's paper and are easily found from the arithmetic statements in this program-segment. Two elements are defined outside the DO-loops however. M1(1) must be zero by definition and a value for N2(1) must be supplied. Jolly suggests that it is put equal to N2(2) if the population seems fairly stable at first and this suggestion is followed in the program. If, however, the results indicate a strongly defined trend in N2 then an approximate value for N2(1) may be guessed and will have been read in as P1. The value adopted for N2(1) is required only in order to compute the standard error of N2 later: it is not an estimate in the sense that the other elements of N2 are estimated.

The computation of standard errors for the estimates of N2, PH1 and B is carried out in the next sections from the very complicated algebraic expressions given by Jolly. One of these, the standard error of B, involves

Outline flowchart for Program 47

Population estimates by marking, release and recapture (Jolly's stochastic method).

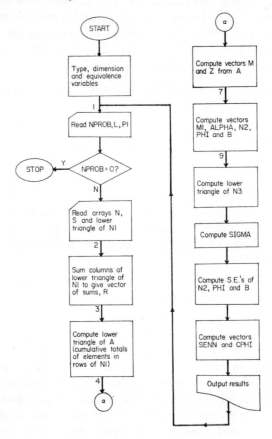

a Fortran arithmetic statement that extends over 5 lines, even after some components in it have been evaluated previously! The main complication in computing the standard errors, however, is the occurrence of a summation term in the expression for the standard errors of the population-sizes (SEN in Fortran). This term—denoted by SIGMA—has to be evaluated before-hand and it requires the setting up of yet another lower triangular matrix, denoted by N3. This, which corresponds to Jolly's $N_i(j)$, denotes the expected number in the population at time i which first joined the population between times j and $(j+1)$. The elements of N3 are quite easy to calculate but the reader should remember when comparing the Fortran coding with Jolly's account that his terminology involves zero subscripts and that these are avoided in the program by arranging for summation over $j = 1$ to $j = i$ instead of $j = 0$ to $j = (i - 1)$. The logical structure of these two program-segments (the calculation of N3 and the summation terms SIGMA) are probably the most difficult parts of the program to follow and illustrate again the way in which subscript manipulation and the use of computed indexing parameters add greatly to the power of the DO-statements.

The computation of two further quantities completes the main part of this program. They are SENN(I), equivalent to $\sqrt{V(\hat{N}_i|N_i)}$ of Jolly's account, and CPHI(I), equivalent to the expression

$$\left\{ V(\hat{\phi}_i) - \frac{\hat{\phi}_i^2(1-\hat{\phi}_i)}{\hat{M}_{i+1}} \right\}^{\frac{1}{2}}$$

The output follows very closely the lay-out of Table 4 in Jolly's paper with brief descriptions given at the head of each column. The symbols placed in parentheses in the output headings—(ALPHA), (M), (N), etc. are intended to represent Jolly's symbols in a simplified way; they are not necessarily the Fortran variable names used in the program. Only one other point need be made concerning the output. Not all the parameters can be estimated for the first and the last two occasions. These three rows are therefore dealt with by separate WRITE statements, while the main body of the table is printed by the DO-loop ending on statement 16.

Program 47

Population estimates by marking, release and recapture (Jolly's stochastic method). Sample data and output are also listed.

```
C
C..............................................................
C
C   THIS PROGRAM COMPUTES POPULATION PARAMETERS FROM MARK AND RECAPTURE
C   DATA USING JOLLY'S STOCHASTIC MODEL WITH DEATH AND IMMIGRATION. FOR
```

Program 47 (*continued*)

```
C    DETAILS SEE JOLLY, G.M. (1965) BIOMETRIKA, VOL.52, PP.225-247. INPUT
C    PARAMETERS ARE AS FOLLOWS -
C       NPROB = PROBLEM REFERENCE NUMBER (ZERO ON LAST DATA CARD)
C       L     = NUMBER OF SAMPLES
C       P1    = ESTIMATE OF POPULATION SIZE AT TIME OF FIRST SAMPLING
C               (LEAVE BLANK IF NO ESTIMATE IS AVAILABLE)
C       N     = VECTOR OF LENGTH L GIVING NUMBER OF ANIMALS CAPTURED
C               IN I-TH SAMPLE
C       S     = VECTOR OF LENGTH (L-1) GIVING NUMBER OF ANIMALS RELEASED
C               FROM I-TH SAMPLE AFTER MARKING
C       N1    = NUMBER OF ANIMALS IN I-TH SAMPLE LAST CAPTURED IN J-TH
C               SAMPLE (ONLY LOWER TRIANGLE OF L BY L MATRIX N1 IS USED)
C
C............................................................................
C
      REAL N, N1, M, M1, N2, N3
      DIMENSION N(25), S(25), N1(25,25), A(25,25), R(25), Z(25),
     A   M(25), M1(25), ALPHA(25), N2(25), PHI(25), B(25), N3(25,25),
     B   SENN(25), SEN(25), SEPHI(25), SEB(25), CPHI(25)
      EQUIVALENCE (N1(1,1), N3(1,1))
C
C    READ PROBLEM SPECIFICATION AND DATA.
    1 READ (5,100) NPROB, L, P1
  100 FORMAT (2I5, F10.0)
      IF (NPROB .EQ. 0) STOP
      LM = L - 1
      READ (5,101) (N(I), I = 1,L)
      READ (5,101) (S(I), I = 1,LM)
  101 FORMAT (12F6.0)
      DO 2 I = 2,L
      IM = I - 1
    2 READ (5,101) (N1(I,J), J = 1,IM)
C
C    COMPUTE R(I).
      DO 3 I = 2,LM
      SUM = 0.0
      IP = I + 1
      DO 3 J = IP,L
      SUM = SUM + N1(J,I)
    3 R(I) = SUM
C
C    COMPUTE LOWER TRIANGLE OF MATRIX A.
      DO 4 I = 2,L
      SUM = 0.0
      IM = I - 1
      DO 4 J = 1,IM
      SUM = SUM + N1(I,J)
    4 A(I,J) = SUM
C
C    COMPUTE Z(I+1) AND M(I).
      LM2 = L - 2
      DO 6 I = 1,LM2
      SUM = 0.0
      IP2 = I + 2
      DO 5 J = IP2,L
    5 SUM = SUM + A(J,I)
    6 Z(I+1) = SUM
      M(1) = 0.0
      DO 7 I = 2,L
    7 M(I) = A(I,I-1)
```

Program 47 (continued)

```
C
C   ESTIMATE REMAINING PARAMETERS.
        DO 8 I = 2,LM
        M1(I) = S(I)*Z(I)/R(I) + M(I)
        ALPHA(I) = M(I) / N(I)
      8 N2(I) = M1(I) / ALPHA(I)
        ALPHA(L) = M(L) / N(L)
        M1(1) = 0.0
        N2(1) = N2(2)
        IF (P1 .GT. 0.0) N2(1) = P1
        DO 9 I = 1,LM2
        PHI(I) = M1(I+1)/(M1(I)-M(I)+S(I))
      9 B(I) = N2(I+1) - PHI(I)*(N2(I)-N(I)+S(I))
C
C   COMPUTE LOWER TRIANGLE OF MATRIX N3.
        DO 10 I = 2,LM
        N3(I,I) = B(I-1)
        IP = I + 1
        DO 10 J = IP,LM
     10 N3(J,I) = (N2(J)-B(J-1)) / N2(J-1) * N3(J-1,I)
        N3(1,1) = N2(1)
        DO 11 J = 2,LM2
     11 N3(J,1) = (N2(J)-B(J-1)) / N2(J-1) * N3(J-1,1)
C
C   COMPUTE STANDARD ERRORS.
        DO 13 I = 2,LM
        SIGMA = N3(I,1) * N3(I,1) / N2(1)
        DO 12 J = 2,I
     12 SIGMA = SIGMA + N3(I,J)*N3(I,J)/B(J-1)
        Q = N2(I) * (N2(I)-N(I)) * ((M1(I)-M(I)+S(I))/M1(I)*(1.0/R(I)-
      A  1.0/S(I)) + (1.0-ALPHA(I))/M(I))
        SENN(I) = SQRT(Q)
     13 SEN(I) = SQRT(Q+N2(I)-SIGMA)
        DO 14 I = 2,LM2
        T = (M1(I+1)-M(I+1)) * (M1(I+1)-M(I+1)+S(I+1)) / M1(I+1)**2 *
      A  (1.0/R(I+1) - 1.0/S(I+1))
        U = (M1(I)-M(I)) / (M1(I)-M(I)+S(I)) * (1.0/R(I) - 1.0/S(I))
        SEPHI(I) = SQRT(PHI(I)**2 * (T + U + (1.0-PHI(I))/M1(I+1)))
     14 SEB(I) = SQRT(B(I)**2 * T + U * (PHI(I)*S(I)*(1.0-ALPHA(I))/
      A  ALPHA(I))**2 + (N2(I)-N(I))*(N2(I+1)-B(I))*(1.0-ALPHA(I)) *
      B  (1.0-PHI(I))/(M1(I)-M(I)+S(I)) + N2(I+1)*(N2(I+1)-N(I+1))*
      C  (1.0-ALPHA(I+1))/M(I+1) + PHI(I)**2 * N2(I)*(N2(I)-N(I)) *
      D  (1.0-ALPHA(I))/M(I))
        SEPHI(1) = SQRT(PHI(1)**2 * ((M1(2)-M(2)) * (M1(2)-M(2)+S(2))/
      A  M1(2)**2 * (1.0/R(2)-1.0/S(2)) + (M1(1)-M(1))/(M1(1)-M(1)+
      B  S(1)) * (1.0/R(1)-1.0/S(1)) + (1.0-PHI(1))/M1(2)))
        DO 15 I = 1,LM2
     15 CPHI(I) = SQRT(SEPHI(I)**2 - PHI(I)**2 * (1.0-PHI(I))/M1(I+1))
C
C   OUTPUT RESULTS.
        WRITE (6,102) NPROB
    102 FORMAT (47H1POPULATION ESTIMATES (JOLLY'S METHOD). PROBLEM, I4///
      A  7X, 10HPROPORTION, 3X, 5HTOTAL, 6X, 5HTOTAL, 3X, 11HPROBABILITY,
      B  2X, 6HNUMBER, 37X, 9HCOMPONENT, 2X, 9HCOMPONENT/ 1X, 4HTIME,
      C  4X, 6HMARKED, 5X, 6HMARKED, 5X, 6HNUMBER, 2X, 11HOF SURVIVAL,
      D  2X, 7HJOINING, 4X, 7HS.E.(N), 3X, 9HS.E.(PHI), 3X, 7HS.E.(B),
      E  3X, 9HOF S.E(N), 1X, 11HOF S.E(PHI)// 2X, 3H(I), 3X, 7H(ALPHA),
      F  6X, 3H(M), 8X, 3H(N), 7X, 5H(PHI), 7X, 3H(B))
        WRITE (0,100) M1(1), PHI(1), SEPHI(1), CPHI(1)
    103 FORMAT (/3X, 1H1, 13X, F9.2,16X, F6.4, 27X, F6.4, 27X, F6.4)
        DO 16 I = 2,LM2
```

Program 47 (continued)

```
 16   WRITE (6,104) I, ALPHA(I), M1(I), N2(I), PHI(I), B(I), SEN(I),
     A   SEPHI(I), SEB(I), SENN(I), CPHI(I)
104   FORMAT (/I4, F11.4, 2F11.2, F11.4, 2F11.2, F11.4, 2F11.2, F11.4)
      WRITE (6,105) LM, ALPHA(L-1),M1(L-1), N2(L-1), SEN(L-1),
     A   SENN(L-1), L, ALPHA(L)
105   FORMAT (/I4, F11.4, 2F11.2, 22X, F11.2, 22X, F11.2//I4, F11.4)
      GO TO 1
      END
```

DATA

```
  1   13         500
 54  146   169   209   220   209   250   176   172   127   123   120
142
 54  143   164   202   214   207   243   175   169   126   120   120
 10
  3   34
  5   18    33
  2    8    13    30
  2    4     8    20    43
  1    6     5    10    34    56
  0    4     0     3    14    19    46
  0    2     4     2    11    12    28    51
  0    0     1     2     3     5    17    22    34
  1    2     3     1     0     4     8    12    16    30
  0    1     3     1     1     2     7     4    11    16    26
  0    1     0     2     3     3     2    10     9    12    18    35
  0
```

Program 48 (*continued*)

OUTPUT

POPULATION ESTIMATES (JOLLY'S METHOD). PROBLEM 1

TIME (I)	PROPORTION MARKED (ALPHA)	TOTAL MARKED (M)	TOTAL NUMBER (N)	PROBABILITY OF SURVIVAL (PHI)	NUMBER JOINING (B)	S.E.(N)	S.E.(PHI)	S.E.(B)	COMPONENT OF S.E.(N)	COMPONENT OF S.E.(PHI)
1				0.6486			0.1103			0.0891
2	0.0685	35.02	511.36	1.0150	262.98	149.85	0.1108	178.64	149.46	0.1112
3	0.2189	170.54	778.97	0.8671	291.79	130.11	0.1076	138.10	129.68	0.1057
4	0.2679	258.00	962.89	0.5637	406.48	141.59	0.0627	120.72	140.97	0.0577
5	0.2409	227.73	945.31	0.8360	96.83	124.02	0.0745	112.29	122.84	0.0721
6	0.3684	324.99	882.12	0.7901	107.05	97.05	0.0715	75.06	95.31	0.0689
7	0.4480	359.50	802.46	0.6510	135.65	75.18	0.0575	55.39	72.78	0.0533
8	0.4886	319.33	653.52	0.9848	−13.82	62.75	0.0961	53.19	60.05	0.0959
9	0.6395	402.13	628.78	0.6862	49.00	63.47	0.0813	34.79	60.81	0.0784
10	0.6614	316.45	478.44	0.8844	84.14	52.35	0.1212	39.96	49.50	0.1200
11	0.6260	317.00	506.38	0.7714	74.54	66.27	0.1291	40.06	64.16	0.1272
12	0.6000	277.71	462.86			70.02			68.18	
13	0.6690									

Chapter 15

Efficiency in Programming

Enough has already been said for the reader to appreciate that some methods of programming a mathematical or statistical calculation are more satisfactory than others. The concept of efficiency is, of course, central to any serious use of computer programs, but it is difficult to define rigorously. For example, the more general a program is the more complicated it sometimes becomes so that a simpler special-purpose version may be more efficient for some restricted range of application. On the other hand, several such special purpose programs may be needed within the given field of research, so that the apparent advantages of any one program must be set off against the need for more programs. Again, a program taking very little computer time may demand large amounts of core-storage and therefore be totally impracticable when the data exceed quite modest dimensions. A slower program capable of dealing with larger bodies of data may then be more efficient in the long run. Furthermore, what is most efficient for an individual research-worker may be unsatisfactory in terms of a large computer installation with its many other users. Thus, it may be quicker for the individual to use a slow or otherwise inefficient program which happens to be available rather than to write or adapt a better one, though if all users did the same the installation would rapidly accumulate a back-log of work. Optimum efficiency usually requires a compromise among many conflicting demands.

Nevertheless, despite difficulties in formulating simple criteria of over-all efficiency, it is not difficult to see that in a given situation certain procedures are clearly preferable to others. Some of these desirable techniques are therefore discussed in this chapter under the following headings:

1. Choice of algorithms
2. Economy in computer time

3. Reducing demand for core storage
4. Accuracy considerations
5. Ease of testing
6. Use of available package programs or subroutines

Choice of Efficient Algorithms

This is perhaps the most important factor determining the efficiency of a program but also the most difficult on which to make general recommendations since any algorithm—i.e. procedure for carrying out a calculation—may have implications for time and space requirements, accuracy considerations and so on. A very clear indication of the importance of selecting an efficient algorithm comes from considering the possible ways of computing the determinant of a matrix in terms of the number of multiplications needed. With very small matrices the determinant can easily be evaluated by the method of cofactors, as when we write the determinant of a 2×2 matrix

$$\begin{vmatrix} a & b \\ c & d \end{vmatrix}$$

as $ad - bc$ or that of a 3×3 matrix as

$$\begin{vmatrix} a & b & c \\ d & e & f \\ g & h & i \end{vmatrix} = a\begin{vmatrix} e & f \\ h & i \end{vmatrix} - b\begin{vmatrix} d & f \\ g & i \end{vmatrix} + c\begin{vmatrix} d & e \\ g & h \end{vmatrix}$$

$$= a(ei - fh) - b(di - fg) + c(dh - eg)$$

The number of multiplications needed to evaluate the determinant of a matrix of order n by this method is about $2n!$—i.e. $2n(n-1)(n-2)\ldots 1$. This means that to evaluate the determinant of a 20×20 matrix on a computer with a multiplication time of 10 μsec (i.e. a quite fast computer) would take millions of years! On the other hand the algorithm used in the determinant program on p. 246 requires around n^3 multiplications and is therefore a perfectly practicable method for large matrices. The important point is that no amount of attention to programming detail would make the first algorithm satisfactory: the whole strategy of approach needed to be altered. The biologist with limited mathematical resources will obviously find it difficult to know of or to evaluate alternative algorithms. Fortunately there are many works available which assess numerical methods in the light of computer requirements. A selection is listed among the references at the end of this book and it is perhaps worth mentioning especially the elementary account by Pennington (1970), which includes many FORTRAN II programs, and the advanced series edited by Ralston and Wilf (1960→). These provide valuable guidance on the choice of algorithms and should save the biologist from many pitfalls.

How widespread are the dangers may be seen by considering what seems a very simple task, viz., the evaluation of a polynomial such as

$$y = 2x^3 + 3x^2 + 2x + 4.$$

If this is programmed as

$$Y = 2.*X**3 + 3.*X*X + 2.*X + 4.$$

it requires a total of 6 multiplications, but if recast in the form

$$Y = X * (X * (2.*X + 3.) + 2.) + 4.$$

it needs only 3 multiplications. Not only has this reduced the time required to about a half, but it also reduces the error which may arise when real numbers—which cannot in general by represented exactly in binary form—are multiplied repeatedly.

It is therefore not only through the choice of suitable large-scale 'strategic' algorithms that one may hope to achieve greater efficiency, but also by the employment of suitable small-scale 'tactical' algorithms at all stages of the program. Almost any program will benefit from careful scrutiny of the choice of algorithms, and though this often demands more mathematical ingenuity than most biologists possess, they have everything to gain by developing some understanding of the principles involved in selecting efficient algorithms. These, one should remember, are not only a question of mathematics, but also demand a knowledge of programming techniques *per se* and even of particular computer systems. Some of the problems involved are discussed below in connection with accuracy control or economy in computer time. The important thing is to attempt an assessment of the alternatives rather than rushing ahead by the first method which one has always used on a desk-calculator. One might also remember that in many elementary accounts of FORTRAN—including the present book—the need to explain the FORTRAN coding with the minimum of other complications has often led to simple, straightforward but not necessarily efficient algorithms being used.

Economy in Computer Time

The time required to run a program on a computer is, of course, the sum of the times needed to carry out all the operations involved. This is not easy to estimate accurately but a major part of it is due to the slower operations. So far as the mathematical processes are concerned, this means multiplication, division and the calling of mathematical functions such as logarithms, trigonometric ratios and so on. Any attempt to reduce the time taken for a program to run must therefore concentrate on reducing the frequency of these operations—hence the insistence above that algorithms requiring a smaller number of multiplications were more efficient. Some of the steps to achieve these ends have been discussed in previous chapters as they arose in particular programs. Here we bring together a number of them for convenient reference.

(i) *Avoid repeated computation of the same expression.* Many programs use the same mathematical expression repeatedly, and in such cases it should always be evaluated once and for all near the start of the program. For example, the program segment

$$Y1 = EXP(X*X + 2.0) - COS(X)/(X*X + 2.0)$$
$$Y2 = (COS(X)*COS(X))/EXP(COS(X))$$
$$Y3 = (X*X + 2.0)**2$$

requires 6 multiplications

        2 divisions

        2 calls to EXP

   and 4 calls to COS

By rewriting it as

$$X2 = X*X + 2.0$$
$$CX = COS(X)$$
$$Y1 = EXP(X2) - CX/X2$$
$$Y2 = CX*CX/EXP(CX)$$
$$Y3 = CX*CX$$

we have 3 fewer multiplications and 3 fewer calls to COS. True we have two extra statements in the program, but the gain is nevertheless worthwhile and it becomes progressively more economical in computer time if $COS(X)$ and $X*X + 2.0$ are needed subsequently in the program.

(ii) *Exclude from DO-loops all arithmetical statements involving only constants.* Consider the following program segment intended to compute N values of Y from N values of X:

$$DO\ 1\ I = 1,N$$
$$1\ \ Y(I) = A*A/((1.0 + A)*COS(THETA))$$
$$\quad\quad\quad\quad\quad\quad\quad + A*ALOG(X(I))/(X(I)*X(I))$$

The first part of the right-hand side of this statement—i.e.

$$A * A/((1.0 + A) * COS(THETA))$$

—remains unchanged in value throughout, yet it has to be re-evaluated afresh on each run through the loop—perhaps a hundred or more times. If, however, the segment is rewritten as:

```
    AFUNCT = A * A/((1.0 + A) * COS(THETA))
    DO 1 I = 1,N
  1 Y(I) = AFUNCT + A * ALOG(X(I))/(X(I) * X(I))
```

the constant expression is evaluated only once, a very considerable saving of time.

(iii) *Exclude from DO-loops expressions in which the subscripts do not vary.* As a special case of the preceding rule we may note the following feature. Where a set of nested DO-loops occurs, some of the subscripted terms can often be moved into the outer DO-loop so that they are not repeatedly evaluated, as they would be in the inner loops. For example,

```
    DO 1 I = 1,M
    P = X(I) * X(I)/ALOG(X(I))
    DO 1 J = 1,N
  1 Y(I,J) = P * X(J)
```

is much more efficient than

```
    DO 1 I = 1,M
    DO 1 J = 1,N
  1 Y(I,J) = X(J) * (X(I) * X(I)/ALOG(X(I)))
```

In the second segment the expression within brackets changes only M times yet it is evaluated M × N times. The first segment evaluates it only the minimum necessary number of times.

A second example of this kind concerns the very frequent operation of matrix multiplication, which is commonly coded as follows:

```
    DO 1 I = 1,N1
    DO 1 J = 1,N3
    C(I,J) = 0.0
    DO 1 K = 1,N2
  1 C(I,J) = C(I,J) + A(I,K) * B(K,J)
```

Although the value stored as C(I,J) changes with each run through the innermost loop, the subscripts themselves do not alter in value throughout the N2 cycles, having been set by the two outer loops. It is therefore more efficient to rewrite the segment as:

```
    DO 2 I = 1,N1
    DO 2 J = 1,N3
    S = 0.0
    DO 1 K = 1,N2
  1 S = S + A(I,K) * B(K,J)
  2 C(I,J) = S
```

The gain in efficiency may not be very great in this example—on one computer such a modification required two-thirds of the time needed for the program as originally coded. Nevertheless matrix multiplications often account for an appreciable part of the time needed in programs of, say, multivariate statistical analysis, and the economy is worth while. The reader might review all the examples given in Chapter 8 which employ the more direct but less efficient method of multiplication.

(iv) *Compute only one triangle of symmetric matrices.* In a symmetric matrix of order n, not all of the n^2 elements are different. If the main diagonal and *either* the upper *or* lower off-diagonal elements are computed, the other off-diagonal elements can then be inserted by a simple assignment statement. This means that only $n(n+1)/2$ elements are evaluated by a more or less complicated calculation—a saving of almost 50% in computer time for large matrices. As an example we take the case of matrix A multiplied by matrix B to give the symmetric matrix C. The first 6 statements perform this multiplication for the upper triangle while the last statement inserts the appropriate values in the lower triangle.

```
    DO 2 I = 1,N1
    DO 2 J = I,N1
    S = 0.0
    DO 1 K = 1,N2
  1 S = S + A(I,K) * B(K,J)
    C(I,J) = S
  2 C(J,I) = S
```

Note especially the second statement, which ensures that subscript J takes values which run from I to N1 not, as is more usual, from 1 to N1. As we shall see below, symmetric matrices offer further possibilities for economy by a device which conserves storage space as well as time.

(v) *Economy in input/output routines.* So far we have considered only economies resulting from re-arrangement of mathematical operations within the computer registers. There is, however, much to be gained from greater efficiency of input–output operations, whether these are conducted by on-line card reader and printer or by off-line equipment. The main savings are to be obtained from efficient input and output of large matrices and the principles involved have already been outlined in Chapter 8: extensive use of implied DO-loops will reduce appreciably the number of calls needed to READ or WRITE routines provided by the system. Thus, to read a data matrix by the instructions

    DO 10 I = 1,N
    DO 10 J = 1,M
 10 READ (5,100) X(I,J)
100 FORMAT (F7.0)

where N and M are both 50 would require 2500 READ instructions, each operating on a single card bearing a single punched value. Card readers can process only a limited number of cards per minute so the greater efficiency of the following is obvious:

    READ (5,100) ((X(I,J), J = 1,M), I = 1,N)
100 FORMAT (10F7.0)

One needs, however, to bear in mind the way in which data are most conveniently collected and punched. In biological work it often happens that an individual record or case comprises a set of observations conveniently punched on a single card or a group of cards, each new case starting at the beginning of a fresh card. Under such circumstances the following input scheme may be more generally useful:

    DO 10 I = 1,N
 10 READ (5,100) (X(I,J), J = 1,M)
100 FORMAT (10F7.0)

So far as output is concerned, it is important to remember that the time taken depends less on the total amount of data than on the number of lines they occupy, printers working at so many lines per minute. Thus an output format containing 5 numbers per line would take twice as long to print a given matrix as one containing 10 numbers per line. Certain types of problem can generate vast quantities of output very easily and the programmer should consider seriously how much of it needs to be printed. Options which allow part of the output to be suppressed are well worth including in such programs and large symmetric matrices need never be printed in full: the upper triangle, with row and column numbers, is enough and can be produced by a coding sequence such as

```
    DO 10 I = 1,N
 10 WRITE (6,100) I,(J,X(I,J), J = I,M)
100 FORMAT (///4H ROW, I4/(10(I4,F8.3)))
```

Note again the coding of the implied DO-loop which ensures that J varies from I to M rather than 1 to M.

Conservation of Core Storage

Many biological problems, especially those in multivariate analysis, make considerable demands on storage space so that efficient use of core-storage and the backing stores provided by magnetic tapes or disks becomes very important. One must, however, accept that economies in storage space are often attained at the expense of rapid execution. Many devices for coping with large volumes of data or large intermediate matrices have already been discussed as they arose in particular programs, but again it is convenient to bring together much of this information and to illustrate it where necessary. The following are some of the devices that may be used:

(i) *Use of EQUIVALENCE statements.* This is a simple and very valuable technique. When two or more arrays are used only in different sections of a program they can be declared in an EQUIVALENCE statement. The compiler will then allot them to the same set of storage registers, which is perfectly satisfactory as they are never required to be in store simultaneously. Examples of this procedure are given in programs on pp. 297 and 305 and the rules governing its use are discussed in all general accounts of FORTRAN. Two points are especially worth noting: (a) it is desirable (or even essential in many cases) for the variables named as equivalent to be of the same type, and (b) different computers differ in the manner of handling subscripts in equivalent variables. Some require all subscripts to be mentioned in a two or three-dimensional array, while in others the linear subscript alone is required. For example, in the first case one would write, say,

```
DIMENSION A(10,10), B(100)
EQUIVALENCE (A(1,1), B(1))
```

while in the second it would be necessary to write

    DIMENSION A(10,10), B(100)
    EQUIVALENCE (A(1), B(1))

In still other cases the processor will accept either of the above forms.

(ii) *Input and/or output of matrices line by line.* In this way one requires only the locations needed for a working row or column vector. An example from matrix multiplication will make this clear. Let us suppose we need to multiply a matrix **A** by another one **B** to give a product matrix **C**:

$$\mathbf{C} = \mathbf{AB}$$

This can, of course, be done in the ordinary way, with all three matrices in core-storage. It may be, however, that matrix A is needed only once for the purposes of this multiplication and that there is therefore no need to preserve it in core. Similarly it may be that we do not require to perform further operations on C, in which case it, too, need not be preserved *in toto*. We are therefore free to read in A one row at a time from punched cards or magnetic tape and to print out C one row at a time. In this case we require core-storage only for B, together with the 'working vectors' AL and CL which will store at any one time a line of A and a line of C. A suitable coding sequence would therefore be as follows, assuming A to have N1 rows and N2 columns, while B has N2 rows and N3 columns, so that C would have N1 rows and N3 columns.

```
         DIMENSION AL(50), B(50,50), CL(50)
         DO 3 I = 1,N1
         READ (5,100) (AL(J), J = 1, N2)
 100     FORMAT (10F7.0)
         DO 2 J = 1,N3
         S = 0.0
         DO 1 K = 1,N2
   1     S = S + AL(K) * B(K,J)
   2     CL(J) = S
   3     WRITE (6,101) I,(J,CL(J), J = 1,N3)
 101     FORMAT (//4H ROW, I4/(5(I5,E16.8)))
```

It should be noted that N1 does not feature in the dimensions of any of the arrays in core storage and may be as large as one likes. Thus a matrix of order, say, 1000×50 can be post-multiplied by one of 50×50 to give a product matrix of 1000×50 yet to do this we require only $50 \times 50 + 50 + 50 = 2,600$

locations in core storage. Of course, not all matrix operations can be conducted line by line—inversion or the evaluation of a determinant, for example, can usually only be carried out on a matrix all of which is in core. Further, the operations described above are much slower than those in a purely 'in-core' multiplication. Nevertheless the method given may be essential in a program requiring large amounts of core storage.

(iii) *Use of magnetic tape or disk backing stores.* The operations involved in the use of magnetic tape as an input–output device differ from those where it is employed as an internal storage medium auxiliary to core-storage. Information is normally stored on input–output tapes in binary-coded decimal form (BCD) and is handled under the usual FORMAT specifications. In this section, however, we are concerned with the internal storage of numerical information. Since numbers are represented within the computer by binary digits, the information on these tapes is in binary, non-formatted form. The operations involved in programming with such tapes are as follows:

READ (i) *list*
WRITE (i) *list*
REWIND i
BACKSPACE i
END FILE i

in all of which i is the logical tape unit number. These numbers are allocated by each computer installation and differ accordingly. They are not normally 5 or 6—which are reserved respectively for the input and output units carrying BCD information between either the card reader or printer and the central processing unit of the computer installation. Since the logical designation i may differ from one installation to another it may be represented in programs by a non-subscripted integer variable whose value in a particular program must be read in as part of the data. In the program on p. 297 the variable NTAPE plays this role.

The effects of the tape-control statements are as follows:

READ (i) *list* reads one *logical record*, (i.e. what is read by a single ordinary READ statement) from tape i, placing the elements of this record in the locations specified by list. For example READ (2) N,X,(Y(I), I = 1,10) would cause a logical record comprising twelve successive binary numbers on tape 2 to be placed respectively in the core-locations reserved for N,X,Y(1), Y(2), Y(3) ... Y(9), Y(10). The values will also remain stored on the tape until something is done to change them.

WRITE (i) *list* causes numbers existing in binary form in core locations specified in the list to be written on to tape i in the form of a single logical record. For example, WRITE (3) A,B,C(10), (D(J), J = 2,10,2) causes eight numbers stored in the locations reserved for A,B,C(10), D(2), D(4), D(6), D(8) and D(10) to be written on to tape 3 to form one logical record.

These statements would be of little value unless the information on the tape were available in an ordered, recoverable sequence. This is the function of

REWIND i

which is always used before a tape is employed for the first time in a program. It causes the tape to be re-wound to its load-point (i.e. to the 'beginning') so that logical records can be written on it or read from it in strict sequence. After each logical record there is a small gap and when the complete set of logical records for the program—making up the 'file'—have been placed on tape a special mark, the end-of-file mark, is made by the instruction

END FILE i

The last tape-manipulation instruction BACKSPACE i causes the tape i to be moved backwards (i.e. towards load-point) exactly one logical record.

It will be clear that correct use of tapes requires careful checking of the number of logical records read or written during a program. It will also be seen that it is quite straightforward to use a WRITE statement repeatedly and then, after rewinding, to use a similar succession of READ statements to bring the same set of logical records back into core storage at a later stage in the program.

For example, at some point in a program we may wish to store a matrix by rows on tape and then later re-read it back row by row into another set of core locations. The first part of this could be coded as

```
      DIMENSION AMAT (50,50)
      REWIND 3
      DO 5 I = 1,N
    5 WRITE (3) (AMAT(I,J), J = 1,M)
      END FILE 3
```

where each of the N rows of the N × M matrix AMAT constitutes a single logical record of M numbers and the whole set of N rows makes up the file. Later we might use the information in this file to make up the matrix BMAT, which is an exact duplicate of AMAT:

```
      DIMENSION BMAT (50,50)
      REWIND 3
      DO 10 I = 1,N
   10 READ (3) (BMAT(I,J), J = 1,M)
```

Note that we rewind before opening the file (i.e. starting to write the numbers which will make it up), we write an end-of-file mark when the file is complete, and we rewind again before reading the file into core. The fact that we rewind to the same load-point on each occasion ensures that the same N logical records (rows of the matrix in this example) are written out and then read in again. The unit operation, so to speak, is the reading or writing (or backspacing) of one logical record.

It will readily be appreciated that if we wrote the above array **AMAT** on tape row by row and then later wished to read it back *column by column* we would be in considerable difficulty. Such operations can be programmed but they require a complicated set of tape manipulations since we have at our disposal only the commands READ i, REWIND i and BACKSPACE i. A better procedure is to avoid the difficulty altogether. In this case, for example, if we wished to use a matrix first row by row and then column by column it would be wisest to store two copies on tape—either on the same or separate tapes—one recorded row by row and the other column by column.

So far it has been assumed that only magnetic tapes have been used as an auxiliary store. In Standard Fortran the use of disk- and drum-storage is programmed as though the storage units were magnetic tapes. This has the great advantage that a program may be used without any alteration when it is transferred from a computer employing tapes to one with disk storage. In other cases, however, separate instructions may be used for disk or drum manipulation. These will not be discussed here: a short account is given by Golden (1965) and further details can be found in manuals for machines using these special conditions.

One final point regarding the use of tape manipulation statements may be made. It is often recommended that programs employing a magnetic tape unit should end:

REWIND i
STOP
END

so that the tape is ready for removal or use in a subsequent job. Tidy though this may be, it has the disadvantage that the end of the job is delayed until a tape has been rewound—an appreciable time if a large body of data is involved —during which the computer is doing nothing else. It is better to omit the final rewind and allow the operating system to rewind the tape at a time when the computer can be employed usefully on another task.

(iv) *Vector storage of matrices.* Though matrices are commonly visualised by the programmer as a set of locations arranged in rows and columns—a picture encouraged by the system of doubly subscripted Fortran variables—

the elements of the matrix can in fact be stored in successive locations, forming a linear array. Any matrix may therefore be stored as a vector and the replacement of double subscripts by a single subscript makes indexing simpler and therefore more efficient. Figure 42(a) shows how the elements

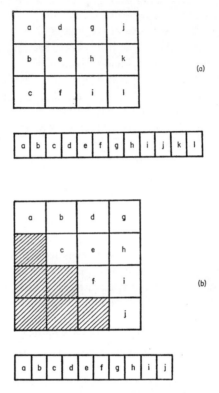

FIG. 42. Vector storage of matrices: (a) Vector storage of a general matrix, arranged columnwise; (b) compressed vector storage of a symmetric matrix, storing only the upper triangular elements; the shaded elements of the lower triangle are not stored.

of a matrix A(I,J) correspond to those of a vector, say VA(IJ), where we use IJ to denote the *single* subscript corresponding to the *two* separate subscripts I and J of matrix A.

Matrix	Vector
A(1,1)	VA(1)
A(2,1)	VA(2)
A(3,1)	VA(3)
A(1,2)	VA(4)

Matrix	Vector
A(2,2)	VA(5)
A(3,2)	VA(6)
A(1,3)	VA(7)
A(2,3)	VA(8)
A(3,3)	VA(9)
A(1,4)	VA(10)
A(2,4)	VA(11)
A(3,4)	VA(12)

In this case, no saving of core-storage would result from storing the matrix in vector form, but if one thinks of a *symmetric* matrix (such as the dispersion or correlation matrices so commonly encountered in multivariate analysis) it is easy to see that a considerable saving of space would follow if one stored as a vector only the main diagonal elements and *one* of the two identical sets of off-diagonal elements. Fig. 42(b) shows how the required elements of a symmetric matrix A(I,J) might be stored as a vector VA(IJ).

Matrix	Vector
A(1,1)	VA(1)
A(1,2)	VA(2)
A(2,2)	VA(3)
A(1,3)	VA(4)
A(2,3)	VA(5)
A(3,3)	VA(6)
A(1,4)	VA(7)
A(2,4)	VA(8)
A(3,4)	VA(9)
A(4,4)	VA(10)

The important feature to notice is that of the 16 elements in the matrix of Fig. 42(b), the 6 shaded elements need not be stored. Thus, instead of using 16 storage locations we can, with the vector form, store all the essential information in 10 locations. In general, a symmetric matrix of order m can be accommodated in vector form in $m(m + 1)/2$ locations instead of the m^2 required for doubly-subscripted matrix storage. For a large matrix, say of order 100, this means almost halving the space needed—5,050 locations instead of 10,000.

The problem is, then, how can we map the elements of matrix A(I,J) listed above on to the corresponding elements of the vector VA(IJ). This must be done by expressing IJ as a function of I and J and a little thought will show that for the symmetric matrix illustrated the appropriate mapping function is

$$IJ = J * (J - 1)/2 + I$$

where the division to the right of the = sign is done in fixed-point arithmetic. In other words, whenever we wish to refer to the matrix element A(I,J) we can now do so as VA(IJ), bearing in mind the above relationship between I, J and IJ. It ought also to be clear that the above method of establishing a correspondence between the elements of a symmetric matrix and a vector is not the only one. We might, for example, have taken the ten successive elements A(1,1), A(1,2), A(1,3), A(1,4), A(2,2), A(2,3), A(2,4), A(3,3), A(3,4), A(4,4) and placed *them* in the vector VA(1) to VA(10), though in this case a different mapping function would be needed to establish the necessary correspondence. There are, however, practical advantages to be gained from adhering to a single type of correspondence and that shown in Fig. 42(b) is advocated because it is employed in the subroutines of the 'Scientific Subroutines Package' obtainable from IBM.

Having established that a correspondence can be set up between matrix and vector we must, of course, rewrite any programs which are to employ vector storage. As an example, we give a subroutine for computing a dispersion or correlation matrix. This performs essentially the same function as the subroutine CORMAT (p. 232) but through using vector storage it reduces the number of locations needed very considerably. Before reproducing the program we may, however, point out that the use of the mapping function is not the only method of generating the vector we require. When the successive elements of the matrix are computed systematically we can place them sequentially in the vector by a set of statements like the following:

```
      ⋮
      IJ = 0
      DO 5 J = 1,M
      DO 5 I = 1,J
      IJ = IJ + 1
    5 VA(IJ) = X(I) * X(J)
      ⋮
```

If the reader cares to follow these loops through for the matrix of Fig. 42(b) where M = 4 he will find that for each value of I and J the appropriate value of IJ is generated.

In the program which follows both methods of computing the appropriate values of IJ are employed: the second method is used repeatedly in subroutine CORVEC while the mapping function is used in the calling program to print out the elements of the vector in a form exactly similar to that used previously in the double-subscripted program CORMAT. The 'occasional programmer' may find it unnecessarily troublesome to acquire some skill in writing

Flowchart for Program 48

Dispersion and correlation matrices with compressed vector storage.

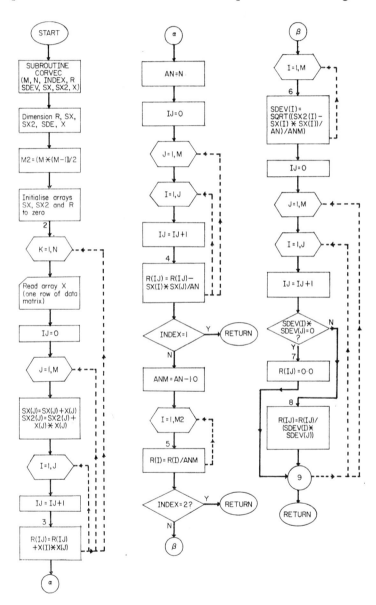

programs with vector storage of matrices. For anyone who contemplates any serious programming, however, the effort spent in mastering the technique is well worth while, especially as it provides the programmer with the most efficient access to the many invaluable programs of the IBM Scientific Subroutine Package (p. 472).

Program 48: Dispersion and Correlation Matrices with compressed vector storage

This program is not discussed further. It may be understood from the above account and by comparison with Program 23 (p. 236).

Program 48

Dispersion and correlation matrices with compressed vector storage. The data are read and computations carried out by SUBROUTINE CORVEC. A suitable calling program (such as that listed) must be supplied to print out the matrix in row and column form.

```
C
C   CALLING PROGRAM FOR SUBROUTINE CORVEC.
C
      DIMENSION R(1275), RR(50), SX(50), SX2(50), SDEV(50), X(50)
      EQUIVALENCE (RR(1), SX(1))
C
    1 READ (5,100) M, N, INDEX
  100 FORMAT (3I5)
      IF (M .EQ. 0) STOP
      CALL CORVEC(M, N, INDEX, R, SDEV, SX, SX2, X)
      GO TO (2, 3, 4), INDEX
    2 WRITE (6,101)
  101 FORMAT (42H1SUMS OF SQUARES AND CROSS-PRODUCTS MATRIX //)
      GO TO 5
    3 WRITE (6,102)
  102 FORMAT (18H1DISPERSION MATRIX //)
      GO TO 5
    4 WRITE (6,103)
  103 FORMAT (19H1CORRELATION MATRIX //)
    5 DO 7 I = 1,M
      DO 6 J = I,M
      IJ = J * (J-1)/2 + I
    6 RR(J) = R(IJ)
    7 WRITE (6,104) I, (J, RR(J), J = I, M)
  104 FORMAT (/4H ROW, I4 / (5(I4,E15.7)))
      GO TO 1
      END


C
      SUBROUTINE CORVEC (M, N, INDEX, R, SDEV, SX, SX2, X)
C
C.........................................................................
C
C   THIS SUBROUTINE COMPUTES AND STORES IN COMPRESSED VECTOR FORM
C   A DEVIATION SUMS OF SQUARES AND CROSS-PRODUCTS MATRIX OR A
C   DISPERSION MATRIX OR A CORRELATION MATRIX. THESE ARE DERIVED FROM
```

Program 48 (*continued*)

```
C     A PRIMARY DATA MATRIX CONTAINING ANY NUMBER OF ROWS, CORRELATIONS
C     BEING COMPUTED BETWEEN COLUMNS. PARAMETERS ARE AS FOLLOWS -
C        M     = NUMBER OF COLUMNS IN PRIMARY DATA MATRIX
C        N     = NUMBER OF ROWS IN DATA MATRIX, READ A ROW AT A TIME
C        INDEX = 1 FOR DEVIATION SQUARES AND CROSS-PRODUCTS MATRIX
C                2 FOR DISPERSION MATRIX
C                3 FOR CORRELATION MATRIX
C        R     = OUTPUT MATRIX STORED AS VECTOR OF LENGTH (M*(M+1))/2
C        SDEV  = VECTOR OF STANDARD DEVIATIONS (LENGTH M)
C        SX    = WORKING VECTOR (LENGTH M)
C        SX2   = WORKING VECTOR (LENGTH M)
C        X     = VECTOR REPRESENTING ROW OF DATA MATRIX (LENGTH M)
C
C.............................................................
C
      DIMENSION R(1275), SX(50), SX2(50), SDEV(50), X(50)
C
C     INITIALISE.
      M2 = (M * (M+1)) / 2
      DO 1 I = 1,M
      SX(I) = 0.0
    1 SX2(I) = 0.0
      DO 2 I = 1,M2
    2 R(I) = 0.0
C
C     READ DATA MATRIX ONE ROW AT A TIME AND ACCUMULATE SUMS.
      DO 3 K = 1,N
      READ (5,100) (X(I), I = 1,M)
  100 FORMAT (10F7.0)
      IJ = 0
      DO 3 J = 1,M
      SX(J) = SX(J) + X(J)
      SX2(J) = SX2(J) + X(J) * X(J)
      DO 3 I = 1,J
      IJ = IJ + 1
    3 R(IJ) = R(IJ) + X(I) * X(J)
C
C     COMPUTE DEVIATION SQUARES AND CROSS-PRODUCTS MATRIX.
      AN = N
      IJ = 0
      DO 4 J = 1,M
      DO 4 I = 1,J
      IJ = IJ + 1
    4 R(IJ) = R(IJ) - SX(I) * SX(J) / AN
      IF (INDEX .EQ. 1) RETURN
C
C     COMPUTE DISPERSION MATRIX.
      ANM = AN - 1.0
      DO 5 I = 1,M2
    5 R(I) = R(I) / ANM
      IF (INDEX .EQ. 2) RETURN
C
C     COMPUTE CORRELATION MATRIX.
      DO 6 I = 1,M
    6 SDEV(I) = SQRT((SX2(I) - SX(I)*SX(I)/AN) / ANM)
      IJ = 0
      DO 8 J = 1,M
      DO 8 I = 1,J
      IJ = IJ + 1
      IF(SDEV(I)*SDEV(J) .EQ. 0.0) GO TO 7
```

Program 48 *(continued)*

```
        R(IJ) = R(IJ) / (SDEV(I)*SDEV(J))
        GO TO 8
    7   R(IJ) = 0.0
    8   CONTINUE
        RETURN
        END
```

Accuracy Considerations

Though biologists are accustomed to the numerical variation found in all living organisms, they are usually much less familiar with the mathematical problems of numerical accuracy and the assessment of error. These form part of the subject matter of numerical analysis, a large branch of mathematics, and they cannot possibly be dealt with in a few pages. A brief introduction is needed, however, if only to warn the reader of its importance and of the need to explore the subject in greater detail through the many available text-books (e.g. Pennington, 1970; Conte, 1965; Fox and Mayers, 1968).

A few elementary points may first be noted. Though there are exact numbers, most of those occurring in scientific work are approximate. A measurement is normally accurate only to some definite tolerance, depending on the sensitivity of the measuring instrument and the number of figures in which the observer expresses his measurement. Thus a measurement recorded as 25.7 mms may be assumed to be accurate to the first decimal place or, in this case, to consist of three significant figures. This means that the true value of the quantity being measured lies somewhere between 25.65 and 25.75. If the measurement had been made more accurately one might have recorded it as, say, 25.73 mms, i.e. with four significant figures, implying that the true value lay between 25.725 and 25.735. Conversely, one might have recorded the measurement correct only to the nearest millimetre, as 26 mms, and therefore to be regarded as lying somewhere between 25.5 and 26.5. Any attempt to convey a spurious impression of accuracy by quoting the length as 26.0000 mms when it had been measured only to the nearest millimetre is, of course, totally unjustifiable.

While it is usually easy to decide how many significant figures should appear in a direct measurement, it is not so obvious how many should be retained in stating the result of an arithmetical calculation. We cannot explore the subject properly here, but it is enough to indicate the kind of conclusions to be borne in mind, summarised as follows:

(i) In adding or subtracting two numbers, the result is not more accurate than the number of significant decimal places in the least accurately determined number. Thus, adding 2.731 and 3.4, when each is accurate to the number of figures given, produces the sum 6.131 but the last two decimal places are not significant and even the first may be in error. This is obvious if

ACCURACY CONSIDERATIONS 459

one remembers that the first number may lie between 2.7305 and 2.7315 while the second lies between 3.35 and 3.45. The sum therefore lies somewhere between 3.35 + 2.7305 = 6.0805 and 3.45 + 2.7315 = 6.1815.

(ii) In multiplication, division and the extraction of roots, the result contains no more significant figures than the operand with the least number of significant figures. Thus $2.318 \times 1.0 = 2.318$ but the product has not more than two significant figures and should be written as 2.3. Again this is clear if one recognises that its true value may lie anywhere between 2.3175×0.95 and 2.3185×1.05, i.e. between 2.201625 and 2.434425.

The above very elementary considerations have been stressed only because so many biologists seem unaware of them. They must be taken into account whenever one carries out a calculation and are in no way peculiar to computer arithmetic. They remind one, however, that if a computer is programmed to print out, say, the arithmetic mean of 29 numbers (each accurate to one decimal place) in the format F10.4, the last three and perhaps even all four decimal places will have no significance. Since one cannot know in advance the error in the input, one can only arrange the output format in what seems a reasonable way; how many figures should be retained when the results are subsequently presented remains to be estimated by the user of the program.

Other more complicated questions of numerical accuracy arise during computer operations (a) because the computer representation of real numbers itself introduces error; (b) because errors, of whatever origin, tend to propagate themselves; and (c) because certain operations bring about a considerable loss of significant digits. The combined effect of these processes may deprive the results of all significance. The computer, it must be recalled, may repeat certain operations hundreds of times in a single calculation and the user generally has no indication of intermediate results such as are available in a hand calculation and often enable accuracy loss to be detected as soon as it begins to occur. In some situations it is possible to minimise accuracy loss by a suitable algorithm; in others the data and the problem may prevent effective action, in which case the programmer should at least be aware of the difficulty and able to incorporate warnings when serious accuracy loss occurs. The subject can conveniently be introduced under the three aspects mentioned above.

(a) *Error in computer representation of numbers*

So long as we consider only integers, any number we think of or write in decimal notation can be represented exactly, i.e. without error, in the binary system of the computer. The same is not generally true of fractions, however. Thus 0.1 (decimal) is represented to 10 binary places as 0.0001100110 and continuing this sequence would show that it is in fact a non-terminating binary fraction which may be written as 0.00011001100011.... Clearly a computer with a finite word-length will only be able to represent this fraction approximately: the longer the word-length (and this may be, say, 36 bits) the less the error, but *some* error there must always be. The same will also hold for most other attempts to represent an exact decimal fraction in binary form. Thus the mere act of representing a real number internally in the computer introduces an element of error into every calculation. The size of the error depends on whether the number is represented in fixed-point or floating-point form and on the physical characteristics of the computer. On the IBM 7090, for example, the absolute difference between a real number and its machine representation is $\leq 2^{-36}$ in fixed-point operations.

The error in converting a real number into a machine number is sometimes referred to as a round-off error. Further round-off errors may be introduced when arithmetical operations are performed on floating-point numbers, even if the numbers themselves are exactly represented. Thus, if two n-digit numbers are multiplied the product is a number with $2n$ digits, but if this operation is performed by a computer with a word-length of m digits ($m < 2n$), the result is truncated or otherwise rounded-off to m digits, thus generating an error. As a rule these errors will be small in relation to the numbers being handled, but the possibility of the errors being propagated may eventually lead to their assuming serious proportions.

(b) *Propagation of errors*

Since each arithmetical operation introduces error, the repetition of operations might be expected to result in increasingly serious errors. This is not necessarily so as some errors may decay in the course of a computation while others may grow slowly and so remain below an acceptable level. In order to appreciate some of the more elementary points involved it is necessary to introduce the concepts of relative error and error accumulation in a few simple arithmetical operations.

If an exact number q is represented by an approximate one q_1 then the error is $|q_1 - q|$ which we may suppose does not exceed a quantity we shall call Δq. That is,

$$|q_1 - q| \leq \Delta q \qquad (1)$$

Δq is therefore the upper limit of the magnitude of the absolute error or, for short, the *absolute error*. The *relative error*—or, strictly speaking, the upper limit of the magnitude of the relative error—is then given by

$$\frac{\Delta q}{|q|} = \frac{\Delta q}{|q_1| - \Delta q} \qquad (2)$$

which, if Δq is small in relation to q_1, reduces to approximately $\Delta q/|q_1|$, thus giving an expression for relative error in terms of the approximate number q_1 and the upper limit of magnitude of the absolute error.

When two approximate positive numbers q_1 and q_2, each with absolute error Δq_1 and Δq_2, are added, the sum $(q_1 + q_2)$ has an absolute error of $(\Delta q_1 + \Delta q_2)$ and therefore a relative error of

$$\frac{\Delta q_1 + \Delta q_2}{q_1 + q_2 - (\Delta q_1 + \Delta q_2)} \tag{3}$$

i.e. the relative error of the sum lies somewhere between the relative errors of the two numbers added. If two approximate positive numbers are subtracted the relative error of the difference is given by

$$\frac{\Delta q_1 + \Delta q_2}{q_1 - q_2 - (\Delta q_1 + \Delta q_2)}. \tag{4}$$

This relation is very important since it indicates that when q_1 and q_2 are very similar in size, the relative error becomes large and may even exceed the difference itself! The 'difference' is then a number devoid of any significance whatsoever. For example, suppose we subtract 9.00 from 9.02, both numbers being correct to two decimal places and therefore having absolute errors of 0.005. The difference is 0.02 and its relative error is

$$\frac{0.005 + 0.005}{0.02 - (0.005 + 0.005)} = \frac{0.01}{0.01} = 1.$$

Comparable relations may be developed for multiplication and division. Thus, assuming the absolute error is small in relation to the numbers multiplied or divided, we have for the product $q_1 q_2$ or the quotient q_1/q_2 a relative error equal to

$$\frac{\Delta q_1}{q_1} + \frac{\Delta q_2}{q_2}$$

i.e. in both cases to the sum of the relative errors of the original numbers.

A detailed analysis of more complicated functions can be made on similar lines, leading to a general expression for error accumulation which states that for any function $f(q)$ in which q is represented by the approximate number q_1 with an absolute error Δq, the difference between the real value of the function and the calculated value $f(q_1)$ is given by

$$f(q) - f(q_1) = f'(q_1)\Delta q \tag{5}$$

where $f'(q_1)$ denotes the derivative of $f(q_1)$. The difficulties of carrying out a full error analysis for a realistic problem are enormous but one can note that the propagation of error may occur either in a linear fashion—when it is generally not serious—or in an exponential fashion, when it may have disastrous consequences. An example is the error introduced in computing $y = e^x$ when x is subject to error. Since the derivative of e^x is also e^x it follows from the general expression for error accumulation (5) that the *relative* error in the computed value of e^x is equal to the absolute error in x. Thus, if $x = 100$ with an absolute error of 0.5, e^x will have a relative error of 0.5 or 50% and thus have no significant figures! In some cases exponential loss of accuracy may be due to the particular algorithm employed—a numerically unstable algorithm as it is called—and the position can be saved only by reformulating the method of computation.

(c) *Loss of significant digits*

The previous discussion has revealed two important sources of error, arising when two almost equal numbers are subtracted or when a division is performed by a number very small in relation to the dividend. The resulting significance error is important not only in itself but also because it can cause serious loss of significance in quantities subsequently computed from the number in error. It is therefore desirable that warning of such a situation should be given by a program at the point where it occurs. This may be done very simply as in:

```
        ⋮
   10   Y = A - B
        IF (ABS(Y) .GT. 0.001 * ABS(A)) GO TO 11
        WRITE (6,101)
   101  FORMAT (30H ACCURACY LOSS AT STATEMENT 10)
   11   Y = Y * Y
        ⋮
```

The size of the decimal fraction given above as 0.001 is, of course, a matter for decision. In the present case it implies that a warning is required if the result Y is 1/1000 or less of the size of A, i.e. that in computing Y there has been a loss of at least three significant figures. If this were too severe a criterion one might rewrite the IF statement as

IF (ABS(Y) .GT. .00001 * ABS(A)) GO TO 11

so that the accuracy loss warning would be printed only if five or more significant figures were lost. Judgement here must be based on how much error one is prepared to tolerate and on the kinds of consideration discussed previously.

A comparable error warning to guard against division by a very small number might be:

⋮

```
10  Y = A/(B * COS(THETA))
    IF (ABS(B * COS(THETA)).GT. 0.001 * A) GO TO 11
    WRITE (6,101)
101 FORMAT (30H ACCURACY LOSS AT STATEMENT 10)
11  Y2 = Y * Y
```

⋮

(d) *Practical considerations concerning error*

Warning messages may be desirable when the results of a computation are being interpreted but do nothing to prevent undesirable accuracy loss. A few comments are therefore needed on steps which enable loss to be avoided or assessed. This is best done by the choice of suitable algorithms. Those which minimise the number of arithmetical operations will, other things being equal, offer fewer opportunities for the propagation of error: the simple example of polynomial evaluation by nesting (p. 442) is efficient not only because it lessens the computer time required but also because it reduces the number of operations in which error propagation occurs. At a more advanced level, the selection of maximum elements as pivots in computing determinants and inverses (pp. 243, 252) tends to reduce the growth of significance errors caused by the use of small divisors. Recursion formulae, i.e. those in which the $(n + 1)$-th term of a series is derived from the n-th term, may cause serious accuracy loss if the series is at all long and the error of the first term subsequently grows exponentially. The choice of stable algorithms is one of the tasks of the numerical analyst whose advice should always be sought by the biologist when he embarks on elaborate computational procedures. Some procedures are known to lead to great inaccuracy when the matrices on which they depend are 'ill-conditioned'—i.e. when small changes in the size of the matrix elements greatly affects the results. Two measures of ill-conditioning are worth bearing in mind as they can often be computed with little extra effort. One is the so-called *condition number* $P(A)$, defined as the ratio of the maximum and minimum latent roots of the matrix. The other is the normalised determinant of the matrix, defined as

$$\text{Norm } A = \frac{|A|}{\alpha_1 \alpha_2 \alpha_3 \ldots \alpha_n}$$

where α_i is the square root of the sum of the squares of the elements in the i-th row. If the condition number is large, say 10^3, or the normalised determinant is small, say 10^{-3}, then the matrix is ill-conditioned. When such matrices are used, for example, in the solution of a system of linear equations, the results may be of little significance.

A more encouraging aspect of error accumulation is provided by the DOUBLE PRECISION facilities available with most of the larger FORTRAN compilers. Since round-off error is due to the representation of a non-terminating fraction by a computer word of finite length, it can obviously be reduced by increasing this word-length. Thus whereas the normal word-length of a computer might be 36 bits, which corresponds to about 8 significant decimal digits in floating point arithmetic, this word-length could be doubled in the DOUBLE PRECISION mode. Numbers could then be represented to about 16 significant decimal digits with a corresponding reduction in round-off error. Computations which encounter serious accuracy loss in the normal single precision mode may then often be conducted without trouble. There is therefore much to be said for employing double precision arithmetic whenever accuracy loss due to round-off error might be feared, as in the solution of differential equations, polynomial curve-fitting or any procedures involving the inversion of large matrices.

The coding and programming techniques used in the double precision mode are described in all general Fortran manuals. Briefly, they involve specifying in a type declaration all the variables for which double precision representation is needed, declaring all functions and function subprograms as double precision, and ensuring that constants are written out to the full number of extra digits required or specified by the D-conversion. Thus, such a program might begin with a declaration like:

DOUBLE PRECISION X, DELTA, SUM

Constants—including data if these are to be in double precision mode—may be written as, say, 3.141592653589793 (for π) or 2.78D − 12 ($=2.78 \times 10^{-12}$), where the letter D has the same significance as the E in the single precision E-specification. Similarly, FORMAT statements employ the D-conversion in place of the E-format, e.g.

FORMAT (4D20.12)

Library functions such as SIN, SQRT, ALOG, etc., must be replaced by the corresponding double precision functions DSIN, DSQRT, DLOG, etc., and the arguments must be double precision quantities. If special FUNCTION subprograms are employed the first statement for each subprogram must indicate its double precision nature, e.g.

DOUBLE PRECISION FUNCTION DET(X,N)

Double precision variables may be combined with single precision real variables according to normal Fortran usage; the type of an expression containing single and double precision quantities is always double precision.

The major drawback of DOUBLE PRECISION is that each variable and constant represented in this mode requires twice as much core-storage space. Thus a 50×50 matrix whose elements are stored in double precision form needs 5,000 locations in place of the 2,500 normally required. The extra space is automatically allocated by the compiler so that DIMENSION statements can remain unchanged, though the additional requirement means that programs that could be accommodated on a given computer in the single precision mode may be too large when in double precision form. Nothing can be done about this except to utilise as fully as possible all the devices for economic use of core-storage (p. 447) and to scrutinise the program carefully to ensure that only those variables are in double precision for which such representation is essential. Double precision operations are also much slower than normal computations—they may take up to 8 times longer according to the computer in use.

A few words may finally be devoted to the error criteria used in stopping an iterative procedure. Let us suppose that some quantity P is being computed and that the 5th, 6th and 7th iterations yield values—we may call them P_5, P_6 and P_7—of 0.00418, 0.00414, 0.00412. When should we stop the procedure? Two methods are generally available. The first is an *absolute error test* which stops the iteration when two successive values differ by less than a predetermined error, i.e.

$$|P_{n+1} - P_n| < \varepsilon.$$

The second employs a *relative error* test:

$$\frac{|P_{n+1} - P_n|}{|P_{n+1}|} < \varepsilon.$$

The size of ε is left to the programmer, depending on how much error he is prepared to tolerate in his results. If accuracy to n decimal places is desired then the ε in the absolute error test should be set at 0.5×10^{-n}. If P is of the order of unity, the two tests are very similar, but if P is expected to be very large or very small the relative error test should be used. It can, of course, be arranged for an appropriate value of ε to be chosen by the user and read in as data to the program. Provision should always be made for a situation where convergence does not occur to within specified error limits after a predetermined number of cycles. The iteration can then be stopped and an appropriate message printed.

This chapter may lead the biologist to wonder whether he can ever believe a result printed out by a computer. There is no need for an unnecessarily pessimistic attitude, however. Many programs raise no accuracy problems at all, while others do so only when the input data or intermediate variables are of an unusual kind. Provided the user does not trust every result blindly, and provided he is aware of some of the obvious pitfalls, most of his programs will run in an entirely satisfactory manner.

Testing of programs

It is very rare for a newly-written program to run successfully when first used, so that an appreciable part of the time involved in programming is spent in correcting one's mistakes. This process—known inelegantly but universally as debugging—is much easier if carried out systematically along the following lines, beginning long before the first trial run.

(a) *Construction of program*: Long, rambling programs without carefully kept notes and devoid of comment statements are nearly always troublesome to correct. Ideally the program should be conceived and flow-charted as a sequence of functional units, best treated as subroutines of not more than, say, 100 statements (excluding comments). Considerable thought may be necessary as to the best way of dividing the total program in this way. Within each subroutine comment statements should define the logical components from which it is built—not less than one comment per 10 instructions might be a general rule with additional ones where the logic becomes at all tortuous. Input–output operations are best reserved for the main program or special I/O subroutines, leaving the other subroutines free of READ or WRITE statements. An exception may perhaps be made for error and warning statements, though even these can be dealt with by allowing the subroutine to return an error code that can be used to initiate a WRITE statement in the calling program. Or, in a large program, one should build up a special error subroutine, calls to which are made as occasion requires and which contains full provisions for warning statements or for terminating the calculations and passing to the next problem. A consistent pattern of numbering statements is helpful, as is the use of mnemonic variable-names and a full list of all input, output and intermediate variables. The program should be written on properly prepared Fortran coding forms with numbered columns, and is best written in a fairly 'open' style with plenty of spaces between variable names and operator symbols.

Programs written in this way are easy to read and understand and can be tested piece by piece before finally being put together for testing as a whole. It is not too much to suggest that a strategy for testing the program should decide much of the form in which it is written.

(b) *Desk-checking the written program*: Before the program is key-punched, it and the accompanying flow-charts and notes should be checked carefully. This is best done by several separate and systematic checks such as:

(1) for clerical mistakes, including spelling, punctuation, unbalanced parentheses, characters in wrong columns or the confusion of symbols such as I, O and Z with 1, zero and 2.

(2) for detailed errors in programming logic, of which the following are only a selection:

1. Incorrectly nested DO-loops. To check these draw lines from the first to the last statement of each DO-loop. If the lines cross there is an error.

2. DO-loops ending on an illegal statement.

3. Undimensioned or incorrectly dimensioned arrays.

4. Inaccessible statements or the repeated use of the same statement number.

5. Wrong character-counts in H-formats.

6. Mixed mode operations—even if the compiler you use allows these, they take longer to execute and reduce the compatibility of the program with other compilers.

7. Errors in the number, type and sequence of arguments in calls to subprograms and in the subprograms themselves.

8. Misrepresentation of your intentions in IF statements.

9. Failure to initialise correctly.

10. Errors in co-ordinating your tape manipulation statements or omission of REWIND before using a tape or disk area.

11. Check also that all variables have been defined before use, either (i) through being read in as data, or (ii) through their appearance on the left-hand side of an arithmetic or assignment statement, or (iii) through transmission via COMMON or the parameter list of a subprogram call.

12. Correct correspondence of DIMENSION and type declarations between calling program and subprograms.

13. Correct control cards if the program is being run under a monitor system.

(3) Check also for strategic logical errors, depending on the purpose of the program and its overall structure.

Despite your diligence in desk-checking a program for errors like the above, some will almost certainly remain in the untested program and the above list may be referred to again when correcting a program that has failed to run correctly.

(c) *Punching and deck composition.* After key-punching the cards should be checked against the written program. Ideally this should be done by the key-punch operator using a verifying machine. If cards have to be checked visually it helps to ensure that the punch produces a legible record along the top edge of the card. The newly-punched cards can also be 'listed'—i.e. run through a listing machine which prints out the contents of each card a row at a time. Check especially for spelling and syntax errors, and for off-punched cards (i.e. those where a fault in positioning the card has led to the holes being punched slightly out of place). It is usually worth identifying each card by symbols or numbers in columns 73–80. These numbers will not be read and are invaluable if a deck gets out of order. Always ensure that the cards are in correct sequence and protect the deck by enclosing it between backing-boards held by rubber bands or keeping it in a properly constructed card-tray. Check that the appropriate control cards have been incorporated in the right places and that the data cards are punched in the required format without numerical errors, illegal characters or misplaced decimal points. Gross errors in the data cards usually give themselves away when the results of a program are examined, but small errors in a small proportion of the cards may never be detected in the output.

(d) *Test runs with small body of data.* No program should be used until it has been tested numerically: the mere fact that results look plausible is no guarantee of the correctness of the program. Simple sets of data can be used for the test runs, the results having previously been worked out on a desk-calculator, or one may use worked examples from textbooks of statistics or numerical analysis. When a program incorporates several options or branches, all of these must be tested by using appropriate data. If necessary the program should be checked with data containing zeros, negative numbers or numerical values known or thought to raise accuracy problems. In practice it is almost certain that the program will at first fail on some of these tests. These failures may concern compilation and/or execution.

(e) *Errors in compilation.* Failure of the program to compile—i.e. to produce a satisfactory object program from the punched source program—is usually accompanied by a set of error diagnostic messages from the compiler. These vary according to the compiler in use—they may be in the form of a simple letter or numerical code, which must be interpreted from the computer manual, or they comprise a brief verbal description of the error and the location at which it occurs—usually given in octal. From this message and the listing produced by the computer, which gives the location(s) of each line, it is sometimes very easy to correct the error. In other cases, however, the diagnostic message simply states trouble caused by a prior defect in logic or program construction which is not itself identified. For example, the diagnostic may report the use of an illegal subscript at a certain point. This, however, may have been caused through an error in a DIMENSION declaration at the start of the program or through incorrect subscript manipulation elsewhere. To trace the real cause of the symptoms reported by the compiler may be difficult and often requires a systematic re-checking of the whole program. The list of possible errors on p. 467 may help in this process. It must also be remembered that subroutines and function subprograms are compiled independently of each other and of the main program, and that difficulties may subsequently arise through incorrect interactions between the main program and the subprograms which could not have been detected by the compiler. Even within a program or subprogram there may be fatal flaws which the compiler is not competent to diagnose. Thus, while it is always a step forward to have a program which compiles, the mere fact that it does so is no guarantee of its correctness.

(f) *Errors in execution.* If a program compiles but produces incorrect or no results, one must locate the causes of faulty execution. Again, the computer system may help by producing a set of diagnostics for execution errors. No computer can detect errors in the logical construction of a program, however, so long as these errors are self-consistent. A program containing such faults is often unusually difficult to correct, especially since a programmer who has already committed a logical fault is unlikely to recognise his mistake on sight.

The simplest way to begin correcting such a program is therefore to incorporate in it a set of WRITE statements (each with an appropriate FORMAT statement) which will print out values of all the variables and other parameters as they are used during the running of the program. The entire course of the arithmetical calculations (beginning with the data read in) then becomes clear and the places at which errors arise can be identified rapidly. Note, however, that it may be undesirable to incorporate WRITE statements within all DO-loops at first, though once a loop has been identified as causing trouble a statement that prints out values of variables and indexing parameters on each pass through the loop may be necessary. Inserting the WRITE statements requires, of course, that the programmer fully understands the object of each section of the program (which he may not have written in its entirety). If suitable comment cards have been included and if the whole program has been systematically organised into functional units, with adequate documentation, the task of correction is very much easier. Once the trouble has been corrected the diagnostic WRITE statements and associated FORMAT declarations are removed from the deck and the program run once again to check that this has been done correctly.

Some of the larger compilers have associated with them an elaborate 'debugging language', in which has been written a set of statements making up a 'debug packet' that can be run with the defective program and causes it to print out variables, thus enabling faults to be diagnosed. It is often simpler, however, to create one's own debugging procedure by using WRITE statements as indicated above.

An especially irritating situation arises when a program that has run successfully when tested, may fail during subsequent use. There are many reasons why this could occur, perhaps the simplest being a mistake in the new set of data. For this reason it is often desirable to include one or more 'echo-checks' in the program, i.e. instructions ensuring that all data are printed out immediately after being read in. Visual checking of data cards is no substitute for this since it cannot in itself always detect errors or inconsistencies in format, nor can it recognise more subtle failures to store the data correctly or make sure that they are properly available for any subroutines that may operate on them. For example, incorrect application of the method of adjustable dimensions may result in operations being performed on a matrix numerically different from that read in by the main program. Other causes of program failure may be at work, however, and a few of these are listed below. In all cases they occur through the data or intermediate variables having values which the programmer had not envisaged and for which he did not make provision when writing or testing the program.

(i) The SQRT function will not normally operate on a negative argument: if this occurs in the data or at intermediate points a statement incorporating SQRT will usually terminate execution. The logic of the program should be re-examined if this occurs. A common fault is to forget that what should be a very small positive number may, through round-off or other errors, actually be represented in the computer by a very small negative value. Thus the programmer who wishes to use the square-root of a variance may feel he is safe in writing SQRT (VAR) because, by definition, a variance cannot be negative. It may, however, happen that small and in themselves not unacceptable errors cause a very small or zero variance to be represented as a very small negative number. If this situation is not to stop execution the programmer must write SQRT(ABS(VAR)).

(ii) Similar difficulties arise over logarithmic functions (for which the argument must always exceed zero) or for other functions operating only on arguments within a given interval, e.g. in ARCOS(X) and ARSIN(X), X must lie between -1 and $+1$. It is also necessary to bear in mind that some compilers will, on encountering an inappropriate argument, carry out a 'recovery' and print a warning message (e.g. that in ARCOS(X), X was greater than 1 and has been set equal to zero). While this may ensure that execution proceeds to the bitter end, the results may be quite unacceptable. In reality the program should have been rewritten to cater for the possibility of an inappropriate value turning up. Warning messages of a recovery operation are always to be treated seriously and here—as also in cases where numerical accuracy loss may be involved (p. 462)—the mere fact that execution is completed does not guarantee the correctness of the results.

(iii) The use of the logical operators .GE., .LE., .GT., .LT., etc., with real numbers may also produce unexpected results if round-off error affects the real number slightly. For example, IF(X.EQ.0.0) GO TO 10 may cause trouble if X instead of being the exact zero expected is, say, 0.0000000000001. Small though this inaccuracy is, the logic of the computer inexorably treats X as a positive number. It may be better in such cases to write, say, IF (ABS(X).LT.1.0E–10) GO TO 10 or to re-organise the program to allow the IF statement to operate on a number represented without error.

(iv) Subscripts, or indices (as in computed GO TO statements), or the limits of DO-loops, all of which have not simply been read in as data but which are derived in the course of a calculation, may exceed the legitimate values. Some system of checking those computed parameters before they are used is therefore very desirable.

Needless to say there are limits to the extent to which unlikely contingencies need be guarded against. In many cases a programmer who discovers that execution has been stopped by errors of the above kind may be content to recognise that his data were aberrant. Programs, after all, are intended to be used realistically and it is hardly necessary to precede every use of an ARCOS statement by qualifying conditions when one knows perfectly well they are never going to apply. Here, as in most things, a reasonable person knows how far to go.

Use of available package programs or subroutines

Anyone wishing to develop some skill as a programmer will obviously want to write many programs himself. Once the necessary facility has been acquired, however, it is clearly wasteful to continue writing every program one needs when such a great variety of tested programs is readily available. This chapter is intended to guide the reader with a working knowledge of Fortran in his search for reliable and efficient programs.

Many textbooks of statistics or numerical analysis now include Fortran programs and are a convenient first resort when they can be located. Among the works that could be cited are Pennington (1970), Conte (1965), Carnahan et al. (1969), James et al. (1967) and Ralston & Wilf (1960) for numerical methods and Cooley & Lohnes (1962), Veldman (1967), Sokal & Rohlf (1969) and the present book for statistical techniques. Beyond textbooks one turns to the specialist periodical literature and especially to journals such as *Applied Statistics*, the *Computer Journal*, the *Journal of the Association for Computing Machinery*, *Computer Programs in Biomedicine* or the *International Journal of Biomedical Computing* which may contain programs or the flow-charts and algorithms from which programs can be written relatively easily.

A second major source of programs are the libraries organised by many of the larger computer manufacturers, some of which contain many hundreds of programs written by users and others in the course of their research. Details of these programs are available from the particular computer installation at which one works and they exist in a variety of forms—as FORTRAN decks (the most convenient form if one wishes to amend them or adapt them), as relocatable binary decks, or as part of a tape or disk library from which they can be called by the appropriate control cards. Even assuming they are available as FORTRAN listings, the documentation associated with these programs is rather variable both in quality and quantity. Ideally the documentation should indicate the mathematical method used, provide an outline flow-chart of the program, discuss accuracy considerations if these are critical, indicate clearly the dimensions and how they may be varied and summarise the input variables and formats used. The biologist may be disappointed when faced with a newly located program in which documentation is not up to standard or is written in a jargon intelligible only to the mathematician and the professional programmer. Perseverance and the assistance of other users will generally overcome difficulties, but there are always dangers in using a program one does not fully understand, and anyone who makes repeated use of a relatively complex program in a research project should certainly seek expert guidance if he cannot interpret most of it himself.

A particularly important set of mathematical and statistical subroutines has been gathered together by IBM to form the so-called System/360 Scientific Subroutine Package. This is a collection of over 250 subroutines (some 200 of which are available both in single and double precision modes) forming input/output-free building blocks, from which the user can readily construct a large number of programs by providing I/O routines and the instructions needed to link the necessary subroutines. The collection is discussed in detail in the programmer's manual (IBM, 1968), often available for consultation from the program librarian of any large computer installation. This manual includes a complete listing of every subroutine with details of computational methods where these are in any way unusual. Further, there are given a number of complete sample programs in which several subroutines are employed to carry out some relatively large scale statistical or mathematical operation. Though intended primarily for the System/360 computers these programs can in most cases be run without change on other machines—even on computers not manufactured by IBM. The subroutines are liberally provided with comment statements and are of the very greatest value to anyone programming scientific computations. They can, if necessary, be punched direct from the listings given in the programmer's manual though many installations have them available in BCD tape form, from which FORTRAN decks may be run off immediately and made available to the user. At large installations they may be called directly from a library stored on disk.

Among the SSP subroutines of special interest to the biologist are those concerned with the following operations:

data-screening and tabulating;
correlation and regression techniques, including multiple regression, probit analysis and canonical correlation;
time-series analysis;
non-parametric statistics;
generation of random variates;
matrix algebra, including inversion and extraction of latent roots and vectors;
numerical integration and differentiation and the solution of first-order ordinary differential equations.

The sample programs include those for ordinary and step-wise multiple regressions, polynomial regression, analysis of variance, factor analysis, discriminant analysis and the solution of simultaneous equations. Even when they are not immediately applicable the SSP subroutines and programs can often be adapted to one's requirements or provide the basis on which a suitable program can be written.

Three special features of these subroutines need to be discussed briefly. Examination of an SSP subroutine shows that the arrays are all dimensioned to contain a single location, e.g. X(1), MAT(1), etc. The calling program provided by the user must, however, be fully dimensioned, e.g. X(100), MAT(500). The fact that dimensioned areas in the subroutines are then not the same as in the calling program may not matter. It is the calling program which reserves space and provided the compiler does not check the upper limit of the dimensioned area in the subroutine no difficulties arise. If, however, one were working with a compiler where subscript checking occurred, then the DIMENSION statements in the calling program and subroutines would need to be made the same or the device of adjustable dimensions employed for the subroutines.

A second feature of the SSP subroutines causes greater difficulties when first encountered. This is that all matrices are stored in vector fashion, so that they are referred to by a single subscript. Thus a matrix A, normally dealt with by two subscripts so that an element would be referred to as A(I,J), is in these subroutines always handled as a vector, say VA(IJ). We have already encountered something like this when dealing with economical storage of symmetric matrices, but the SSP subroutines employ vector storage for *all* matrices. The user who wishes to employ a subroutine in a program containing the usual double-subscripted general matrices can, however, do so by a simple device. So long as the matrix occupies the *whole* of its dimensioned area the storage locations occupied by its elements are the same, no matter whether it is represented in the program as A(I,J) or VA(IJ). By using statements such as the following, therefore, one can introduce subroutines employing vector storage of matrices with very little difficulty:

DIMENSION A(10,6), VA(60)
EQUIVALENCE (A(1,1), VA(1))
⋮
*CALL SUB (VA, M, N)
⋮

Provided the array A is never overdimensioned and that it and VA are mentioned in an EQUIVALENCE statement one can then refer to what is really the same set of locations as VA in the call to the SSP subroutines but as A elsewhere. If, however, one wishes to work with a dimensioned area *greater* than the matrix actually being processed then it is necessary to establish a correspondence between the elements of VA(IJ) and those of A(I,J). This can be done by using special subroutines provided in the SSP collection or—with some thought—by mapping functions based on the storage relationships illustrated in Fig. 43. Alternatively the programmer can take the more radical step of writing his calling programs using vector storage of matrices throughout.

The third feature of SSP subroutines is that symmetric and diagonal matrices are handled by compressed storage in vector form. Compressed vector storage of symmetric matrices has been touched on above (p. 451). That for diagonal matrices simply stores the main diagonal elements in vector form while the off-diagonal zeros are not stored. To link up such subroutines to a main program in which all matrices are stored in the usual non-compressed double-subscripted form is not very difficult and is again helped by the use of special subroutines included in the SSP package for converting from one storage mode to another. The programmer can himself, however, incorporate a few statements in his main program to do the work or, once more, consider the more far-reaching alternative of using compressed vector storage for all symmetric and diagonal matrices. The general principles

---

* SUB is a fictitious name; it is not a SSP subroutine.

involved will be gathered from the discussion on p. 451 and from a consideration of the program CORVEC. Further details are outside the scope of this book.

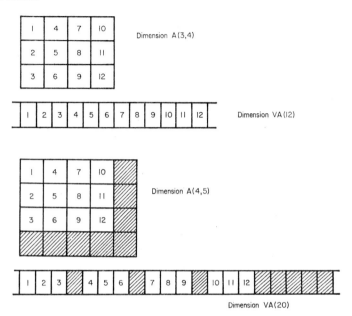

FIG. 43. Relationships between symbolic and physical storage locations of matrices and vectors. When a matrix A of size, say, 3×4 is stored in a dimensioned symbolic area A(3,4) the physical storage arrangement is in the form of twelve successive locations as shown in the upper two diagrams. The twelve successive locations may equally well be referred to as the twelve elements of a vector VA(12). If, however, the same 3×4 matrix is stored in a symbolic area dimensioned as A(4,5), the physical storage arrangement is as shown below, in the twelve numbered locations out of twenty successive ones. The crosshatched elements or locations are empty.

In both cases, to locate VA(IJ) in terms of A(I,J) we can use the mapping function:

$$IJ = (J - 1) * M + I$$

where IJ is the required vector subscript for VA, I and J are the row and column subscripts of the matrix A, and M is the number of rows in the DIMENSION statement for A. Conversely, to locate A(I,J) in terms of VA(IJ) we can use the two functions:

$$J = (IJ - 1)/M + 1$$
$$I = IJ - M * (J - 1).$$

Only in the first arrangement (upper two diagrams) do the elements of arrays A and VA correspond in storage in such a way that they can be made the subject of an EQUIVALENCE statement.

A different approach to the question of ready-made programs is provided by the Biomedical Data Programs sponsored by the University of California and described in detail in the BMD manuals edited by Dixon (1969, 1970).

These are a set of over 40 complete programs which read data in a relatively uniform standard format, carry out a more or less elaborate statistical analysis and write out the results. Apart from making up the deck correctly and specifying one or two formats, no programming skills are required to run the programs. The details of deck set-up, computational procedures and options available are set out at length in the BMD manual.

The programs available cover the following main types of statistical analysis:

> Description and tabulation;
> Correlation and regression analysis;
> Multivariate analysis;
> Life tables and survival rates;
> Probit analysis;
> Guttman scale analysis;
> Time series analysis (including power spectral analysis);
> Analysis of variance and covariance.

A special feature of the BMD programs is the use of transgeneration procedures, whereby one or more variables can be transformed and used as well as or instead of the original variables in the subsequent statistical analyses.

Conclusion

One may perhaps end this chapter by mentioning a few features which improve the usefulness of a program likely to be employed often by a variety of users. These include:

(i) Incorporation of instructions which allow the program to deal with a series of problems in succession. In most of the programs in this book we have achieved this object by causing a problem reference number to be read first (usually denoted as NPROB) and stopping only if this has a value like zero. The last data card before the end-of-file control card should then be a blank card or preferably, to avoid misunderstanding, one which has a zero punched in the appropriate column. Many similar procedures can readily be devised.

(ii) Use of object-time FORMAT declarations (p. 56). These allow the programmer the opportunity of choosing his own input format, a freedom which may be absolutely essential if several sets of data are to be dealt with in a single run.

(iii) Use of adjustable dimensions (p. 76) in subroutines. These have the advantage that when the dimensions of a program have to be altered, it is only the main program that needs to be changed. The subroutines can be left unchanged or can be used permanently as relocatable binary decks (which would otherwise have to be prepared afresh).

(iv) Use of relocatable binary decks for all or part of the program so that the source program does not have to be compiled anew each time the program is run. Naturally this method of saving computer time is most profitable when compilation time is relatively long in relation to execution time; this is often the case in programs run for scientific purposes (as compared with simple but very large data-processing operations).

(v) The use of A-formats, enabling the user to read in a heading or two with his numerical data, is often very desirable. It provides the best method of identifying output when many similar problems are being run and when the data themselves are not printed out.

(vi) The inclusion of several options in a program is often desirable since it may be very little extra trouble to extend certain parts of a program in ways that are occasionally more useful than the original purpose for which the program was written. There should, however, be limits to this approach: it is relatively easy to write programs with dozens of options but wasteful if, as often happens, a very large program has to be compiled in order that one small branch of it can be used.

(vii) Finally, a point which has not perhaps been stressed sufficiently in this book is the extent to which a large library of subroutines can be built up to provide the units from which hundreds of programs can be constructed. This is the philosophy underlying the Scientific Subroutines Package discussed on p. 472 and it can very readily be applied in the more restricted fields likely to interest individual research workers. Whenever a programmer devotes more than a few lines to some specific computation known to be of wider applicability, there are good arguments for writing it as a subroutine which thereafter is permanently available for incorporation into many other programs.

References

A bibliography of computer science has been published by the National Computer Centre (1968). Two very useful sources of general information on many of the topics covered in the present book are provided by Klerer and Korn (1967) and by Sterling and Pollack (1968).

J. Aitchison and J. A. C. Brown (1957). 'The Lognormal Distribution'. University Press, Cambridge.

T. W. Anderson (1958). 'An Introduction to Multivariate Statistical Analysis'. Wiley, New York.

W. R. Ashby (1956). 'Introduction to Cybernetics'. Methuen, London.

E. H. Ashton, M. J. R. Healy, C. E. Oxnard and T. E. Spence (1965). The combination of locomotor features of the primate shoulder-girdle by canonical analysis. *J. Zool.* **147**, 406–29.

N. T. J. Bailey (1951). On estimating the size of mobile populations from recapture data. *Biometrika* **38**, 293–306.

N. T. J. Bailey (1952). Improvements in the interpretation of recapture data. *J. anim. Ecol.* **21**, 120–27.

N. T. J. Bailey (1957). 'The Mathematical Theory of Epidemics'. Griffin, London.

N. T. J. Bailey (1959). 'Statistical Methods in Biology'. English Universities Press, London.

N. T. J. Bailey (1967). 'The Mathematical Approach to Biology and Medicine'. Wiley, New York.

C. A. Barlow, J. A. Graham and S. Adisoemarto (1969). A numerical factor for the taxonomic separation of *Pterostichus pensylvanicus* and *P. adstrictus* (Coleoptera: Carabidae). *Canad. Ent.* **101**, 1315–19.

T. C. Bartee (1966). 'Digital Computer Fundamentals', 2nd Edn. McGraw-Hill, New York.

M. S. Bartlett (1947). The use of transformations. *Biometrics* **3**, 39–52.

M. S. Bartlett (1949). Fitting a straight line when both variables are subject to error. *Ibid.* **5**, 207–12.

M. S. Bartlett, (1960). 'Stochastic Population Models in Ecology and Epidemiology'. Methuen, London.

R. Bellman (1960). 'Introduction to Matrix Analysis'. McGraw-Hill, New York.

C. I. Bliss (1953). Fitting the negative binomial distribution to biological data. *Biometrics* **9**, 176–200.

K. A. Brownlee (1951). 'Industrial Experimentation', 4th Edn. H.M.S.O., London.

R. C. Campbell (1967). 'Statistics for Biologists'. University Press, Cambridge.

B. Carnahan, M. A. Luther and J. O. Wilkes (1969). 'Applied Numerical Methods'. Wiley, New York.

W. G. Cochran (1954). Some methods of strengthening the common χ^2 tests. *Biometrics* **10**, 417–51.

W. G. Cochran and G. M. Cox (1957). 'Experimental Designs', 2nd Edn. Wiley, New York.

REFERENCES

A. J. Cole (Ed.) (1969). 'Numerical Taxonomy'. Academic Press, London and New York.

S. D. Conte (1965). 'Elementary Numerical Analysis'. McGraw-Hill, New York.

W. W. Cooley and P. R. Lohnes (1962). 'Multivariate Procedures for the Behavioural Sciences'. Wiley, New York.

D. R. Cox (1958). 'Planning of Experiments'. Wiley, New York.

D. R. Cox and P. A. W. Lewis (1966). 'The Statistical Analysis of Series of Events'. Methuen, London.

J. M. Craddock (1968). 'Statistics in the Computer Age'. Hodder, London.

G. Dahlberg (1947). 'Mathematical Methods for Population Genetics'. Karger, Basel.

D. Dimitry and T. Mott (1966). 'Introduction to Fortran IV Programming'. Holt, Rinehart & Winston, New York.

W. J. Dixon (Ed.) (1969). 'BMD Biomedical Computer Programs. X-Series Supplement'. University of California Press, Berkeley and Los Angeles.

W. J. Dixon (Ed.) (1970). 'BMD Biomedical Computer Programs'. University of California Press, Berkeley and Los Angeles.

W. J. Dixon and F. J. Massey (1969). 'Introduction to Statistical Analysis'. 3rd Edn. McGraw-Hill, New York.

M. A. Efroymson (1960). Multiple regression analysis. In: A. Ralston and H. S. Wilf (1960). 'Mathematical Methods for Digital Computers'. Chap. 7. Wiley, New York.

D. A. Evans (1953). Experimental evidence concerning contagious distributions in ecology. *Biometrika* **40**, 186–211.

W. J. Ewens (1969). 'Population Genetics'. Methuen, London.

W. I. Ferrar (1957). 'Algebra: a Textbook of Determinants, Matrices and Algebraic Forms', 2nd Edn. University Press, Oxford.

D. J. Finney (1952). 'Probit Analysis'. 2nd Edn. University Press, Cambridge.

R. A. Fisher (1930). 'The Genetical Theory of Natural Selection'. Clarendon Press, Oxford.

R. A. Fisher (1953). Note on the efficient fitting of the negative binomial. *Biometrics* **9**, 197–200.

R. A. Fisher and F. Yates (1963). 'Statistical Tables for Biological, Agricultural and Medical Research', 6th Edn. Oliver & Boyd, Edinburgh.

L. Fox (1964). 'An Introduction to Numerical Linear Algebra'. Clarendon Press, Oxford.

L. Fox and D. F. Mayers (1968). 'Computing Methods for Scientists and Engineers'. Clarendon Press, Oxford.

J. T. Golden (1965). 'Fortran IV Programming and Computing'. Prentice–Hall, Englewood Cliffs, N.J.

B. C. Goodwin (1963). 'Temporal Organisation in Cells'. Academic Press, New York and London.

J. C. Gower (1966). Some distance properties of latent root and vector methods used in multivariate analysis. *Biometrika* **53**, 325–28.

J. C. Gower and G. J. S. Ross (1969). Minimum spanning trees and single linkage cluster analysis. *Appl. Stat.,* **18**, 54–64.

A. Greenstadt (1960). The determination of the characteristic roots of a matrix by the Jacobi method. In A. Ralston and H. S. Wilf (1960): 'Mathematical Methods for Digital Computers'. Chap. 7. Wiley, New York.

P. Greig-Smith (1964). 'Quantitative Plant Ecology', 2nd Edn. Butterworth, London.
F. S. Grodins (1963). 'Control Theory and Biological Systems'. Columbia University Press, New York.
J. Gurland (Ed.) (1964). 'Stochastic Models in Medicine and Biology'. University of Wisconsin Press, Madison.
J. B. S. Haldane (1932). 'The Causes of Evolution'. Longman, London.
J. M. Hammersley and D. C. Handscomb (1964). 'Monte Carlo Methods'. Methuen London.
H. O. Hartley (1960). Analysis of variance. In A. Ralston and H. S. Wilf, 'Mathematical Methods for Digital Computers'. Chap. 20. Wiley, New York.
C. Hastings (1955). 'Approximations for Digital Computers'. University Press, Princeton.
P. Hoel (1971). 'Introduction to Mathematical Statistics', 4th Edn. Wiley, New York.
K. Hope (1968). 'Methods of Multivariate Analysis'. University of London Press, London.
H. Hotelling (1931). The generalisation of Student's 'ratio'. *Ann. math. Stat.* **2,** 360–378.
H. Hotelling (1933). Analysis of a complex of statistical variables into principal components. *J. Educ. Psychol.,* **24,** 417–41; 498–520.
A. Huitson (1966). 'Analysis of Variance'. Griffin, London.
IBM (1968). 'System/360 Scientific Subroutine Package (360A-CM-03X) Version III. Programmer's Manual'. IBM, New York.
M. L. James, G. M. Smith and J. C. Wolford (1967). 'Applied Numerical Methods for Digital Computation with Fortran'. International Textbook Co., New York.
N. Jaspen (1965). The calculation of probabilities corresponding to values of z, t, F and chi-square. *Educ. Psychol. Measurement* **25,** 877–80.
G. M. Jolly (1963). Estimates of population parameters from multiple recapture data with both death and dilution—deterministic model. *Biometrika* **50,** 113–28.
G. M. Jolly (1965). Explicit estimates from capture–recapture data with both death and immigration—stochastic model. *Biometrika* **52,** 225–57.
R. Jones (1964). A review of methods of estimating population size from marking experiments. *Rapp. Proc. Verbaux J. Conseil Perm. int. Expl. Mer.* **155,** 202–9.
O. Kempthorne (1957). 'An Introduction to Genetic Statistics'. Wiley, New York.
M. G. Kendall (1957). 'A Course in Multivariate Analysis'. Griffin, London.
M. Klerer and G. A. Korn (1967). 'Digital Computer User's Handbook'. McGraw-Hill, New York.
D. N. Lawley and A. E. Maxwell (1963). 'Factor Analysis as a Statistical Method'. Butterworth, London.
R. S. Ledley (1965). 'Uses of Computers in Biology and Medicine'. McGraw-Hill, New York.
C. C. Li (1955). 'Population Genetics'. University Press, Chicago.
C. C. Li (1964). 'Introduction to Experimental Statistics'. McGraw-Hill, New York.
A. J. Lotka (1924). 'Elements of Physical Biology'. Williams & Wilkins, Baltimore, (Reprinted in 1956 as 'Elements of Mathematical Biology'. Dover, New York.)
R. Margalef (1969). 'Perspectives in Ecological Theory'. University Press, Chicago.

REFERENCES

F. F. Martin (1968). 'Computer Models and Simulation'. Wiley, New York.
A. E. Maxwell (1968). 'Analysing Qualitative Data'. Methuen, London.
J. Maynard Smith (1968). 'Mathematical Ideas in Biology'. University Press, Cambridge.
D. D. McCracken (1965). 'A Guide to Fortran IV Programming'. Wiley, New York.
J. H. Milsum (1966). 'Biological Control Systems Analysis'. McGraw-Hill, New York.
L. Mirsky (1955). 'An Introduction to Linear Algebra'. Clarendon Press, Oxford.
D. F. Morrison (1967). 'Multivariate Statistical Methods'. McGraw-Hill, New York.
N. E. Morton (Ed.) (1969). 'Computer Applications in Genetics'. University of Hawaii Press, Honolulu.
J. Moshman (1967). Random number generation. In A. Ralston and H. S. Wilf, 'Mathematical Methods for Digital Computers', Vol. 2. Wiley, New York.
National Computer Centre (1968). 'International Computer Bibliography'. N.C.C., Manchester.
T. H. Naylor, J. L. Balintfy, D. S. Burdick and K. Chu (1966). 'Computer Simulation Techniques'. Wiley, New York.
A. Orden, (1960). Matrix inversion and related topics by direct methods. In A. Ralston and H. S. Wilf, 'Mathematical Methods for Digital Computers', Vol. 1, Chap. 2. Wiley, New York.
E. I. Organick (1966). 'A Fortran IV Primer'. Addison-Wesley, Reading, Mass.
J. Ortega (1967). The Givens–Householder method for symmetric matrices. In A. Ralston and H. S. Wilf, 'Mathematical Methods for Digital Computers', Vol. 2. Wiley, New York.
R. H. Pennington (1970). 'Introductory Computer Methods and Numerical Analysis', 2nd Edn. Macmillan, New York.
E. C. Pielou (1970). 'Introduction to Mathematical Ecology'. Wiley, New York.
M. H. Quenouille (1968). 'The Analysis of Multiple Time Series'. Griffin, London.
A. Ralston and H. S. Wilf (1960, 1967). 'Mathematical Methods for Digital Computers'. Wiley, New York.
C. R. Rao (1952). 'Advanced Statistical Methods in Biometric Research'. Wiley, New York.
N. Rashevsky (1960). 'Mathematical Biophysics', 2 vols. Dover, New York.
H. Scheffé (1959). 'The Analysis of Variance'. Wiley, New York.
H. Seal (1964). 'Multivariate Statistical Analysis for Biologists'. Methuen, London.
S. R. Searle (1966). 'Matrix Algebra for the Biological Sciences'. Wiley, New York.
S. Siegel (1956). 'Non-Parametric Statistics for the Behavioral Sciences'. McGraw-Hill, New York.
G. G. Simpson, A. Roe and R. C. Lewontin (1960). 'Quantitative Zoology" 2nd Edn. Harcourt Brace Jovanovitch, New York.
R. R. Sokal and F. J. Rohlf (1969). 'Biometry'. Freeman, San Francisco.
R. R. Sokal and P. H. A. Sneath (1963). 'Principles of Numerical Taxonomy'. Freeman, San Francisco.
T. R. E. Southwood (1966). 'Ecological Methods with Particular Reference to the Study of Insect Populations'. Methuen, London.
R. W. Stacey and B. D. Waxman (Eds.) (1965). 'Computers in Biomedical Research', 2 vols. Academic Press, New York and London.
T. D. Sterling and S. V. Pollack (1966). 'Computers and the Life Sciences'.

Columbia University Press, New York.
T. D. Sterling and S. V. Pollack (1968). 'Statistical Data Processing'. Addison-Wesley, Englewood Cliffs.
D'Arcy W. Thompson (1917). 'On Growth and Form', 2nd Edn. (1942). University Press, Cambridge.
K. D. Tocher (1963). 'The Art of Simulation'. English Universities Press, London.
D. J. Veldman (1967). 'Fortran Programming for the Behavioural Sciences'. Holt, Rinehart & Winston, New York.
K. E. F. Watt (1968). 'Ecology and Resource Management'. McGraw-Hill, New York.
P. A. White (1958). The computation of eigenvalues and eigenvectors of a matrix. *J. Soc. ind. appl. Maths.* **6**, 393–437.
N. Wiener (1961). 'Cybernetics: or Control and Communication in the Animal and the Machine', 2nd Edn. M.I.T. Press, Cambridge, Mass.
J. H. Wilkinson (1965). 'The Algebraic Eigenvalue Problem'. Clarendon Press, Oxford.
C. B. Williams (1964). 'Patterns in the Balance of Nature and Related Problems in Quantitative Ecology'. Academic Press, London and New York.
S. Wright (1931). Evolution in Mendelian populations. *Genetics* **16**, 97–159.
F. Yates (1936). Incomplete randomised blocks. *Ann. Eug.* **7**, 121–40.
F. Yates and A. Anderson (1966). A general computer program for the analysis of factorial experiments. *Biometrics* **22**, 503–24.
H. P. Yockey (Ed.) (1958). 'Symposium on Information Theory in Biology'. Pergamon Press, London.

Author Index

Numbers in italics refer to pages on which references are listed at the end of the book.

A

Adisoemarto, S., 291, *478*
Aitchison, J., 346, *478*
Anderson, A., 148, *482*
Anderson, T. W., 274, *478*
Ashby, W. R., 377, *478*
Ashton, E. H., 311, *478*

B

Bailey, N. T. J., 1, 82, 89, 92, 185, 187, 377, 431, *478*
Barlow, C. A., 291, *478*
Bartee, T. C., 23, *478*
Bartlett, M. S., 208, 210, 213, 346, 377, *478*
Bellman, R., 216, *478*
Balintfy, D. S., 378, 394, *481*
Bliss, C. I., 366, 367, *478*
Brown, J. A. C., 346, *478*
Brownlee, K. A., 148, 174, *478*
Burdick, D. S., 378, 394, *481*

C

Campbell, R. C., 82, 313, *478*
Carnahan, B., 471, *478*
Chu, K., 378, 394, *481*
Cochran, W. G., 149, 314, *478*
Cole, A. J., 421, *478*
Conte, S. D., 386, 458, 471, *479*
Cooley, W. W., 264, 471, *479*
Cox, D. R., 149, 198, *479*
Cox, G. M., 149, *478*
Craddock, J. M., 201, *479*

D

Dahlberg, G., 378, *479*

Dimitry, D., 32, *479*
Dixon, W. J., 82, 148, 475, *479*

E

Efroymson, M. A., 283, *479*
Evans, D. A., 366, *479*
Ewens, W. J., 2, *479*

F

Ferrar, W. I., 216, *479*
Finney, D. J., 414, 416, 417, *479*
Fisher, R. A., 154, 167, 173, 174, 367, 377, 403, *479*
Fox, L., 216, 264, 458, *479*

G

Golden, J. T., 32, 403, 451, *479*
Goodwin, B. C., 377, *479*
Gower, J. C., 310, 421, *479*
Graham, J. A., 291, *478*
Greenstadt, A., 264, *479*
Greig-Smith, P., 2, *479*
Grodins, F. S., 377, *479*
Gurland, J., 394, *479*

H

Haldane, J. B. S., 377, *479*
Hammersley, J. M., 394, *479*
Handscomb, D. C., 394, *479*
Hartley, H. O., 148, *480*
Hastings, C., 348, *480*
Healy, M. J. R., 311, *478*
Hoel, P., 185, *480*
Hope, K., 274, 301, *480*
Hotelling, H., 261, 284, 291, *480*
Huitson, A., 145, 147, 154, 158, *480*

AUTHOR INDEX

I

IBM, 148, 348, 472, *480*

J

James, M. L., 471, *480*
Jaspen, N., 404, *480*
Jolly, G. M., 431, 436, *480*
Jones, R., 431, *480*

K

Kempthorne, O., 2, *480*
Kendall, M. G., 264, *480*
Klerer, M., *480*
Korn, G. A., *480*

L

Lawley, D. N., 261, 265, 295, *480*
Ledley, R. S., 1, *480*
Lewis, P. A. W., 198, *479*
Lewontin, R. C., 147, 208, 210, 213, *481*
Li, C. C., 2, 145, *480*
Lohnes, P. R., 264, 471, *479*
Lotka, A. J., 377, *480*
Luther, M. A., 471, *478*

M

McCracken, D. D., 32, *480*
Margalef, R., 377, *480*
Martin, F. F., 378, 394, *480*
Massey, F. J., 82, *479*
Maxwell, A. E., 261, 265, 295, 361, *480*
Mayers, D. F., 458, *479*
Maynard-Smith, J., 377, 387, *480*
Milsum, J. H., 377, *480*
Mirsky, L., 216, *480*
Morrison, D. F., 274, 287, 297, *480*
Morton, N. E., 377, *480*
Moshman, J., 394, *480*
Mott, T., 32, *479*

N

National Computer Centre, *480*
Naylor, T. H., 378, 394, *481*

O

Orden, A., 255, *481*
Organick, E. I., 32, *481*
Ortega, J., 264, *481*
Oxnard, C. E., 311, *478*

P

Pennington, R. H., 244, 386, 441, 458, 471, *481*
Pielou, E. C., 2, *481*
Pollack, S. V., 1, *481*

Q

Quenouille, M. H., 198, *481*

R

Ralston, A., 441, 471, *481*
Rao, C. R., 274, *481*
Rashevsky, N., 377, *481*
Roe, A., 147, 208, 210, 213, *481*
Rohlf, F. J., 471, *481*
Ross, G. J. S., 421, *479*

S

Scheffé, H., 145, *481*
Seal, H., 274, *481*
Searle, S. R., 216, 270, *481*
Siegel, S., 313, 326, 332, 338, 340, *481*
Simpson, G. G., 147, 208, 210, 213, *481*
Smith, G. M., 471, *480*
Sneath, P. H. A., 2, 421, *481*
Sokal, R. R., 2, 421, 471, *481*
Southwood, T. R. E., 2, *481*
Spence, T. E., 311, *478*
Stacey, R. W., 1, *481*
Sterling, T. D., 1, *481*

T

Thompson, D'Arcy W., 377, *481*
Tocher, K. D., 394, 402, 403, *481*

V

Veldman, D. J., 471, *481*

W

Watt, K. E. F., 2, 377, *481*
Waxman, B. D., 1, *481*
White, P. A., 264, 270, *481*
Wiener, N., 377, *481*
Wilf, H. S., 441, 471, *481*
Wilkes, J. O., 471, *478*
Wilkinson, J. H., *481*

Williams, C. B., 2, *481*
Wolford, J. C., 471, *480*
Wright, S., 377, *482*

Y

Yates, F., 148, 154, 167, 173, 174, 178, 403, *479*, *482*
Yockey, H. P., 377, *482*

Subject Index

A

Abbott's formula, 414
Absolute error, 460, 465
Access-time, 20
Accumulator, 24
Accuracy, 458ff;
  of numerical integration, 385
Adder, 25
Address, 20, 27
Adjustable dimensions, 73, 77, 476
Algol, 31
Algorithm, 32, 441ff;
  for sorting, 107ff;
  for tabulation, 117, 125
Alphameric character, 36
Analog computer, 5
Analysis of variance, 145ff;
  balanced incomplete blocks, 172ff;
  complete 3-factor, 161ff;
  incomplete 3-factor, 154ff;
  Latin square, 166ff;
  programs, 148;
  and regression, 189ff, 193ff;
  two-factor, 149ff, 154ff
ANOVA, see Analysis of variance
Applications programmer, 31
Argument,
  of intrinsic functions, 44;
  of mathematical functions, 43;
  of subprograms, 67, 70ff;
  transmission of, 70ff
Arithmetic expression, 40ff
Arithmetic IF statement, 58
Arithmetic mean, 82, 86, 87
Arithmetic statement, 45
Arithmetic statement function, 66
Arithmetic unit, 23ff
Array, 61ff, 77ff;
  storage of, 77ff;
  three-dimensional, 122, 155, 161

Array declarator, see DIMENSION
Assembler, 29
Assembly language, 29
Autocorrelation, 198ff

B

Balanced incomplete block, 172ff
Basic external functions, 43, 75
Batch processing, 17
BCD, see Binary coded decimal
Binary coded decimal, 13
Binary digit, see Bit
Binary notation, 6ff
Binary number conversion, 10
Binomial distribution, 360ff, 400ff
Biomedical Data Programs, 148, 475ff
Bit, 7, 21, 22, 55
Bivariate normal distribution, 181, 182
Blank COMMON, 77
BMD programs, 148, 475ff
Buffer, 16

C

CALL, 71
CALL EXIT, 46
Canonical analysis, see Multiple discriminant analysis
Canonical variate, 303
Card reader, 16
Carriage control, 52
Central store, 17ff
Character set, 36
Characteristic equation, 258
Chi-squared distribution, 402, 403ff
Chi-squared test,
  of association, 313ff;
  of goodness-of-fit, 318ff, 350, 358
Class-interval, 115
Class mark, 116

SUBJECT INDEX

Cluster analysis, 293, 420ff
Cobol, 31
Coefficient,
 of contingency, 314, 316;
 of variation, 83, 86, 87
Cofactor, 240
Comment, 38
COMMON, 70, 72ff, 74, 77
Common storage, 77
Compilation errors, 468
Compiler, 30
Compiler language, 30
Complement, 24
COMPLEX, 74, 75
Component score, 292
Compressed vector storage, 452ff, 474
Computed GO TO, 58
Computer,
 structure and function, 14ff;
 uses in biology, 1ff
Computer register, 7, 8, 19, 24, 28
Computer storage, 17ff
Computer word-structure, 7, 19, 55, 107
Condition number, 463
Contingency table, 121ff, 126, 127, 184, 313ff
Continuation card, 36
Continuation column, 36
CONTINUE, 65
Continued product, 225
Constant, *see* Fortran constant
Control cards, 80
Control unit, 27
Core-memory plane, 19, 20
Core-storage, 17ff, 447ff
Correlation, 88ff, 181ff
Correlation coefficient, 89, 93, 94, 183, 185
Correlation matrix, 219, 232ff, 456ff
Correlogram, 201
Covariance, 88, 93, 94
Covariance matrix, 219, 232ff, 285, 456ff
Cross-correlation, 202
Crossed classification, 156
Cumulative frequency, 116

D

DATA, 74
Data cards, 80
Data processing, 105

Data summarising programs, 129ff
Decimal notation, 5ff
Decimal number conversion, 10
Decision box, 34
Deck, 80
Degrees of freedom, 146ff, 350, 358, 404
Dependent variable, 183, 274
Destructive read-in, 18
Determinant, 239ff, 251, 441, 463
Deterministic model, 376
Differential equations, 387ff
DIMENSION, 63, 73, 74, 76ff;
 in subprograms, 76
Dimensions,
 in COMMON statement, 76, 77
Discriminant coefficients, 284, 285, 301
Discriminant function,
 for two groups, 284ff;
 for several groups, 300ff
Discriminant score, 287, 290, 303
Disk, *see* Magnetic disk
Dispersion matrix, 232ff; *see also* Covariance matrix
Dispersion test, 361
Distribution,
 binomial, 360ff, 400ff;
 chi-squared, 402, 403ff;
 exponential, 402ff;
 F, 403ff;
 negative binomial, 366ff;
 normal, 344ff, 399ff, 403ff;
 Poisson, 355ff, 401ff;
 Student's t, 403ff;
 uniform, 394ff
Distribution-free statistics, 312ff
Division by zero, 96
DO statement, 64ff, 443ff
Doolittle method, 278
Dosage-response curve, 411ff
DOUBLE PRECISION, 74, 75, 464ff
Dummy argument, 67, 70

E

Eigenvalue, *see* Latent root
Eigenvector, *see* Latent vector
Electronic digital computer, 5
END, 46
END FILE, 450
End-of-file card, 80
EQUIVALENCE, 74, 77ff, 447

Error, 459ff, 463ff, 465;
    propagation of, 460ff
Error mean square, 150, 176
Error subroutine, 466
Error sum of squares, 149, 176
Errors,
    in programs, 467
Exchange sort, 106, 107ff
Execution cycle, 28
Execution errors, 469
Explicit type, 39
Exponential distribution, 402ff
Exponentiation, 40, 42
EXTERNAL, 70, 74, 75
External functions, 43, 75

F

F-distribution, 403ff
F-test, 98ff; *see also* Variance ratio test
False position, 369
Ferrite core, 18
Field descriptor, *see* Format specification
Field width, 47
File protection, 23
Fixed effects, 147
Floating-point number, 8, 61
Flowchart, 33ff;
    symbols for, 35
Formal parameter, 70
FORMAT, 46, 47ff, 74
Format specification, 47ff;
    A-specification, 54, 477;
    E-specification, 48;
    F-specification, 48;
    H-specification, 51;
    I-specification, 47;
    X-specification, 51
Format,
    variable, 54, 56, 477
Formatted READ, 46ff
Formatted WRITE, 47ff
Fortran, 32, 36ff
Fortran coding form, 37
Fortran constant, 38ff;
    exponent form, 38;
    fixed-point, 38;
    floating-point, 38;
    integer, 38;
    real, 38
Fortran functions, 42ff

Fortran mathematical functions, 42ff, 470
Fortran variable, 39;
    subscripted, 40, 61
Frequency distribution, 115, 132ff, 347ff
Frequency table, 115, 118, 119, 132ff, 347ff
Function, 42ff, 470
FUNCTION subprogram, 67ff

G

Generalised distance, 291, 310
Givens–Householder technique, 264, 310
GO TO statement, 57

H

Hexadecimal notation, 6
High-speed printer, 25, 26
Hotelling's T^2, 284ff
Hysteresis curve, 18

I

IBM, 472
Identity matrix, 248
IF statement, 58ff
Implicit type, 39
Independent variable, 183
Indexing parameters,
    computed, 108, 143
Initialization, 34
Input,
    of matrices, 219ff, 446;
    statements, 46ff
Input devices, 15ff
Instruction, 29;
    storage of, 14, 20ff
Instruction cycle, 28
Instruction format, 21
INTEGER, 74, 75
INTEGER FUNCTION, 74, 75
Integration, 380ff
Interaction, 146, 154, 161, 168, 174
Intrinsic functions, 44, 75
Inverse hyperbolic sine transformation, 347
Inverse matrix, 247ff
Inverse sine transformation, 346
Iterative methods, 261, 267, 368, 465

J

Jacobi technique, 264, 310
Job control, *see* Control cards
Jolly's method, 430ff

K

Key-punch, 15
Kruskal–Wallis test, 338ff

L

Labelled COMMON, 77
Latent roots, 258ff, 295;
  of real non-symmetric matrix, 267ff, 301;
  of real symmetric matrix, 260ff, 264
Latent vectors, 258ff, 267ff, 295;
  normalization of, 260, 268;
  of real non-symmetric matrix, 267ff, 301;
  of real symmetric matrix, 260ff, 264;
  right and left, 267
Latin square, 166ff
Library, 69
Line printer, *see* High-speed printer
Linear functional relationship, 208ff;
  significance tests, 209ff
Listing, 80
Logarithmic transformation, 346
LOGICAL, 74, 75
Logical IF statement, 59
Logical operators, 60, 471
Loop, 34, 59, 64

M

Machine language, 29
Macro-instruction, 30
Main effects, 146
Magnetic core memory, 17
Magnetic disk, 17, 22, 449ff
Magnetic drum, 17, 22, 451
Magnetic tape, 17, 22, 449ff;
  manipulation, 123ff, 449ff;
  output, 27
Mark-recapture data, 430ff
Matching coefficient, 420, 421
Mathematical functions, 42ff, 470

Matrix, 216;
  addition, 226;
  input–output of, 219ff;
  interchange of rows, 225;
  main diagonal, 217;
  manipulation of, 224ff;
  multiplication, 226ff;
  order of, 217;
  real non-symmetric, 267;
  real symmetric, 218, 445;
  trace of, 217, 225;
  vector storage of, 451ff
Matrix algebra, 216ff
Maximum,
  computation of, 124, 132, 138
Mean, 82, 86, 87
Means,
  comparison of, 97, 101, 102
Mean square, 146ff
Memory, *see* Core-storage
Minimum,
  computation of, 124, 132, 138
Minor, 240
Mixed modes, 41, 42, 467
Models,
  in analysis of variance, 146, 149, 154, 156, 161, 168, 174;
  mathematical, 376ff
Monitor program, 80
Multiple correlation coefficient, 275, 278
Multiple discriminant analysis, 300ff
Multiple regression, 273, 274ff
Multiplicative congruential method, 394ff
Multivariate analysis, 273ff

N

NAMELIST, 74
Negative binomial distribution, 366ff
Nested classification, 156, 158
Nested DO-loops, 65, 66
Nested parentheses, 41
Nominal scale, 312
Non-destructive read-out, 19
Non-executable statement, 74
Non-formatted READ, 449
Non-formatted WRITE, 124, 449
Nonparametric statistics, 312ff
Normal deviate, 97, 101, 102, 347
Normal distribution, 344ff, 403ff

SUBJECT INDEX

Normalised determinant, 463
Normalised latent vector, 260, 268
Number systems, 5ff;
  interconversion, 10ff
Numerical analysis, 458
Numerical integration, 380ff
Numerical solution of differential
  equations, 387ff

O

Object program, 29
Object-time format, 56, 476
Octal notation, 8ff
Octal number conversion, 12
Off-line, 16
On-line, 16
OP-code, *see* Operation code
Operation code, 27
Order,
  of statements, 74
Ordinal scale, 312
Output,
  of matrices, 221ff, 446;
  paper-tape, 26;
  punched card, 26
Output devices, 25ff
Output statements, 46ff
Over-dispersion, 366
Overflow, 25

P

Packages, 471ff
Paper-tape, 15;
  output, 26;
  reader, 16
Parameter studies, 378ff
Parentheses, 41, 42
PAUSE, 46
Pivot, 243
PL/1, 31
Poisson distribution, 355ff, 401ff
Polynomial,
  evaluation of, 348, 442
Polynomial approximations, 348, 406, 414
Population-size estimates, 430ff
Power-efficiency, 313
Power residue method, 394
Power spectral analysis, 201

Principal component, 292
Principal component analysis, 291ff
Printer, 25, 26
Probability, 403ff
Probability switch, 396
Probit, 413
Probit analysis, 410ff
Product-moment correlation coefficient, 89; *see also* Correlation coefficient
Program construction, 466
Program planning, 32ff, 440ff, 476
Program preparation, 80
Program testing, 466ff
Programming languages, 28
Proportional frequency, 116
Pseudorandom numbers, 394ff
Punched card, 15;
  output, 26

Q

Quadratic form, 231, 251

R

Random access, 17
Random effects, 147
Random numbers, 394ff
Random variables, 394ff
Randomised block, 172
Randomness,
  tests for, 403
Range, 123, 132
Range,
  of DO-loop, 64
Rank correlation coefficient, 322ff
Ranking, 106, 111ff, 322ff
READ, 46ff, 449
REAL, 74, 75
REAL FUNCTION, 68
Reciprocal transformation, 347
Register, 7, 8, 19, 24, 28
Regression, 88ff, 181ff;
  coefficient, 91;
  comparisons, 193ff;
  equation, 91, 93, 94;
  line, 90, 185;
  multiple, 273ff;
  parameters, 91;
  significance tests, 185, 189ff;
  test of linearity, 189ff
Relational operators, 59

Relative error, 460, 465
Relocatable binary, 17, 26, 69, 477
Replication, in analysis of variance, 154
residual, 146
Residual sum of squares, 149ff
RETURN, 68, 70
REWIND, 450, 451
Round-off error, 13, 460
Runge–Kutta method, 388ff
Running mean, 114ff

S

Scalar, 229, 231, 239, 251
Scatter diagram, 89, 90, 181
Scientific Subroutines Package, 148, 472
Scratch tape, 23
Sense-winding, 18
Sheppard's correction, 135, 184
Shift register, 24
Sign bit, 7
Signed ranks test, 330ff
Significant digits, 458, 462
Similarity level, 421
Similarity matrix, 421
Simpson's rule, 380ff
Simultaneous equations, 248ff
Single-linkage clustering, 420ff
Slash format, 53
Sorting, 106, 109, 110
Source program, 29
Spearman's rank correlation coefficient, 322ff
Square-root transformation, 347
SQRT function, 470
Standard deviation, 83, 86, 87
Standard error,
  in balanced incomplete blocks, 176, 178;
  of mean, 83, 86, 87;
  of regression coefficient, 92
Standard Fortran, 32
Standard partial regression coefficient, 275
Standardised normal deviate, 348
Statement, 36
Statement function, 66, 74
Statement label, see Statement number
Statement number, 36
Stepwise multiple regression, 283

Stochastic model, 376
Stochastic population estimates, 430ff
STOP, 46
Storage, 17ff, 63;
  conservation of, 77
Student's t-distribution, 403ff
Subprogram, 67ff
SUBROUTINE subprogram, 67ff
Subscript, 61
Subscripted variable, 40, 61ff
Sum of squares, 146ff, 175
Supervisory program, 80
System control cards, 80
Systems programmer, 31

T

t-test, 92, 97, 98ff, 101, 102
Tabulation, 115ff;
  algorithm, 117, 125
Tabulation of functions, 378ff
Tape, see Magnetic tape, paper tape
Tape-drive unit, 23
Thin-film memory, 20
Three-way ANOVA, 154ff, 161ff
Tied ranks, 323ff
Time,
  economies in, 442ff
Time-series, 141, 198, 202
Trace, 217, 225
Transfer statements, 57ff
Transformations, 346, 347
Transpose, 230
two-way ANOVA, 149ff, 154ff
Type-declaration, 39, 68, 74

U

Unconditional GO TO, 57
Underflow, 25
Unformatted READ, 449
Unformatted WRITE, 449

V

Variable, see Fortran variable
Variable format, 54, 56, 105
Variance, 83, 86, 87; see also Analysis of variance

Variance-covariance matrix, 219, 232ff, 285, 456ff
Variance ratio, 98;
  test, 146, 154ff, 164
Vector, 61, 112, 217;
  for matrix storage, 451ff, 473;
  multiplication of, 229
Verifier, 80
Visual display, 27

W

Weighting coefficients, 416
Wilcoxon's signed ranks test, 330ff
Word-structure, 7, 19, 55
WRITE, 47ff, 449

Y

Yates' correction, 314